Communication Techniques for Digital and Analog Signals

Morton Kanefsky
University of Pittsburgh

CO-AKQ-799

1817

HARPER & ROW, PUBLISHERS, New York
Cambridge, Philadelphia, San Francisco, London,
Mexico City, São Paulo, Singapore, Sydney

Sponsoring Editor: Peter Richardson
Project Editor: Jonathan Haber
Cover Design: Michel Craig
Text Art: Reproduction Drawings, Ltd.
Production: Delia Tedoff/Marion Palen
Compositor: Black Dot, Inc.
Printer and Binder: The Maple Press Company

Communication Techniques for Digital and Analog Signals

Library of Congress Cataloging in Publication Data

Kanefsky, Morton.

  Communication techniques for digital and analog signals.

   Includes index.
    1. Digital communications.   I. Title.
TK5103.7.K36 1985   621.38'0413   84-22389
ISBN 0-06-043475-9

84 85 86 87 9 8 7 6 5 4 3 2 1

Contents

Chapter 3 Overview of Discrete Communications 59

Chapter 4 Communicating Analog Signals 106

Chapter 5 Statistical Modeling of Random Signals 136

Chapter 6 Analysis of Binary Communications 176

Chapter 7   Discrete Communication Theory   211

Chapter 8   Continuous Waveform and Other
             Suboptimal Techniques   254

Chapter 9   Introduction to Error-Correcting Codes   288

Chapter 10 Cyclic and Convolution Codes   315

Preface

Many excellent texts on modern communication theory have become available in recent years. My motivation is not to write a "better" one, which I cannot do, but to write one that contains some substantial differences. I shall attempt to explain and justify these differences.

While most new texts intended primarily for undergraduates emphasize discrete communications, there is still a strong tendency to base the development on continuous waveform techniques. My intention is to reverse this procedure—to introduce the basic concepts in terms of discrete communication techniques. I believe that with the growth of computer technology in our curricula this approach should now be natural. Also, since many schools require only one semester of communications, this approach has the advantage of guaranteeing a strong modern discrete communications emphasis. In addition, I have attempted to make this text as brief as possible by not including some specialized techniques. These details are unnecessary in a text intended for undergraduates, and their omission should make the book useful as a compact reference on communications techniques.

The organization of this book permits considerable flexibility on the part of the instructor. If the text is followed in order, one should be able to cover the first six chapters in one semester. If, however, after Chapter 5 one covers the first half of Chapter 8 followed by a few topics in Chapter 6, considerable emphasis can be given to continuous waveform communications. If, on the other hand, one skips Chapter 4 and includes Chapter 7, the resulting treatment will be fully oriented toward discrete communications. The last two chapters are an introduction to error-correcting coding that is more extensive than those found in other undergraduate texts. It is my intention that this material could be included in a second course in communications and should be a helpful reference.

A significant feature of this text is the presentation of the theory of spectral analysis in Chapter 2. A measurement-oriented definition of the power spectrum permits an early and rigorous spectral analysis without the need for probability concepts. Other undergraduate texts do not discuss the very basic concept of the power spectrum because of the complexity of the more traditional definition. For their spectral analysis, they rely on the Fourier transform only and ignore the random nature of the signals. Because of the rigorous spectral analysis in this text, a number of important results and concepts can be presented. Examples include a closed-form solution for the FSK spectrum, proper insight into the underlying assumptions of the sampling theorem, and the ability to discuss such random processes as shot noise, impulse noise, and thermal noise.

Other definitions pertinent to random processes, which are given in Chapter 5, differ somewhat from those of even graduate-oriented engineering texts. Traditional texts follow the ensemble approach to random processes, which was first introduced by Wiener in the early 1940s. The ensemble approach has several drawbacks, however. It in fact offers little insight into the processes, and it is difficult to present to undergraduates. Indeed, most elementary texts either omit the ideas or include them superficially. In contrast, the approach to time series analysis of such theoretical mathematicians as Birkhoff, Von Neumann, and Rosenblatt has great conceptual advantages. With the omission of some complex details, these concepts can be presented to undergraduates. I hope that Chapters 2 and 5 prove this point. One clear advantage to this approach is the ability to discuss the power spectrum prior to the introduction of probability concepts.

Chapter 7 includes material on decision theory that exploits the 1956 text *Principles of Communication Engineering* by Wozencraft and Jacobs. The clear, geometric visualizations of this early text are beginning to reappear in recent textbooks as the emphasis on discrete communications grows.

I wish to recognize the many students who have helped over the years to clarify these ideas and to improve my understanding. I want to thank Jay Sherfey for editing the manuscript and Annette Earl, Kathy Tamenne, and Erika Van Sickel for their efforts in typing various versions. Finally, I wish to thank the staff of Harper & Row and my editor, Carl McNair, for his encouragement and suggestions, as well as the reviewers for their many helpful improvements and corrections.

Morton Kanefsky

Chapter 1

An Introduction to Communications

1.1 INTRODUCTION

When most of us think of communications, we focus on commercial broadcast communications: radio (both AM and FM) and television and their associated electronic hardware. In reality these are a relatively minor aspect of communications and represent only the start of the efforts that led to our vast array of communication capabilities. The techniques and hardware of broadcast communications are quite different from those of modern communications. Undoubtedly, the enormous impact on our lives of the news and entertainment media is responsible for these views. If we focus briefly, however, on commercial, scientific, and military communication needs and their historical development, we should obtain a better understanding of modern communication techniques.

1.2 A VERY BRIEF HISTORY OF COMMUNICATIONS

The introduction of telegraphy by Morse in the 1840s can be thought of as the beginning of communications. The next significant pioneering effort was Bell's invention of the telephone in 1876 and its subsequent rapid development, which enabled speech to be transmitted over wires. Almost simultaneously, Hertz was developing techniques to prove that electric sparks could be transmitted over long distances by electromagnetic waves. His ideas were quickly commercialized by Marconi, whose development of radio telegraphy enabled information to be transmitted without wires using a discrete code, the

Morse code. By the turn of the century, the goal was radiotelephony, which was to permit speech communication to moving receivers such as ships and planes. It took only two more decades of amazing progress to bring this goal to fruition, no doubt spurred on by the military needs of World War I.

Early in the century came the crucial invention of the vacuum tube, by Flemming and deForest. At the same time, techniques and circuits for using electromagnetic waves to transmit speech were devised by researchers such as Fessenden and Armstrong, who also invented the radio amplifier. The remaining hardware was developed early in the second decade, which saw the introduction of the heterodyne receiver, by Fessenden, and of oscillators, by Nicholson, Colpits, and others. Radiotelephony became possible, and the first commercial broadcast transmitters were constructed by 1920.

The next two decades saw the extensive development of continuous waveform (cw) modulation techniques. At the beginning of this period, efforts shifted from increasing signal fidelity to reducing the ultimate limitation, the susceptibility to noise. This led to new modulation techniques —single sideband (SSB) and frequency modulation (FM). There was a great expansion of commercial mobile applications and an introduction of the new broadcast bands of FM radio, television, and citizens band radio. The field of cw communications was nearly fully realized by the early 1940s, but the impetus toward new directions had already begun.

The development of radar was motivated by World War II but had a major impact on communications. It was now clear that discrete decision-oriented techniques, such as those already used in telegraphy, are necessary to combat the effects of noise. Because of this, pulse communication techniques were developed. At the same time researchers including Wiener, Shannon, and Rice presented theories to explain the true capabilities and limitations of communication channels. Nevertheless, it took two technological revolutions in the late 1940s to bring modern discrete communication techniques to fruition. These were the development of the electronic digital computer at the Universities of Pennsylvania and Illinois in 1944 and the invention of the transistor at Bell Telephone Laboratory in 1947, which paved the way to modern integrated circuit technology. It was not just the possibility of developing cheap, simple hardware, but also the remarkable synergistic effect of these three technologies that resulted in ballistic missiles, satellites for deep-space exploration, communication satellites, high-speed computer networks, and other important applications. Each of these examples requires nearly noise-free communications. By the early 1960s the underlying theory of modern discrete communication systems was fully realized.

These techniques have, of course, made their way into the older technologies of telephone and radio. Present telephone systems accommodate discrete computer-to-computer links. Push-button dialing techniques are inherently discrete and are being used generally as a means of communicating with computers. In all likelihood, when the new broadband optical fiber channels are fully in place, telephony will become totally discrete. On the other hand, although commercial mobile radio hardware now accommodates discrete signals for such applications as environmental monitoring, most radio usage is likely to remain cw because of the large bandwidths generally required by discrete signaling techniques.

1.3 WHAT ARE DISCRETE COMMUNICATIONS?

Let us consider the two signals shown in Figures 1.1. The signal in Figure 1.1a might represent a measurement of some kind or a speech signal and is called an analog signal. The second (Figure 1.1b) is a series of pulses intended to transmit the binary language of computers (1s and 0s) from one computer to another. If these signals are transmitted over any distance, they are inevitably distorted by noise, as in Figures 1.2a and 1.2b, respectively. Since we can have no a priori information about the shape of analog signals, the noise cannot be distinguished from the analog signal at the receiver. On the other hand, for the discrete signal we expect only positive and negative pulses, and thus we can easily determine from Figure 1.2b the relevant "information" in the transmitted signal. This ability to decide which of a finite number of pulse types was transmitted can be accomplished with hardware and represents one advantage of discrete communications. Errors occur occasionally, however, and noise cannot be completely removed, but discrete information is more immune to noise than is analog information.

Usually we are not transmitting a series of binary messages but rather messages that take on one of M values. For example, the dialing information from the headset of a push-button telephone is a series of messages from a finite alphabet of 12 symbols. This telephone headset is increasingly used as a home computer terminal in addition to its normal dialing function. A possible signal would be a sinusoidal burst of one of 12 possible frequencies. This signal set, called *frequency position modulation* (FPM), is particularly practical because of the ease of designing an oscillator whose frequency is set by connecting one of 12 tuning capacitors with a button. However, while it is possible to determine which frequency is present, the decision mechanism is

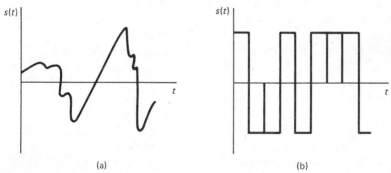

(a)   (b)

Figure 1.1   (a) An analog and (b) a discrete signal.

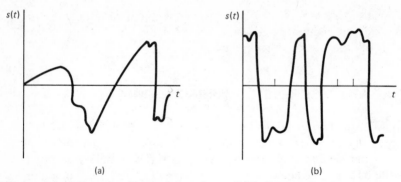

Figure 1.2   Effect of noise on (a) analog and (b) discrete signals.

complicated by the fact that the receiver does not know when to expect the pulses. This is an example of a pulse synchronization problem, which is present to some degree in all discrete communications. Indeed, the main hardware difficulties in discrete communication techniques are the various kinds of timing problems. We must spend some effort learning how to deal with pulse and other kinds of synchronization problems. Determining which signal was sent is relatively easy after we know when to look.

Even though messages rarely occur in binary form, we can use binary codes to represent them. We can encode up to 256 symbols using a series of eight binary pulses since there are $2^8 = 256$ different possible combinations. Thus the signal in Figure 1.1b (10010111) could represent a signal message or word. This technique is called *pulse code modulation* (PCM).

Analog signals can be converted to discrete signals by the process of sampling and quantization. Suppose that, instead of transmitting the analog signal in Figure 1.3, we transmitted the quantization interval in which each of the samples lie, as indicated by the value at the midpoint of the appropriate interval (see the crosses in Figure 1.3). The receiver can then reconstruct a reasonable replica of the analog signal $s(t)$ from this sequence of discrete values. We shall study the small errors inherent in these conversion processes in detail later in the text. Since no detector has infinite resolution and real analog signals are limited in bandwidth and have inherent measurement inaccuracies, these errors can be made insignificant. This ability to convert analog signals to discrete (or digital) ones is another reason for the importance of discrete communications. For example, if speech signals are sampled at a rate of 9000/s and quantized into 32 levels, a human listener cannot distinguish between the reconstructed signal and the original analog signal. If each sample is encoded into a five-pulse binary code, we have converted the analog speech signal into a digital PCM signal of 45,000 pulses/s.

Let us consider the problem of transmitting images from deep space, as in the Mariner and Voyager space probes. The images were originally scanned, as in television, to generate 200 lines of one-dimensional data. Each line was sampled 200 times, and each sample was quantized into 64 intensity

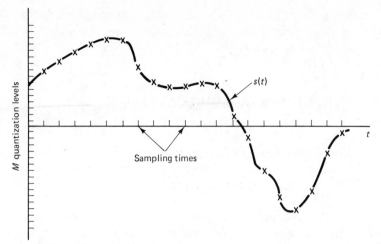

Figure 1.3 Analog to discrete conversion.

levels and then encoded into a six-bit binary code. The total time to convert this analog image into a digital signal of 240,000 binary pulses was approximately 1 min. If this signal were transmitted during this minute—that is, in real time—many errors would be made because of the extremely weak signal levels. For this particular problem, however, the signals were stored and transmitted over a period of many hours. This technique, called *time scaling*, is illustrated by Figure 1.4.

If we know we are looking at two long binary pulses, it is not too difficult to discern that the first pulse is positive and the second negative. Indeed hardware can integrate such signals and make decisions even more easily than human observers. To transmit these signals by electromagnetic waves requires some high-frequency signals, and in this case the binary pulses are sinusoidal bursts. This signal is called *phase shift keying* (PSK).

As a last brief example, let us consider the problem of sending binary digital signals for the purpose of guiding a ballistic missile. Although these signals may be strong and relatively noise free, errors could have drastic consequences and must be virtually eliminated. Because the information must

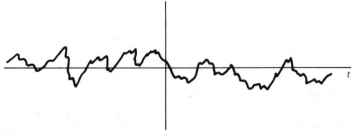

Figure 1.4 Two long, noisy binary pulses.

be sent in real time, time scaling is not available. For such problems, special error-correcting binary codes have been devised. These techniques are discussed in some detail in Chapters 9 and 10, along with all the other examples considered here.

1.4 CONSTRAINTS OF THE COMMUNICATIONS CHANNEL

This text is concerned with various aspects of the problem of transmitting information from a source to a receiver, usually at a considerable distance. Let us divide up the transmission path according to the block diagram of Figure 1.5.

There are many kinds of inputs that we must consider. For example, the input can be a continuous acoustical waveform arising from speech. The first transducer is a device that converts such signals into an electrical signal $m(t)$, and transducer 4 performs the inverse function if needed. The encoder is the transmitter hardware that converts the source signal $m(t)$ into the electrical signal $s(t)$ that is to be transmitted. This hardware is called a *modulator* when cw techniques are employed and an *encoder* when discrete signals are used. The choice of which encoder to use depends surprisingly little on the type of source being considered. What is important is the bandwidth of $m(t)$ if it is an analog signal or the message rate of $m(t)$ if it is discrete. The power or energy available at the transmitter for the encoded signal $s(t)$ is also important. These considerations are called the *source constraints*.

Source Constraints
1. the bandwidth or message rate of source $m(t)$
2. the power or energy of encoded signal $s(t)$

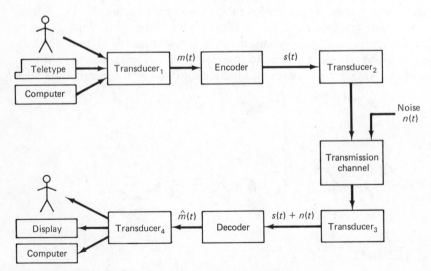

Figure 1.5   Components of communication systems.

If the transmission channel is a wire or cable, the signal $s(t)$ is transmitted directly and transducers 2 and 3 of Figure 1.5 are not needed. Often the channel requires signals that are not electrical but acoustic, electromagnetic, or optical and the transducers are needed to convert the form of energy. For example, if the channel is the atmosphere or a microwave link and electromagnetic signals are required, the transducers are antennas. As another example, if the channel is an optical fiber and optical signals are required, transducer 2 is a laser and transducer 3 could be a diode that is light sensitive. The channel medium does not directly affect the type of encoder that should be used. What is important is the frequency range or bandwidth of the channel and the type of noise that is introduced to the channel. In most cases the noise is a purely additive distortion. While we shall occasionally consider such effects as the "fading" of signal intensity, which can be visualized as a multiplicative distortion, this text will concentrate almost exclusively on the additive noise model. The only variables are the intensity and statistical properties of this noise. Thus there are only two channel constraints:

Channel Constraints
1. the frequency range or bandwidth of the channel
2. the amount and statistical properties of the additive noise

Some of the constraints may be due to regulatory agencies rather than natural causes. Specifically the source constraint of signal power and the channel bandwidth constraint may be imposed by regulatory agencies. The goal of this text is to develop the tools needed to analyze various encoding schemes that meet the constraints in order to determine their susceptibility to noise. Whether the constraints are natural or imposed makes little difference.

For cw techniques the decoder or demodulator of Figure 1.5 is the receiver hardware that performs the inverse operation of the modulator. For discrete techniques the decoder is a decision mechanism whose output $\hat{m}(t)$ is an estimate of the input message $m(t)$. This represents the principal difference between analog and discrete communications. The susceptibility of the encoder to noise depends to a large extent on the decoder used. Thus the analysis of encoding schemes must be based on both the encoder and decoder hardware. Such an analysis should lead to methods for determining the "best" decoder for a given encoder.

Finally, there may be additional important constraints due to practical considerations of the communication system; these can be called system constraints. The existence or absence of a real-time constraint has a major impact. If not constrained to operate in real time, the signals can be time scaled, which requires memory capability in both the encoder and decoder. One of the most important system constraints is that of cost or hardware complexity. Hardware costs cannot usually be quantitatively evaluated. As a result, we generally compare encoding and decoding schemes strictly on the basis of performance in additive noise subject to the constraints of the source and channel. The practical hardware differences are examined separately, and the actual choice is left to the system designer.

REFERENCES

A number of excellent references consider the historical development of communications. A few of these are listed here.

1. A. Still, *Communication Through the Ages*, Holt, NY, 1946.
2. A. B. Carlson, *Communication Systems*, McGraw-Hill, NY, 1975.
3. M. D. Fagen, Ed., *A History of Engineering and Science in the Bell System; the Early Years*, Bell Telephone Laboratories, Murray Hill, NJ, 1975.

Chapter 2

Frequency Domain Concepts

2.1 INTRODUCTION

We are all familiar with visualizing signals in the time domain and with characterizing them as functions of time. For example, we may let $x(t)$ represent the *value* of a signal at time t (perhaps in units of volts) and let $x^2(t)$ represent the instantaneous *power* of a signal at time t (perhaps in units of volts squared). Any instantaneous measurement of a signal (such as power) is then characterized by some $F[x(t)]$. Communication engineers must feel just as comfortable visualizing and characterizing signals in the frequency domain as in the time domain. The goal of this chapter is to develop the analytic techniques necessary for such a visualization.

There are three categories of signals, each of which requires a different characterization in the frequency domain. The first category contains those signals whose energy E, defined by

$$E = \int_{-\infty}^{\infty} x^2(t) \, dt \tag{2.1}$$

is finite. These signals include bounded signals (i.e., signals that satisfy $|x(t)| < A$ for some A) of finite time duration, as well as those that decay to zero sufficiently fast for their energy to be finite. Examples are shown in Figure 2.1a. The frequency characterization that we seek for these signals will be based on their energy.

The second category comprises *periodic* signals that repeat every T s. An example is shown in Figure 2.1b. Important characterizing measurements for these infinite-energy signals are time averages. The average value and average power of periodic signals are defined by

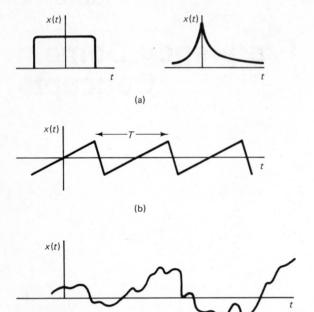

Figure 2.1 Signal categories: (a) finite-energy signals, (b) a periodic signal, and (c) a random signal.

$$\langle x(t) \rangle = \frac{1}{T} \int_a^{T+a} x(t)\, dt \tag{2.2}$$

and

$$\langle x^2(t) \rangle = \frac{1}{T} \int_a^{T+a} x^2(t)\, dt \tag{2.3}$$

where a can be chosen arbitrarily. The interval T is called the period of the signal. The frequency domain characterization for these signals will be based on their average powers.

The final category of signals includes those nondeterministic, infinite energy signals that are random in nature as the example shown in Figure 2.1c. We are most interested in this latter category, which includes speech, measured data of all kinds, all signals used for communication purposes, and noise disturbances. For many of these signals we can still measure time average values and the average power, defined by, respectively,

$$\langle x(t) \rangle = \lim_{T \to \infty} \frac{1}{T} \int_a^{T+a} x(t)\, dt \tag{2.4}$$

and

$$\langle x^2(t)\rangle = \lim_{T\to\infty} \frac{1}{T}\int_a^{T+a} x^2(t)\, dt \tag{2.5}$$

Unfortunately such time averages do not exist for all random signals. The limits may never converge, or they may converge to different values when the measurement is repeated. We shall develop a frequency domain concept for those random signals whose average value and average power exist.

We begin with the first two deterministic signal categories.

2.2 FOURIER TRANSFORMS AND FINITE-ENERGY SIGNALS

The Fourier transform of a signal $x(t)$ is denoted $X(f)$ and defined by

$$X(f) = \int_{-\infty}^{\infty} x(t)e^{-j2\pi ft}\, dt \tag{2.6}$$

The transform may be either a real or a complex function of the frequency f.

EXAMPLE 2.1 ■

Let us determine the Fourier transform of the rectangular pulse shown in Figure 2.2:

$$X(f) = a\int_{-W/2}^{W/2} e^{-j2\pi ft}\, dt = a\frac{e^{j\pi fW} - e^{-j\pi fW}}{j\pi fW}$$

$$= aW\frac{\sin \pi fW}{\pi fW} = aW \operatorname{sinc} fW$$

where sinc x is defined as $\sin(\pi x)/\pi x$. This transform is real and is plotted in Figure 2.3.　　　　　　　　　　　　　　　　　　　■ ■

Let us now set $a = 1/W$ (i.e., we consider a pulse of unit area) and let $W \to 0$. In this way the pulse is made to approach the impulse function $\delta(t)$, which is defined by the following properties:

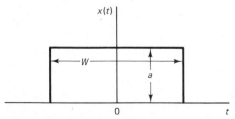

$x(t)$

W

a

0

t

Figure 2.2　Rectangular pulse.

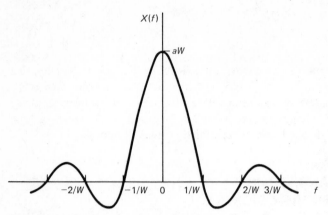

Figure 2.3 Fourier transform of a rectangular pulse.

1. $\delta(t) = \begin{cases} \infty & t = 0 \\ 0 & \text{otherwise} \end{cases}$ (2.7)

2. $\int_{-\infty}^{\infty} \delta(t - t_i)\, dt = 1$ (2.8)

3. $\int_{-\infty}^{\infty} f(t)\delta(t - t_i)\, dt = f(t_i)$ (2.9)

We observe from Figure 2.3 that as $W \to 0$ the transform approaches the constant 1. Alternatively,

$$\int_{-\infty}^{\infty} \delta(t)e^{-j2\pi ft}\, dt = e^{-0} = 1 \tag{2.10}$$

Similarly, if we set $a = 1$ in Example 2.1 and let $W \to \infty$, we find that the pulse approaches the infinite-energy constant of unity. As $W \to \infty$, the $W \operatorname{sinc} fW$ function of Figure 2.3 approaches an impulse function $\delta(f)$. Thus

$$\int_{-\infty}^{\infty} 1e^{-j2\pi ft}\, dt = \lim_{W \to \infty} \int_{-W/2}^{W/2} e^{-j2\pi ft}\, dt = \lim_{W \to \infty} W \operatorname{sinc} fW = \delta(f) \tag{2.11}$$

The impulse function and the constant unity are said to be Fourier transform pairs.

We now show that the inverse Fourier transform is given by

$$x(t) = \int_{-\infty}^{\infty} X(f)e^{j2\pi ft}\, df \tag{2.12}$$

This inverse relation is developed as follows:

$$\int_{-\infty}^{\infty} X(f)e^{j2\pi ft}\, df = \int_{-\infty}^{\infty} e^{j2\pi ft} \left[\int_{-\infty}^{\infty} x(\xi)e^{-j2\pi f\xi}\, d\xi \right] df$$

$$= \int_{-\infty}^{\infty} x(\xi) \left[\int_{-\infty}^{\infty} e^{-j2\pi f(\xi - t)}\, df \right] d\xi$$

$$= \int_{-\infty}^{\infty} x(\xi)\delta(\xi - t)\, d\xi \quad \text{(from Equation 2.11)}$$

$$= x(t) \quad \text{(from Equation 2.9)}$$

Equations 2.6 and 2.12 define the Fourier transform pairs.

2.3 PROPERTIES OF FOURIER TRANSFORMS

We first show that the Fourier transform is linear, which means that the property of *superposition* holds:

Property 2.1. If $F_1(f)$ and $F_2(f)$ are Fourier transforms of $f_1(t)$ and $f_2(t)$, respectively, then $aF_1(f) + bF_2(f)$ is the Fourier transform of $af_1(t) + bf_2(t)$.

    Proof

$$\int_{-\infty}^{\infty} [af_1(t) + bf_2(t)]e^{-j2\pi ft}\, dt = a\int_{-\infty}^{\infty} f_1(t)e^{-j2\pi ft}\, dt + b\int_{-\infty}^{\infty} f_2(t)e^{-j2\pi ft}\, dt$$

$$= aF_1(f) + bF_2(f)$$

Property 2.2. If $x(t)$ and $X(f)$ are Fourier transform pairs, then $x(t - D)$ and $e^{-j2\pi fD}X(f)$ are also Fourier transform pairs.

    Proof

$$\int_{-\infty}^{\infty} x(t - D)e^{-j2\pi ft}\, dt = \int_{-\infty}^{\infty} x(\xi)e^{-j2\pi f(\xi+D)}\, d\xi \quad (\xi = t - D)$$

$$= e^{-j2\pi fD}\int_{-\infty}^{\infty} x(\xi)e^{-j2\pi f\xi}\, d\xi$$

$$= e^{-j2\pi fD}X(f)$$

Property 2.3. If $x(t)$ and $X(f)$ are Fourier transform pairs, then $d^n x(t)/dt^n$ and $(j2\pi f)^n X(f)$ are also Fourier transform pairs.

    Proof

$$\frac{d^n x(t)}{dt^n} = \frac{d^n}{dt^n}\left[\int_{-\infty}^{\infty} X(f)e^{j2\pi ft}\, df\right] \quad \text{(from Equation 2.12)}$$

$$= \int_{-\infty}^{\infty} X(f)\frac{d^n}{dt^n}(e^{j2\pi ft})\, df$$

$$= \int_{-\infty}^{\infty} (j2\pi f)^n X(f)e^{j2\pi ft}\, df$$

EXAMPLE 2.2 ■

The unit step function $u(t)$ is defined to be unity for $t \geq 0$ and 0 for $t < 0$. The derivative of this function is the impulse function $\delta(t)$. If $U(f)$ is the Fourier transform of $u(t)$, it follows from property 2.3 that

$$1 = (j2\pi f)U(f) \quad \text{or} \quad U(f) = \frac{1}{j2\pi f} \tag{2.13}$$

If $r(t)$ is the rectangular pulse defined by $r(t) = u(t) - u(t - W)$, it follows from properties 2.1 and 2.2 that

$$
\begin{aligned}
R(f) &= \frac{1}{j2\pi f} - e^{-j2\pi fW}\frac{1}{j2\pi f} \\
&= \frac{1 - e^{-j2\pi fW}}{j2\pi f} \\
&= \frac{e^{-j\pi fW}(e^{j\pi fW} - e^{-j\pi fW})}{j2\pi f} \\
&= e^{-j\pi fW}\frac{\sin \pi fW}{\pi f} \\
&= e^{-j\pi fW}W \operatorname{sinc} fW
\end{aligned}
$$

■ ■

The next-to-last step utilized Euler's expansion for the complex exponential ($e^{\pm jx} = \cos x \pm j \sin x$). We could have obtained this result directly from property 2.2 and Example 2.1.

The next property concerns even and odd symmetry. A function $x(t)$ is said to have even symmetry and $y(t)$ is said to have odd symmetry if

$$x(t) = x(-t) \quad \text{and} \quad y(t) = -y(-t) \tag{2.14}$$

Examples of functions with even symmetry are the cosine function ($\cos at$) and the square wave function if the time origin is chosen halfway between the discontinuities. Examples of odd functions are the sine function ($\sin at$) and square waves if the time origin is chosen at one of the discontinuities. Some important properties of functions with even or odd symmetry are summarized in the following equations where the subscripts e and o refer to even and odd functions, respectively:

$$
\begin{aligned}
x_e(t)y_e(t) &= z_e(t) \\
x_o(t)y_o(t) &= z_e(t) \\
x_e(t)y_o(t) &= z_o(t)
\end{aligned} \tag{2.15}
$$

Furthermore, it is evident that

$$\int_{-A}^{A} x_e(t)\, dt = 2\int_{0}^{A} x_e(t)\, dt$$

and   (2.16)

$$\int_{-A}^{A} y_o(t) \, dt = 0$$

Property 2.4. If $x(t)$ is real and even, its transform $X(f)$ is real and even. If $x(t)$ is real and odd, its transform $X(f)$ is imaginary and odd.

Proof: Using Euler's expansion, the transform becomes

$$X(f) = \int_{-\infty}^{\infty} x(t) \cos 2\pi ft \, dt - j \int_{-\infty}^{\infty} x(t) \sin 2\pi ft \, dt$$

If $x(t)$ is even, it follows from Equations 2.15 and 2.16 that

$$X(f) = 2\int_{0}^{\infty} x(t) \cos 2\pi ft \, dt$$

and

$$X(-f) = 2\int_{0}^{\infty} x(t) \cos(-2\pi ft) \, dt = X(f)$$

Similarly, if $x(t)$ is odd,

$$X(f) = -2j\int_{0}^{\infty} x(t) \sin 2\pi ft \, dt$$

and

$$X(-f) = -2j\int_{0}^{\infty} x(t) \sin(-2\pi ft) \, dt = -X(f)$$

We shall be particularly concerned with even functions, where the Fourier transform equations become

$$X(f) = 2\int_{0}^{\infty} x(t) \cos 2\pi ft \, dt$$

and   (2.17)

$$x(t) = 2\int_{0}^{\infty} X(f) \cos 2\pi ft \, df$$

Some Fourier transform pairs for real and even functions are given in Table 2.1. The variables ξ and η are used and, because of the complete symmetry of Equations 2.17, either variable can represent the time t or frequency f.

The remaining properties we require deal with the response of filters to signals.

TABLE 2.1   FOURIER TRANSFORM PAIRS FOR REAL EVEN FUNCTIONS

$f(\xi)$

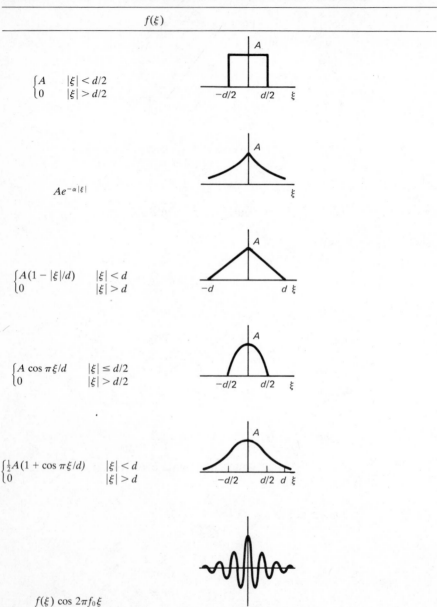

$\begin{cases} A & \lvert\xi\rvert < d/2 \\ 0 & \lvert\xi\rvert > d/2 \end{cases}$	
$Ae^{-\alpha\lvert\xi\rvert}$	
$\begin{cases} A(1 - \lvert\xi\rvert/d) & \lvert\xi\rvert < d \\ 0 & \lvert\xi\rvert > d \end{cases}$	
$\begin{cases} A\cos \pi\xi/d & \lvert\xi\rvert \le d/2 \\ 0 & \lvert\xi\rvert > d/2 \end{cases}$	
$\begin{cases} \frac{1}{2}A(1 + \cos \pi\xi/d) & \lvert\xi\rvert < d \\ 0 & \lvert\xi\rvert > d \end{cases}$	
$f(\xi) \cos 2\pi f_0\xi$	

2.4 FILTER TRANSFER FUNCTIONS

In this section we consider systems that are linear (i.e., satisfy the superposition principle) and time invariant [i.e., if $y(t)$ is the response to $x(t)$, then $y(t - D)$ is the response to $x(t - D)$ for any D]. The most general input–output model for such a system is that of the convolution integral:

$$F(\eta)$$

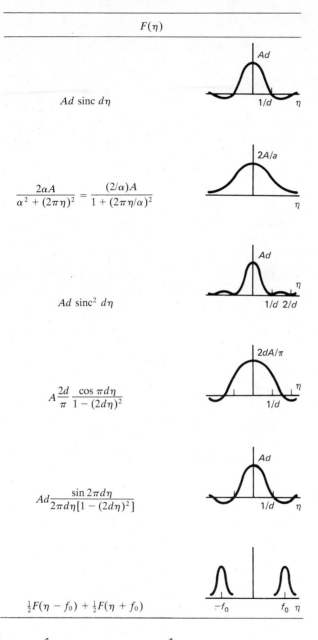

Ad sinc $d\eta$	
$\dfrac{2\alpha A}{\alpha^2 + (2\pi\eta)^2} = \dfrac{(2/\alpha)A}{1 + (2\pi\eta/\alpha)^2}$	
Ad sinc^2 $d\eta$	
$A\dfrac{2d}{\pi}\dfrac{\cos \pi d\eta}{1 - (2d\eta)^2}$	
$Ad\dfrac{\sin 2\pi d\eta}{2\pi d\eta[1 - (2d\eta)^2]}$	
$\tfrac{1}{2}F(\eta - f_0) + \tfrac{1}{2}F(\eta + f_0)$	

$$y(t) = \int_{-\infty}^{\infty} h(\xi)x(t - \xi)\, d\xi = \int_{-\infty}^{\infty} x(\xi)h(t - \xi)\, d\xi \tag{2.18}$$

The weight $h(t)$ is called the impulse response since it is the response to an impulse function input (see Equation 2.9). A system is said to be *causal* if its response does not precede its input or, equivalently, if $h(t) = 0$ for $t < 0$. For causal systems Equation 2.18 can be written

$$y(t) = \int_0^\infty h(\xi)x(t - \xi)\, d\xi = \int_{-\infty}^t x(\xi)h(t - \xi)\, d\xi \tag{2.19}$$

A less general model for linear, time-invariant, causal systems, but one that is often found for systems such as electrical networks, is that of the linear differential equation model with constant coefficients. A typical model of this type is

$$b_n \frac{d^n y(t)}{dt^n} + b_{n-1} \frac{d^{n-1} y(t)}{dt^n} + \cdots + b_0 y(t) = a_m \frac{d^m x(t)}{dt^m} + \cdots + a_0 x(t) \tag{2.20}$$

where the order of the model is the largest value of n or m. Either model leads to the following important property of Fourier transforms.

Property 2.5. If $X(f)$ and $Y(f)$ are the Fourier transforms of the input and output of a linear, time-invariant system, then

$$Y(f) = H(f)X(f) \tag{2.21}$$

where $H(f)$ is the Fourier transform of the impulse response $h(t)$ and is called the *transfer function*.

Proof: We take the Fourier transform of the convolution model (Equation 2.18 or 2.19):

$$Y(f) = \int_{-\infty}^{\infty} \left[\int_{-\infty}^{\infty} h(\xi)x(t - \xi)\, d\xi \right] e^{-j2\pi ft}\, dt$$

$$= \int_{-\infty}^{\infty} h(\xi)e^{-j2\pi f\xi} \left[\int_{-\infty}^{\infty} x(t - \xi)e^{-j2\pi f(t-\xi)}\, dt \right] d\xi$$

Setting $t - \xi = \eta$, we obtain

$$Y(f) = \int_{-\infty}^{\infty} h(\xi)e^{-j2\pi f\xi} \left[\int_{-\infty}^{\infty} x(\eta)e^{-j2\pi f\eta}\, d\eta \right] d\xi$$

$$= X(f)\int_{-\infty}^{\infty} h(\xi)e^{-j2\pi f\xi}\, d\xi = X(f)H(f)$$

Alternatively, taking the Fourier transform of the differential equation model (Equation 2.20) and using property 2.3, we have

$$[b_n(j2\pi f)^n + \cdots + b_0]Y(f) = [a_m(j2\pi f)^m + \cdots + a_0]X(f)$$

Rearranging terms,

$$Y(f) = H(f)X(f)$$

where

$$H(f) = \sum_{k=0}^{m} a_k(j2\pi f)^k \bigg/ \sum_{k=0}^{n} b_k(j2\pi f)^k \tag{2.22}$$

When the input is an impulse function, $X(f) = 1$. Therefore, $Y(f) = H(f)$, which must be the Fourier transform of the impulse response.

EXAMPLE 2.3 ■

A first-order linear system could be modeled by

$$A\frac{dy(t)}{dt} + y(t) = x(t) \tag{2.23}$$

where the coefficient A is called the time constant. The response to a unit step input $[x(t) = u(t)]$ can be seen to be $y(t) = (1 - e^{-t/A})u(t)$ by substituting this answer back into Equation 2.23 and observing that it fits (i.e., it is a legitimate solution). Since the impulse function is the derivative of $u(t)$, we would expect the impulse response to be the derivative of the step response, $h(t) = (1/A)e^{-t/A}u(t)$. Once again we can verify this solution by substituting into Equation 2.23:

$$x(t) = \delta(t) \quad \text{and} \quad y(t) = \frac{1}{A}e^{-t/A}u(t)$$

The transfer function of this system must be

$$H(f) = \int_0^\infty \frac{1}{A}e^{-t/A}e^{-j2\pi ft}\, dt = \frac{-e^{-(1/A+j2\pi f)t}}{A(1/A + j2\pi f)}\bigg|_{t=0}^{t=\infty}$$

$$= \frac{1}{1 + j2\pi fA}$$

Alternatively, taking the Fourier transform of Equation 2.23 directly yields

$$[A(j2\pi f) + 1]Y(f) = X(f)$$

and therefore

$$H(f) = \frac{Y(f)}{X(f)} = \frac{1}{1 + j2\pi fA} \qquad\qquad ■\ ■$$

The significance of the transfer function will begin to become apparent if we concern ourselves with sinusoidal inputs. Let us employ the common simplifying technique of using the complex input $e^{j2\pi ft}$ instead of a sinusoid. Because of the superposition principle, the real part of the complex response will be the response to the real part of the input $(\cos 2\pi ft)$ and the imaginary part of the response is the response to $\sin 2\pi ft$. From the convolution model (Equation 2.18),

$$y(t) = \int_{-\infty}^{\infty} h(\xi)e^{j2\pi f(t-\xi)}\, d\xi$$

$$= e^{j2\pi ft}\int_{-\infty}^{\infty} h(\xi)e^{-j2\pi f\xi}\, d\xi = e^{j2\pi ft}H(f)$$

In polar form,

$$H(f) = |H(f)|\exp(j\angle H(f)) \tag{2.24}$$

where $|H(f)|$ is called the *magnitude* of the transfer function and $\angle H(f)$ is called its *phase*,

$$y(t) = |H(f)| \exp[j(2\pi ft + \angle H(f))] \tag{2.25}$$

We have shown that the response of a linear time-invariant system to a sinusoid is a sinusoid of the same frequency whose amplitude is multiplied by $|H(f)|$ and phase is shifted by $\angle H(f)$.

EXAMPLE 2.4 ■

Let us evaluate the magnitude of the transfer function for the first-order system of Example 2.3:

$$|H(f)| = 1/\sqrt{1 + (2\pi fA)^2}$$

A plot of the transfer function, as magnitude versus frequency, is given in Figure 2.4. This system passes low-frequency inputs $[f < 1/(2\pi A)]$ relatively unattenuated while simultaneously attenuating high-frequency inputs $[f \gg 1/(2\pi A)]$; it is called a low-pass filter. The bandwidth of a filter is defined, arbitrarily, to be the range of frequencies (not including negative frequencies, which are plotted only for analytic convenience) for which the attenuation is no greater than $1/\sqrt{2}$ or in this case $1/2\pi A$ Hz. ■ ■

EXAMPLE 2.5 ■

An ideal low-pass filter $H_1(f)$ has an amplitude given by

$$|H_1(f)| = \begin{cases} 1 & |f| \le B \\ 0 & |f| > B \end{cases} \tag{2.26}$$

If the phase of this filter is zero for all frequencies, the filter impulse response is determined from Table 2.1 to be

$$h(t) = 2B \text{ sinc } 2Bt \tag{2.27}$$

Since $h(t)$ is not zero, for $t < 0$, this filter is not causal and cannot be realized. Suppose the phase were linear $(\angle H(f) = -2\pi Df)$. This would cause a delay of D s for any sinusoid whose frequency is inside the pass band, as well as the impulse response $[h(t) = 2B \text{ sinc } 2B(t - D)]$, which is sketched in Figure 2.5. This filter is still not causal, but it can be

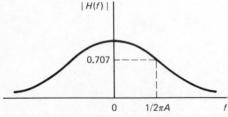

Figure 2.4 Amplitude of a low-pass first-order system.

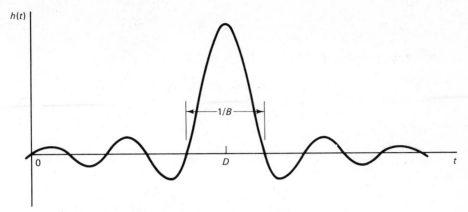

Figure 2.5 Impulse response of an ideal low-pass filter.

closely approximated with a causal system for large D ($D \gg 1/B$) by truncation, so that $h(t) = 0$ for $t < 0$. The transfer function magnitude of such a causal approximation is shown in Figure 2.6. ■ ■

2.5 ENERGY IN THE FREQUENCY DOMAIN

The last property of Fourier transforms that we shall need is *Parseval's theorem*:

Property 2.6. If $x(t)$ is a finite energy signal and $X(f)$ is its Fourier transform, then

$$\int_{-\infty}^{\infty} x^2(t) \, dt = \int_{-\infty}^{\infty} |X(f)|^2 \, df \tag{2.28}$$

Proof: Using the inverse transform (Equation 2.12), we have

$$x^2(t) = \left[\int_{-\infty}^{\infty} X(f)e^{j2\pi ft} \, df \right]^2$$

$$= \int_{-\infty}^{\infty} \int_{-\infty}^{\infty} X(f)X(\xi)e^{j2\pi(f+\xi)t} \, df \, d\xi$$

Calculating the energy of $x(t)$, we obtain

Figure 2.6 Causal approximation to $H_l(f)$.

$$\int_{-\infty}^{\infty} x^2(t)\, dt = \int_{-\infty}^{\infty} \int_{-\infty}^{\infty} \int_{-\infty}^{\infty} X(f) X(\xi) e^{j2\pi(f+\xi)t}\, df\, d\xi\, dt$$

Integrating with respect to t first, we have

$$\int_{-\infty}^{\infty} e^{j2\pi(f+\xi)t}\, dt = \delta(f + \xi) \qquad \text{(see Equation 2.11)}$$

Therefore,

$$\int_{-\infty}^{\infty} x^2(t)\, dt = \int_{-\infty}^{\infty} X(f) \left[\int_{-\infty}^{\infty} X(\xi)\, \delta(f + \xi)\, d\xi \right] df$$

$$= \int_{-\infty}^{\infty} X(f) X(-f)\, df \qquad \text{(see Equation 2.9)}$$

But $X(-f)$ is the conjugate of $X(f)$; hence $X(f)X(-f)$ is equivalent to $|X(f)|^2$, which proves Parseval's theorem.

We shall show that $|X(f)|^2$ characterizes the energy of $x(t)$ in the frequency domain. More specifically, we define $E_x(f)$, the *energy density* of a signal $x(t)$, and show that

$$E_x(f) = |X(f)|^2 \tag{2.29}$$

For a function $E_x(f)$ to be an energy density of a signal $x(t)$, it must satisfy the following conditions:

1. $E_x(f)$ must be real and nonnegative and must have even symmetry about the origin.

2. $\int_{-\infty}^{\infty} E_x(f)\, df = $ total energy $= \int_{-\infty}^{\infty} x^2(t)\, dt$.

3. $2\int_{f_1}^{f_2} E_x(f)\, df = \int_{f_1}^{f_2} E_x(f)\, df + \int_{-f_2}^{-f_1} E_x(f)\, df$ is the energy in the frequency band $[f_1, f_2]$.

The first condition, which is automatically satisfied by $|X(f)|^2$, is needed because energy is real and positive and because negative frequencies, which have no physical meaning, are included only for analytic convenience. We could avoid negative frequencies, if necessary, by defining the energy density as $2|X(f)|^2$ for $f \geq 0$ and 0 for $f < 0$. The second condition holds by Parseval's theorem (property 2.6). The third condition is verified by first recognizing that from property 2.5 (Equation 2.21)

$$|Y(f)|^2 = |H(f)|^2 |X(f)|^2$$

or, from Equation 2.29,

$$E_y(f) = E_x(f) |H(f)|^2 \tag{2.30}$$

Let us now assume that a finite-energy signal $x(t)$ is passed through an ideal bandpass filter that permits energy in the frequency range $[f_1, f_2]$ to be passed and completely attenuates the energy that lies outside this range or

$$|H(f)| = \begin{cases} 1 & f_1 < |f| < f_2 \\ 0 & \text{elsewhere} \end{cases}$$

From Equation 2.30,

$$E_y(f) = \begin{cases} E_x(f) & f_1 < |f| < f_2 \\ 0 & \text{elsewhere} \end{cases}$$

and, from Parseval's theorem,

$$\int_{-\infty}^{\infty} y^2(t)\, dt = \int_{-\infty}^{\infty} E_y(f)\, df = 2\int_{f_1}^{f_2} E_x(f)\, df$$

This is the meaning of the third condition because the energy of a signal in the frequency range $[f_1, f_2]$ is the energy in the output when the signal is passed through an ideal bandpass filter.

EXAMPLE 2.6 ■

Let us consider the finite-energy signal shown in Figure 2.7. We have seen that the Fourier transform of the flat-topped pulse is

$$X(f) = aW\frac{\sin \pi fW}{\pi fW} = aW \operatorname{sinc} fW$$

Hence, from Equation 2.29,

$$E_x(f) = a^2 W^2 \operatorname{sinc}^2 fW \tag{2.31}$$

which is shown in Figure 2.8. Thus the energy of a flat-topped pulse is concentrated in the low-frequency range. It can be shown that 90% of the energy lies in the range $[0, 1/W]$ Hz, 95% in the range $[0, 2/W]$ Hz, and 98% in the range $[0, 3/W]$ Hz. ■ ■

We can define the *bandwidth of a signal* as the range of frequencies in which "most" of the energy lies. With this rather ambiguous definition we may say the bandwidth of a flat-topped pulse is m/W Hz, where $1 \le m \le 3$, depending on how "most" is interpreted. Perhaps a better definition, from a conceptual viewpoint, is the width of the filter or channel that is necessary for

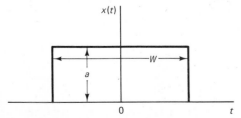

Figure 2.7 Flat-topped pulse.

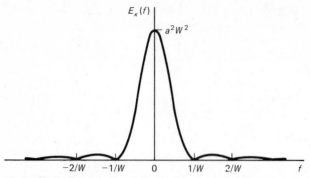

Figure 2.8   Energy density of a flat-topped pulse.

the output to look "similar" to the input. This definition is, unfortunately, even more ambiguous because it depends on the shape of the transfer function of the channel as well as on the interpretation of the word "similar." In particular, Figures 2.9 and 2.10 show the response of an ideal low-pass filter with width m/W Hz and a first-order filter with half-power width m/W Hz to a flat-topped pulse. We see that, regardless of the definition, the bandwidth of a flat-topped pulse is m/W Hz, where $1 \leq m \leq 3$.

Actually, there is no value in defining bandwidth in an unambiguous manner. How similar a received signal should be to the transmitted signal depends, in large part, on how the signal is to be processed. Thus the required channel bandwidth depends on the receiver implementation. It is more informative to say the bandwidth of a flat-topped pulse is m/W Hz, $1 \leq m \leq 3$, where the value of m depends on how the pulse is to be processed.

2.6 PERIODIC SIGNALS AND THE FOURIER SERIES

A signal that repeats every T s is called periodic, and T is called the *period of the signal*. Such a signal can be represented by an infinite sum of sinusoidal components called a *Fourier series*:

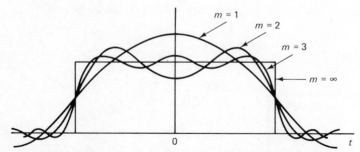

Figure 2.9   Response of an ideal filter with bandwidth m/W H$_z$ to a flat-topped pulse.

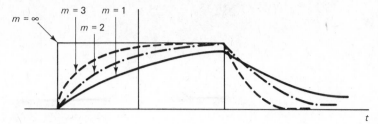

Figure 2.10 Response of a first-order filter with half-power bandwidth of m/W Hz to a flat-topped pulse.

$$x(t) = C_0 + \sum_{n=1}^{\infty} C_n \cos\left(n\frac{2\pi}{T}t + \theta_n\right)$$ (2.32)

or

$$x(t) = \frac{a_0}{2} + \sum_{n=1}^{\infty}\left[a_n \cos\left(n\frac{2\pi}{T}t\right) + b_n \sin\left(n\frac{2\pi}{T}t\right)\right]$$ (2.33)

C_0 (or $a_0/2$) is the average value of the signal and C_n and θ_n are the amplitude and phase, respectively, of the various sinusoidal components: $C_n^2 = a_n^2 + b_n^2$ and $\theta_n = -\tan^{-1}(b_n/a_n)$. The first term ($n = 1$) is called the *fundamental* component, and the others are called *harmonics*. One of the key advantages of the Fourier series representation is that, from the principle of superposition, the response to a linear system with transfer function $H(f)$ is

$$y(t) = C_0|H(0)| + \sum_{n=1}^{\infty} C_n|H(nf_0)| \cos\left[n2\pi f_0 t + \theta_n + \angle H(nf_0)\right]$$ (2.34)

where $f_0 = 1/T$ and is called the fundamental frequency. While the representation of Equation 2.32 is most convenient from a conceptual viewpoint, the equivalent representation of Equation 2.33 is more convenient for determining the unknown coefficients.

Let us first prove the *orthogonality* principle of sinusoids, which is defined by

$$\frac{2}{T}\int_a^{T+a} \sin\left(\frac{2\pi}{T}nt\right) \sin\left(\frac{2\pi}{T}mt\right) dt = \delta_{nm}$$ (2.35)

$$\frac{2}{T}\int_a^{T+a} \cos\left(\frac{2\pi}{T}nt\right) \cos\left(\frac{2\pi}{T}mt\right) dt = \delta_{nm}$$ (2.36)

$$\frac{2}{T}\int_a^{T+a} \sin\left(\frac{2\pi}{T}nt\right) \cos\left(\frac{2\pi}{T}mt\right) dt = 0 \qquad \text{for all } m \text{ and } n$$ (2.37)

where

$$\delta_{nm} = \begin{cases} 1 & n = m \\ 0 & n \neq m \end{cases}$$

and a can be chosen arbitrarily. Substituting the trigonometric identity $2 \sin A \times \sin B = \cos(A - B) - \cos(A + B)$ into Equation 2.35 results in

$$\frac{1}{T}\int_a^{T+a} \cos\frac{2\pi}{T}(n-m)t\,dt + \frac{1}{T}\int_a^{T+a} \cos\frac{2\pi}{T}(n+m)t\,dt = \delta_{nm}$$

With the exception of the case $n = m$, where the first integral is

$$\frac{1}{T}\int_a^{T+a} dt = 1$$

all of the integrands are sinusoids and the integral of a sinusoid over an integral number of cycles is zero. Equation 2.36 can be proven in a similar fashion with the identity $2 \cos A \cos B = \cos(A - B) + \cos(A + B)$. Using the identity $2 \sin A \cos B = \sin(A + B) - \sin(A - B)$, one sees that the integral of Equation 2.37 is always zero.

Consider now the integral

$$\frac{2}{T}\int_a^{T+a} x(t) \cos\frac{2\pi}{T}nt\,dt$$

Substituting the expansion of Equation 2.33 for $x(t)$ into the integral and using the orthogonality properties of sinusoids, we are led (see Problem 2.13) to

$$a_n = \frac{2}{T}\int_a^{T+a} x(t) \cos\frac{2\pi}{T}nt\,dt \qquad (2.38)$$

Similarly,

$$b_n = \frac{2}{T}\int_a^{T+a} x(t) \sin\frac{2\pi}{T}nt\,dt \qquad (2.39)$$

These equations enable us to find the coefficients a_n and b_n (and hence C_n and θ_n). One is free to choose any value for a, but usually one sets $a = 0$ or $-T/2$. Finally, we observe that C_0 is the average or dc value of the signal $x(t)$:

$$C_0 = \frac{a_0}{2} = \frac{1}{T}\int_a^{T+a} x(t)\,dt \qquad (2.40)$$

If we set $a = -T/2$, then it follows from Equations 2.15 and 2.16 (the properties of even and odd functions, respectively) that for even signals all the b_n coefficients vanish and for odd signals all the a_n coefficients vanish. Hence, if the signal has even or odd symmetry, by proper choice of the time origin half the coefficients can be made to vanish, and the Fourier series will have a simple form.

EXAMPLE 2.7: The dc Supply or Envelope Detector ■

Let us examine the problem of constructing a constant (or dc) signal source. Because electrical energy is transmitted generally as a 60-Hz sinusoidal signal, this problem is the same as constructing a device which converts a sinusoidal signal into a dc signal. Such a device must include a nonlinearity. Let us examine the system in Figure 2.11 to determine its

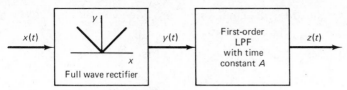

Figure 2.11 A dc supply or envelope detector.

usefulness for this purpose. The output of the rectifier, called a *rectified sine wave*, is a periodic signal that can be represented by a Fourier series where the frequency of the fundamental component is twice that of the input sinusoid (see Figure 2.12). Thus, if the input sinusoid is $\cos 2\pi f_0 t$, the output of the rectifier is

$$C_0 + \sum_{n=1}^{\infty} C_n \cos(4\pi n f_0 t + \theta_n)$$

If this signal is then passed through a low-pass filter which passes the dc, or average, value unattenuated and greatly attenuates the sinusoidal components, then the conversion will be accomplished.

By choosing the time origin as in Figure 2.12, the signal has even symmetry and $b_n = 0$ for all n. Recall that $T_0 = 2T$. Solving for a_n then yields

$$
\begin{aligned}
a_n &= \frac{2}{T}\int_{-T/2}^{T/2} x(t)\cos\frac{2\pi}{T}nt\,dt = \frac{4}{T_0}\int_{-T_0/4}^{T_0/4}\cos\left(\frac{2\pi}{T_0}t\right)\cos\left(\frac{2\pi}{T_0}2nt\right)dt \\
&= \frac{2}{T_0}\int_{-T_0/4}^{T_0/4}\cos\left[\frac{2\pi}{T_0}(2n+1)t\right]dt + \frac{2}{T_0}\int_{-T_0/4}^{T_0/4}\cos\left[\frac{2\pi}{T_0}(2n-1)t\right]dt \\
&= \frac{1}{\pi}\left\{\frac{2\sin[(\pi/2)(2n+1)]}{(2n+1)} + \frac{2\sin[(\pi/2)(2n-1)]}{(2n-1)}\right\} \\
&= \frac{2}{\pi}\frac{-2\cos n\pi}{(2n+1)(2n-1)} = -\frac{2}{\pi}(-1)^n\frac{2}{(2n+1)(2n-1)}
\end{aligned}
\qquad (2.41)
$$

Substituting this result (along with $b_n = 0$) into Equation 2.33 gives the Fourier series for the rectified sinusoid,

$$y(t) = (2/\pi)(1 + \tfrac{2}{3}\cos 4\pi f_0 t - \tfrac{2}{15}\cos 8\pi f_0 t + \tfrac{2}{35}\cos 12\pi f_0 t - \cdots) \quad (2.42)$$

Consider now the first-order low-pass filter of Example 2.4; its transfer function is $H(f) = 1/(1 + j2\pi fA)$. If the rectified sinusoid is

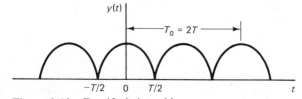

Figure 2.12 Rectified sinusoid.

passed through this filter, its components are attenuated by the factor $|H(2nf_0)|$. Thus

$$z(t) = \frac{2}{\pi} + \frac{2}{\pi}\left[\frac{2}{3\sqrt{1 + (4\pi f_0 A)^2}} \cos 4\pi f_0 t \right.$$
$$\left. - \frac{2}{15\sqrt{1 + (8\pi f_0 A)^2}} \cos 8\pi f_0 t + \cdots\right] \tag{2.43}$$

We want the second term, called the ripple, to be much smaller than the desired dc component, and this is achieved if $(4\pi f_0 A)^2 \gg 1$. ■ ■

2.7 AVERAGE POWER IN THE FREQUENCY DOMAIN FOR PERIODIC SIGNALS

Periodic signals do not have finite energy, and it is therefore more appropriate to consider the average power of such signals. As indicated earlier, average power is a well-defined notion for periodic signals. The average value of a periodic signal $x(t)$ is given by Equation 2.40 and equals C_0. Let us calculate the average power of $x(t)$ using the Fourier series representation of Equation 2.32:

$$\langle x^2(t)\rangle = \frac{1}{T}\int_0^T x^2(t)\,dt = C_0^2 + 2C_0 \sum_{n=1}^{\infty} C_n \frac{1}{T}\int_0^T \cos\left(n\frac{2\pi}{T}t + \theta_n\right)dt$$
$$+ \sum_{n=1}^{\infty}\sum_{m=1}^{\infty} C_n C_m \frac{1}{T}\int_0^T \cos\left(n\frac{2\pi}{T}t + \theta_n\right)\cos\left(m\frac{2\pi}{T}t + \theta_m\right)dt$$

Since $\int_0^T \cos[n(2\pi/T)t + \theta_n]\,dt = 0$ and

$$\frac{1}{T}\int_0^T \cos\left(n\frac{2\pi}{T}t + \theta_n\right)\cos\left(m\frac{2\pi}{T}t + \theta_m\right)dt = \tfrac{1}{2}\delta_{mn}$$

(by Equation 2.36), then

$$\langle x^2(t)\rangle = C_0^2 + \sum_{n=1}^{\infty}\sum_{m=1}^{\infty} C_n C_m \tfrac{1}{2}\delta_{nm}$$
$$= C_0^2 + \sum_{n=1}^{\infty}\frac{C_n^2}{2} = \left(\frac{a_0}{2}\right)^2 + \sum_{n=1}^{\infty}\frac{(a_n^2 + b_n^2)}{2} \tag{2.44}$$

Analogous to Equation 2.30 for finite-energy signals, the average power distribution for the output of a linear filter is given by

$$\langle y^2(t)\rangle = C_0^2|H(0)|^2 + \sum_{n=1}^{\infty}\frac{C_n^2}{2}|H(nf_0)|^2 \tag{2.45}$$

Equation 2.32 (or 2.33) can be interpreted as

$$x_N(t) = C_0 + \sum_{n=1}^{N} C_n \cos(2\pi n f_0 t + \theta_n) \to x(t) \qquad \text{as} \qquad N \to \infty \tag{2.46}$$

which means that the truncated Fourier series $x_N(t)$ converges to $x(t)$ as N increases beyond bound. The mean square value of the error signal $[x(t) - x_N(t)]$ is defined by

$$\langle [x(t) - x_N(t)]^2 \rangle = \left\langle \left[\sum_{n=N+1}^{\infty} C_n \cos(2\pi n f_0 t + \theta_n) \right]^2 \right\rangle = \sum_{n=N+1}^{\infty} \frac{C_n^2}{2} \tag{2.47}$$

Since the average power is finite, that is, $\sum_{n=1}^{\infty} (C_n^2/2) < \infty$, it follows that

$$\sum_{n=N+1}^{\infty} \frac{C_n^2}{2} \to 0 \qquad \text{as} \quad N \to \infty$$

Thus the mean square value of the error signal converges to zero. When this happens, the series is said to *converge in the mean*.

EXAMPLE 2.8 ■

Let us consider the square wave periodic signal of Figure 2.13. The average value C_0 or $a_0/2$ is 1. The origin was chosen so that the signal has even symmetry and the b_n coefficients are all zero:

$$a_n = \frac{2}{T} \int_{-T/2}^{T/2} x(t) \cos\left(\frac{2\pi}{T} nt\right) dt = \frac{2}{T} \int_{-T/4}^{T/4} 2 \cos\left(\frac{2\pi}{T} nt\right) dt$$

$$= \frac{4}{T} \frac{T}{2\pi n} \sin\left(\frac{2\pi}{T} nt\right) \Big|_{-T/4}^{T/4} = \frac{4}{\pi n} \sin\frac{\pi}{2} n = \begin{cases} 2 & n = 0 \\ 0 & n \text{ even} \\ -(4/\pi n)(-1)^{(n+1)/2} & n \text{ odd} \end{cases}$$

Thus the Fourier series for the square wave is

$$x(t) = 1 + \frac{4}{\pi}\left(\cos\frac{2\pi}{T}t - \frac{1}{3}\cos\frac{6\pi}{T}t + \frac{1}{5}\cos\frac{10\pi}{T}t - \cdots\right) \tag{2.48}$$

A plot of the average power distribution (see Equation 2.44) is shown in Figure 2.14. The total average power $(1/T) \int_0^T x^2(t)\, dt = 2$ [i.e., $x^2(t) = 4$ half of the time]. Thus 95% of the average power lies in the frequency range $(0 \leq f \leq 3f_0)$, $(C_0^2 + \frac{1}{2}C_1^2 + \frac{1}{2}C_3^2)/2 = 1.9/2 = 0.95$. As with finite-energy signals, defining the bandwidth of the signal as the range of frequencies over which "most" of the average power lies is somewhat ambiguous. Defining it as the width of the filter or channel that is

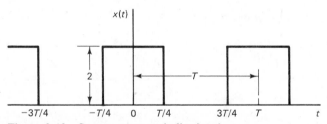

Figure 2.13   Square wave periodic signal.

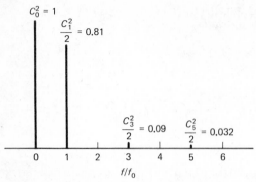

Figure 2.14   Frequency domain average power distribution.

necessary for the output to look "similar" to the input is more appropriate. If we think of this signal as a sequence of positive pulses of width $T/2$, each pulse would be affected by a low-pass filter as shown in Figures 2.9 and 2.10. Thus the bandwidth of the square wave becomes $2m/T$ or $2mf_0$, where $1 \le m \le 3$, depending on how the signal is to be processed. ■ ■

2.8 AVERAGE POWER DENSITIES FOR INFINITE-ENERGY SIGNALS

Although a Fourier series representation is adequate to characterize the manner in which the average power of a periodic signal is distributed in the frequency domain, it is not applicable to nonperiodic or random infinite-energy signals. For random signals the average power is not concentrated at specific frequencies and has to be characterized by a density function.

We seek an average power density $S(f)$, also called a *power spectrum*, that is applicable for both periodic and random infinite-energy signals and that, similar to the energy density, must satisfy the following properties:

1. $S(f)$ must be real and nonnegative and must have even
   symmetry about the origin [i.e., $S(f) = S(-f) \ge 0$].   (2.49)

2. $\int_{-\infty}^{\infty} S(f)\, df$ is the total average power.   (2.50)

3. $\int_{f_1}^{f_2} S(f)\, df + \int_{-f_2}^{-f_1} S(f)\, df = 2\int_{f_1}^{f_2} S(f)\, df$ is the average
   power in the frequency band $[f_1, f_2]$.   (2.51)

As before, we could avoid dealing with negative frequencies by defining a one-sided power spectrum,

$$\hat{S}(f) = \begin{cases} 2S(f) & f \geq 0 \\ 0 & f < 0 \end{cases} \tag{2.52}$$

In order to define a power spectrum for a random process $x(t)$, it is essential that the following time averages exist:

$$\langle x(t) \rangle = \lim_{T \to \infty} \frac{1}{T} \int_a^{T+a} x(t)\, dt \tag{2.53}$$

and

$$R_x(\tau) = \langle x(t)x(t-\tau) \rangle = \lim_{T \to \infty} \frac{1}{T} \int_a^{T+a} x(t)x(t-\tau)\, dt \tag{2.54}$$

The average denoted $R(\tau)$ is called the *autocorrelation function*. We see from the definition of the autocorrelation function that it can be measured in the manner shown in Figure 2.15, where $\tilde{T}$ is large enough to estimate the limit. Of course, if the limit is never reached, $R_x(\tau)$ cannot be measured and does not exist. By inspection of Equation 2.54, $R_x(0) = \langle x^2(t) \rangle$, we see that the existence of the autocorrelation function implies the existence of the average power.

A process for which the time average and the autocorrelation function exists is said to be *wide-sense stationary* (WSS). We can define a power spectrum only for wide-sense stationary processes.

Theorem.   The power spectrum $S_x(f)$ of a WSS random signal $x(t)$ is defined as the Fourier transform of its autocorrelation function:

$$S_x(f) = \int_{-\infty}^{\infty} R_x(\tau)e^{-j2\pi ft}\, d\tau \tag{2.55}$$

$$R_x(\tau) = \int_{-\infty}^{\infty} S_x(f)e^{j2\pi ft}\, df \tag{2.56}$$

In order to prove the theorem, we must first prove that $R_x(\tau)$ is always an even function. From the definition (Equation 2.54),

$$R_x(-\tau) = \lim_{T \to \infty} \frac{1}{T} \int_a^{T+a} x(t)x(t+\tau)\, dt$$

Letting $t + \tau = t'$ and $a + \tau = a'$, we obtain

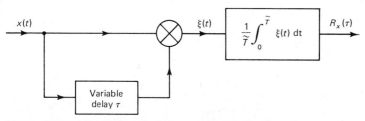

Figure 2.15   Measurement of the autocorrelation function.

$$R_x(-\tau) = \lim_{T \to \infty} \frac{1}{T} \int_{a'}^{T+a'} x(t')x(t' - \tau) \, dt' = R_x(\tau) \tag{2.57}$$

Since we have proven that $R_x(\tau)$ is an even function, it follows from property 2.4 of Fourier transforms and Equation 2.17 that $S_x(f)$ is real, even, and

$$S_x(f) = 2 \int_0^\infty R_x(\tau) \cos 2\pi f \tau \, d\tau \tag{2.58}$$

As a consequence of $S_x(f)$ being real and even,

$$R_x(\tau) = 2 \int_0^\infty S_x(f) \cos 2\pi f \tau \, df \tag{2.59}$$

Equations 2.58 and 2.59 are also used as the defining equations for the power spectrum. The requirement that $S_x(f)$ be nonnegative does not follow automatically from the definition but must nevertheless be true. Thus, for the wide-sense stationary signals that occur in practice, the Fourier transform of their autocorrelation functions must always be nonnegative. All random signals are either *nonstationary*, in which case one cannot define their average power density, or Equation 2.55 (or 2.58) defines a legitimate average power density.

Property 2 (Equation 2.50) is automatically satisfied from the definition. If we set $\tau = 0$ in Equation 2.59, we obtain

$$R_x(0) = 2 \int_0^\infty S_x(f) \, df = \langle x^2(t) \rangle \tag{2.60}$$

Property 3 is harder to prove. Let us first find the important relationship between the spectrum of the output of a linear system and the spectrum of the input. We start with the convolution relationship between the input, $x(t)$, and the output, $y(t)$, of a linear system whose impulse response is $h(t)$:

$$y(t) = \int_{-\infty}^\infty h(\xi)x(t - \xi) \, d\xi \tag{2.61}$$

Substituting this relationship into the definition of the autocorrelation function (Equation 2.54), we obtain

$$R_y(\tau) = \langle y(t)y(t - \tau) \rangle = \left\langle \int_{-\infty}^\infty h(\xi)x(t - \xi) \, d\xi \int_{-\infty}^\infty h(\eta)x(t - \tau - \eta) \, d\eta \right\rangle$$

$$= \int_{-\infty}^\infty \int_{-\infty}^\infty h(\xi)h(\eta)\langle x(t - \xi)x(t - \tau - \eta) \rangle \, d\xi \, d\eta$$

where the symbol $< \; >$ denotes time average. Interchanging the order of averaging and integration is the same as interchanging the order of integration,[1] and, because $h(\xi)$ and $h(\eta)$ are not functions of time, they are

---

[1]Because of the limit operation and the ∞ limits on the integrals, this interchanging requires some mathematical justification, even though it is commonly done.

properly removed from the time average operation. Changing variables ($t' = t - \xi$), we see that

$$\langle x(t - \xi)x(t - \tau - \eta)\rangle = \langle x(t')x(t' + \xi - \eta - \tau)\rangle = R_x(\tau - \xi + \eta)$$

Therefore,

$$R_y(\tau) = \int_{-\infty}^{\infty}\int_{-\infty}^{\infty} h(\xi)h(\eta)R_x(\tau - \xi + \eta)\,d\xi\,d\eta \qquad (2.62)$$

Taking Fourier transforms of both sides, we have

$$S_y(f) = \int_{-\infty}^{\infty} e^{-j2\pi f\tau}\int_{-\infty}^{\infty}\int_{-\infty}^{\infty} h(\xi)h(\eta)R_x(\tau - \xi + \eta)\,d\xi\,d\eta\,d\tau$$

$$= \int_{-\infty}^{\infty} h(\xi)e^{-j2\pi f\xi}\int_{-\infty}^{\infty} h(\eta)e^{j2\pi f\eta}\int_{-\infty}^{\infty} R_x(\tau - \xi + \eta)e^{-j2\pi f(\tau - \xi + \eta)}\,d\xi\,d\eta\,d\tau$$

$$= \int_{-\infty}^{\infty} h(\xi)e^{-j2\pi f\xi}\,d\xi\int_{-\infty}^{\infty} h(\eta)e^{j2\pi f\eta}\,d\eta\int_{-\infty}^{\infty} R_x(\rho)e^{-j2\pi f\rho}\,d\rho$$

$$= H(f)H(-f)S_x(f)$$

$$= |H(f)|^2 S_x(f) \qquad (2.63)$$

$H(f)$ is the transfer function of the linear system. The last step uses the fact that a complex number $H(f)$ multiplied by its conjugate is the square of its magnitude. It should be pointed out that Equation 2.62 is never used. If we seek the autocorrelation of the output of a linear system, we use Equation 2.63 and a table of Fourier transform pairs.

Equation 2.63, besides being important in its own right, is all that is needed to prove the final property of $S(f)$. The argument is identical to that used in Section 2.5. Thus, if $H(f)$ is an ideal bandpass filter,

$$|H(f)| = \begin{cases} 1 & f_1 \le |f| \le f_2 \\ 0 & \text{elsewhere} \end{cases}$$

then the spectrum of the output is

$$S_y(f) = \begin{cases} S_x(f) & f_1 \le |f| \le f_2 \\ 0 & \text{elsewhere} \end{cases}$$

and, from Equation 2.60, the total power in the output is

$$\langle y^2(t)\rangle = R_y(0) = \int_{-\infty}^{\infty} S_y(f)\,df$$

$$= \int_{f_1}^{f_2} S_x(f)\,df + \int_{-f_2}^{-f_1} S_x(f)\,df = 2\int_{f_1}^{f_2} S_x(f)\,df$$

We associate the average power of the signal $x(t)$ in the frequency band $[f_1, f_2]$ with the average power in the output of an ideal bandpass filter.

34   2/FREQUENCY DOMAIN CONCEPTS

The rest of this chapter will be devoted to computing the power spectrum of a number of important random signals.

Consider a way of measuring the power spectrum directly rather than measuring the autocorrelation function first. If we pass a random signal with spectrum $S_x(f)$ through a very-narrow-band filter that is symmetrical about its center frequency f_i, then the average power in the output $\langle y^2(t) \rangle$ is

$$\langle y^2(t) \rangle = \int_{-\infty}^{\infty} S_x(f)|H(f)|^2 \, df \cong S_x(f_i)\int_{-\infty}^{\infty} |H(f)|^2 \, df$$

The approximation is good if the input spectrum can be approximated by a straight line segment over the "width" of the filter. Thus,

$$S_x(f_i) \cong P_y \bigg/ \int_{-\infty}^{\infty} |H(f)|^2 \, df$$

where

$$P_y = \langle y^2(t) \rangle$$

P_y can be measured by squaring the reading of a true rms meter, and the denominator can be calculated from the frequency response of the filter. If we can change the center frequency of the filter, then by sweeping the filter over the frequency range of interest we can measure $S_x(f)$ directly.

Let us verify that the definition for a power spectrum is compatible with our previous concept of the average power for a periodic signal of the type

$$x(t) = \sum_{n=0}^{\infty} C_n \cos(n2\pi f_0 t + \theta_n)$$

Since the signal is periodic, we know that the autocorrelation function exists:

$$R(\tau) = \left\langle \sum_{n=0}^{\infty} \sum_{m=0}^{\infty} C_n C_m \cos(n2\pi f_0 t + \theta_n) \cos[m2\pi f_0(t - \tau) + \theta_m] \right\rangle$$

where the symbol $< \ >$ corresponds to the time average over the fundamental period of the signal $x(t)$. Expand the second cosine term using the identity $\cos(A - B) = \cos A \cos B + \sin A \sin B$. Interchanging the summations and time averages, one obtains

$$R(\tau) = C_0^2 + \sum_{n=1}^{\infty} \sum_{m=1}^{\infty} C_n C_m \langle \cos(n2\pi f_0 t + \theta_n) \cos(m2\pi f_0 t + \theta_m) \rangle \cos m2\pi f_0 \tau$$

$$+ \sum_{n=1}^{\infty} \sum_{m=1}^{\infty} C_n C_m \langle \cos(n2\pi f_0 t + \theta_n) \sin(m2\pi f_0 t + \theta_m) \rangle \sin m2\pi f_0 \tau$$

From the orthogonality principle

$$\langle \cos(n2\pi f_0 t + \theta_n) \cos(m2\pi f_0 t + \theta_m) \rangle = \tfrac{1}{2}\delta_{nm}$$

$$\langle \cos(n2\pi f_0 t + \theta_n) \sin(m2\pi f_0 t + \theta_m) \rangle = 0$$

and

$$R(\tau) = C_0^2 + \sum_{n=1}^{\infty} \frac{C_n^2}{2} \cos n2\pi f_0 \tau$$

The autocorrelation function of a periodic signal is itself periodic. Furthermore,

$$R(0) = C_0^2 + \sum_{n=1}^{\infty} \frac{C_n^2}{2}$$

which is the average power. Taking Fourier transforms, we obtain

$$S(f) = \int_{-\infty}^{\infty} e^{-j2\pi f\tau} \left(C_0^2 + \sum_{n=1}^{\infty} \frac{C_n^2}{2} \cos n2\pi f_0 \tau \right) d\tau$$

$$= C_0^2 \int_{-\infty}^{\infty} e^{-j2\pi f\tau} \, d\tau + \sum_{n=1}^{\infty} \frac{C_n^2}{2} \int_{-\infty}^{\infty} e^{-j2\pi f\tau} \frac{e^{jn2\pi f_0 \tau} + e^{-jn2\pi f_0 \tau}}{2} \, d\tau$$

$$= C_0^2 \delta(f) + \sum_{n=1}^{\infty} \frac{C_n^2}{4} \left(\int_{-\infty}^{\infty} e^{-j2\pi(f-nf_0)\tau} \, d\tau + \int_{-\infty}^{\infty} e^{-j2\pi(f+nf_0)\tau} \, d\tau \right)$$

$$= C_0^2 \, \delta(f) + \sum_{n=1}^{\infty} \frac{C_n^2}{4} [\delta(f - nf_0) + \delta(f + nf_0)] \tag{2.64}$$

The presence of impulse functions indicates that the average power of a periodic signal is concentrated on those frequencies that are harmonics of the signal $x(t)$. The total average power is once again

$$\int_{-\infty}^{\infty} S(f) \, df = C_0^2 + \sum_{n=1}^{\infty} \frac{C_n^2}{2}$$

We have verified that the power spectrum is properly defined for periodic signals.

It is sometimes necessary to indicate the manner in which two random processes are related. A time average that is useful for such characterizations is the *cross-correlation function*, defined analogously to Equation 2.54:

$$R_{xy}(\tau) = \langle x(t)y(t - \tau) \rangle = \lim_{T \to \infty} \frac{1}{T} \int_{a}^{T+a} x(t)y(t - \tau) \, dt \tag{2.65}$$

We define this function here for completeness, but we shall not discuss it further in this chapter. It should be noted, however, that the only property of the cross-correlation function is

$$R_{xy}(\tau) = R_{yx}(-\tau)$$

and its Fourier transform does not have a physical significance analogous to the power spectrum.

2.9 WHITE PROCESSES AND SPECTRAL MODELS

A *white process*, often called *white noise*, is a random process whose spectrum is flat for all frequencies. If the spectral level is $\eta_0/2$ V^2/Hz (i.e., in a band of B Hz, the power is $\eta_0 B$ V^2 because of the two-sided nature of the spectrum), then the autocorrelation function is $\frac{1}{2}\eta_0 \, \delta(\tau)$. This white process appears to be unrealistic or overidealized because of its infinite power, or

$$\langle x^2(t) \rangle = R_x(0) = \int_{-\infty}^{\infty} S_x(f) \, df = \infty$$

This model is quite useful, however, for a process whose spectrum is flat relative to the transfer function of the system to which it is an input. If the input spectrum is flat (level $\eta_0/2$) in the vicinity of the system transfer function $H(f)$, then, from Equation 2.63, the output spectrum is

$$S_y(f) = (\eta_0/2)|H(f)|^2 \tag{2.66}$$

The name "white process" comes from the analogy to white light, whose spectrum is flat over the frequencies of human perception (i.e., flat relative to the system transfer function).

This model is useful for many noise mechanisms whose spectra are flat over a range of frequencies for which the system transfer function is different from zero. An example is thermal noise, which is present in all electric circuits. Electrons tend to move randomly through conducting material with a strong average drift in one direction. The rate of change of charge due to the average drift of electrons is called *current*. The remaining random component is called *thermal noise*. The spectrum of this thermal noise is flat well beyond the useful frequency range of most electric circuits.

EXAMPLE 2.9 ■

Let us determine the spectrum and autocorrelation function of thermal noise, whose input spectral level is 10^{-12} V^2 for the low-pass filter of Example 2.3. The transfer function is $H(f) = 1/(1 + j2\pi fA)$. From Equation 2.66,

$$S_n(f) = \frac{10^{-12}}{1 + (2\pi fA)^2} = \frac{10^{-12}}{2A} \frac{2A}{1 + (2\pi fA)^2} \tag{2.67}$$

From Table 2.1

$$R_n(\tau) = \frac{10^{-12}}{2A} e^{-|\tau|/A} \tag{2.68}$$

The total average power is

$$\int_{-\infty}^{\infty} S_n(f) \, df = R_n(0) = \frac{10^{-12}}{2A} \quad V^2 \qquad\qquad ■ ■$$

When random processes have a nonzero average value, it is common to treat this average or dc component separately. For example, we have discussed the flow of electrons in a circuit in terms of its dc component called the current and its zero-mean random component called thermal noise. If a process $x(t)$ has an average component, $x_{dc} = \langle x(t) \rangle$, and if $x'(t)$ is defined as

$x(t) - x_{dc}$, which has a zero average value, then

$$R_x(\tau) = \langle (x'(t) + x_{dc})(x'(t - \tau) + x_{dc}) \rangle$$
$$= \langle x'(t)x'(t - \tau) \rangle + 2x_{dc}\langle x'(t) \rangle + x_{dc}^2$$
$$= R_{x'}(\tau) + x_{dc}^2 \tag{2.69}$$

$R_{x'}(\tau)$ is called the *autocovariance function* of a process. For a process with zero average value, the autocorrelation function and autocovariance function are the same. Taking Fourier transforms of Equation 2.69, we obtain

$$S_x(f) = S_{x'}(f) + x_{dc}^2 \delta(f) \tag{2.70}$$

When the average value of a process is included in the computation of a power spectrum, it appears as an impulse function at the origin.

It is convenient to have an analytic model for any random process rather than measured values. The task of determining a model from measurements is called the *identification problem*. In many instances, the most convenient analytic model is

$$S(f) = K|H(f)|^2 \tag{2.71}$$

$H(f)$ can be regarded as the transfer function of a system modeled by a differential equation with constant coefficients. Thus the spectrum results from white noise through a linear filter, and the identification problem reduces to that of identifying the filter. Whether there *is* such a filter is immaterial.

So far, all of the processes that we have considered are low-pass processes whose spectra are concentrated at the lower frequencies. If the filter model in Equation 2.71 were a bandpass filter, the spectrum would be a bandpass spectrum. A bandpass process, whose spectrum is symmetric about some center frequency f_c, can be modeled by

$$R(\tau) = \hat{R}(\tau) \cos 2\pi f_c \tau \tag{2.72}$$

where $\hat{R}(\tau)$ is the autocorrelation function of a low-pass process whose spectrum is $\hat{S}(f)$. Taking Fourier transforms of Equation 2.72, we obtain

$$S(f) = \int_{-\infty}^{\infty} \hat{R}(\tau) \cos 2\pi f_c \tau \, e^{-j2\pi f\tau} \, d\tau$$

$$= \int_{-\infty}^{\infty} \hat{R}(\tau) \tfrac{1}{2}(e^{j2\pi f_c \tau} + e^{-j2\pi f_c \tau}) e^{-j2\pi f\tau} \, d\tau$$

$$= \tfrac{1}{2} \int_{-\infty}^{\infty} \hat{R}(\tau) e^{-j2\pi(f-f_c)\tau} \, d\tau + \tfrac{1}{2} \int_{-\infty}^{\infty} \hat{R}(\tau) e^{-j2\pi(f+f_c)\tau} \, d\tau$$

$$= \tfrac{1}{2}\hat{S}(f - f_c) + \tfrac{1}{2}\hat{S}(f + f_c) \tag{2.73}$$

Since $S(f \pm f_c)$ represents a shift of a spectrum centered at zero to one centered at $\pm f_c$, Equation 2.73 defines a bandpass, symmetric spectrum (see Figure 2.16). Equations 2.72 and 2.73 can be used, along with Table 2.1, to treat bandpass spectra.

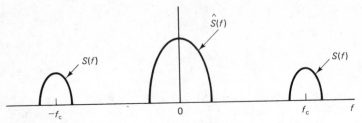

Figure 2.16   Bandpass spectrum and its low-pass equivalent.

EXAMPLE 2.10: Flat Bandpass Spectra ■

Let us determine the autocorrelation function of the spectrum shown in Figure 2.17. The autocorrelation function is related to $\hat{R}(\tau)$, which is the autocorrelation function of $\hat{S}(f)$ shown in Figure 2.18. From the table of Fourier transforms we see that

$$\hat{R}(\tau) = \eta_0 B \text{ sinc } B\tau$$

From Equation 2.72

$$R(\tau) = \eta_0 B \text{ sinc } B\tau \cos 2\pi f_c \tau$$

These autocorrelation functions are shown in Figure 2.19.   ■ ■

2.10 AMPLITUDE MODULATED SIGNALS

The original intention of amplitude modulation techniques was to shift the spectrum of a low-pass process such as speech into a bandpass process which can be transmitted via electromagnetic waves. This procedure, originally called wireless telephony, converts a low-pass speech process $m(t)$ into a signal $s(t)$ according to the relation

$$s(t) = A[1 + m(t)] \cos 2\pi f_c t \tag{2.74}$$

where f_c is some high frequency called the carrier frequency.

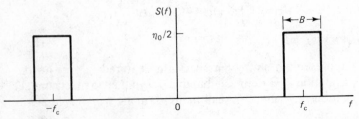

Figure 2.17   Flat bandpass spectrum.

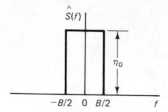

Figure 2.18 Low-pass equivalent.

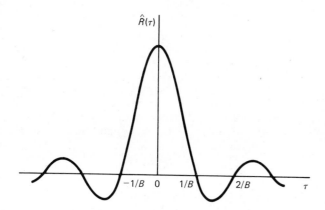

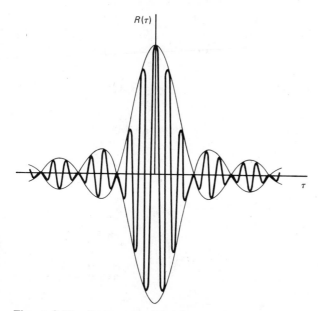

Figure 2.19 Autocorrelation of a process with a flat bandpass spectrum and its low-pass equivalent.

Let us compute the autocorrelation function of the modulated signal $s(t)$:

$$R_s(\tau) = A^2 \langle [1 + m(t)] \cos(2\pi f_c t)[1 + m(t - \tau)] \cos 2\pi f_c(t - \tau) \rangle$$

Using the identity $2 \cos A \cos B = \cos(A - B) + \cos(A + B)$, we obtain

$$R_s(\tau) = \frac{A^2}{2} \langle [1 + m(t) + m(t - \tau) + m(t)m(t - \tau)]$$

$$\times [\cos 2\pi f_c \tau + \cos 2\pi f_c(2t - \tau)] \rangle$$

$$= \frac{A^2}{2} \cos 2\pi f_c \tau + \frac{A^2}{2} \langle m(t)m(t - \tau) \rangle \cos 2\pi f_c \tau$$

$$+ \frac{A^2}{2} [\langle m(t) \rangle + \langle m(t - \tau) \rangle] \cos 2\pi f_c \tau + \frac{A^2}{2} \langle g(t, \tau) \rangle$$

where

$$g(t, \tau) = [1 + m(t) + m(t - \tau) + m(t)m(t - \tau)] \cos 2\pi f_c(2t - \tau)$$

and $\cos 2\pi f_c \tau$ was removed from the average because it is not a function of time. Assuming that $\langle m(t) \rangle = 0$ and $\langle m(t)m(t - \tau) \rangle = R_m(\tau)$ and that it can be proven that $\langle g(t, \tau) \rangle = 0$, we then have

$$R_s(\tau) = \tfrac{1}{2}A^2 \cos 2\pi f_c \tau + \tfrac{1}{2}A^2 R_m(\tau) \cos 2\pi f_c \tau \qquad (2.75)$$

Assume that the carrier frequency f_c is large relative to the bandwidth of $m(t)$, which means that $\cos(4\pi f_c t)$ varies much more rapidly than $m(t)$. If this is the case, then $g(t, \tau)$ is as shown in Figure 2.20. Observe by inspection that $\langle g(t, \tau) \rangle = 0$ for large carrier frequencies, and hence Equation 2.75 is the autocorrelation function for the modulated signal. Taking Fourier transforms of this equation and using the formulas of Equations 2.72 and 2.73, we obtain

$$S_s(f) = \tfrac{1}{4}A^2[\delta(f - f_c) + \delta(f + f_c)] + \tfrac{1}{4}A^2[S_m(f - f_c) + S_m(f - f_c)] \qquad (2.76)$$

This spectrum is plotted in Figure 2.21 along with $S_m(f)$, the spectrum of $m(t)$. Observe that the bandwidth of the modulated signal is twice that of the original process.

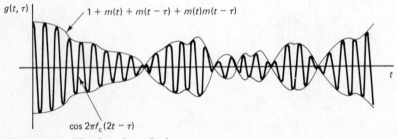

$g(t, \tau)$

$1 + m(t) + m(t - \tau) + m(t)m(t - \tau)$

t

$\cos 2\pi f_c(2t - \tau)$

Figure 2.20   The function $g(t,\tau)$.

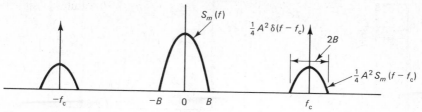

Figure 2.21 Spectrum of an amplitude modulated signal.

We have shown that amplitude modulation (Equation 2.74) achieves the goal of shifting a low-pass audio process into a narrow bandpass process. We shall study this and other continuous waveform modulation techniques in some detail in later chapters.

2.11 RANDOM SEQUENCES OF IMPULSES

In this section we shall consider random signals whose spectrum can be determined analytically but which appear to be unrealistic. When considered as inputs to linear filters, however, they become quite useful.

Consider some periodic sequences of impulses. Although these signals are not random, they provide some necessary insights. We define the *periodic impulse sequence* $[x_a(t)]$ by

$$x_a(t) = \sum_{n=-\infty}^{\infty} \delta(t - nT)$$ (2.77)

Since impulses are difficult to work with, examine the periodic pulse sequence $[\hat{x}_a(t)]$ shown in Figure 2.22. If $\Delta \to 0$, it follows that $\hat{x}_a(t) \to x_a(t)$. The procedure will be to compute $R_{\hat{x}_a}(\tau)$ first, then allow $\Delta \to 0$, whereupon $R_{x_a}(\tau)$, whose Fourier transform is $S_{x_a}(f)$, will be obtained.

$R_{\hat{x}_a}(0)$ is equal to $\langle \hat{x}_a^2(t) \rangle$, where $\hat{x}_a^2(t)$ looks identical to the signal in Figure 2.22 except that the pulse amplitudes are now $(1/\Delta)^2$. The time average of such a signal is $(1/T)(1/\Delta)^2\Delta$ or $1/(T\Delta)$. Similarly, for $|\tau| < \Delta$, $\hat{x}_a(t)\hat{x}_a(t - \tau)$ resembles the signal in Figure 2.22, but in addition to pulse amplitudes of $(1/\Delta)^2$, the width of the pulses are $\Delta - |\tau|$ (see Figure 2.23).

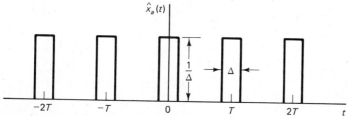

Figure 2.22 Periodic pulse sequence.

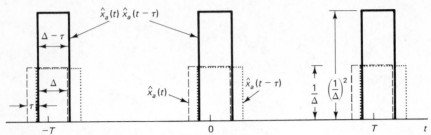

Figure 2.23   Plot of $\hat{x}_a(t)\hat{x}_a(t - \pi)$ for $|\tau| < \Delta$.

The time average of this signal is $(1/T)(1/\Delta)^2(\Delta - |\tau|)$. We have argued that, in the range $|\tau| < \Delta$,

$$R_{\hat{x}_a}(\tau) = \frac{1}{T\Delta}\left(1 - \frac{|\tau|}{\Delta}\right)$$

For delays of nT, where n is an integer, the delayed pulse train is identical to the pulse train itself or $R_{\hat{x}_a}(nT) = R_{\hat{x}_a}(0)$. For τ near nT, $R_{\hat{x}_a}(\tau)$ is a replica of $R_{\hat{x}_a}(\tau)$ evaluated for τ near zero. For the delays not considered, $\Delta < (|\tau| - nT) < T - \Delta$, the delayed pulse train does not overlap the original pulse train anywhere and hence $R_{\hat{x}_a}(\tau)$ must be equal to zero. The resultant autocorrelation function is periodic, as shown in Figure 2.24. Each of the triangle sections of Figure 2.24 approaches $(1/T)\delta(\tau - nT)$ as $\Delta \to 0$. Thus,

$$R_{x_a}(\tau) = \frac{1}{T} \sum_{n=-\infty}^{\infty} \delta(\tau - nT) \tag{2.78}$$

In order to obtain the power spectrum, let us first compute the transform of $R_{\hat{x}_a}(\tau)$ and then take the limit as $\Delta \to 0$. $R_{\hat{x}_a}(\tau)$ is an even periodic signal with period T and can be expanded in the Fourier series (see Problem 2.14)

$$R_{\hat{x}_a}(\tau) = C_0 + \sum_{n=1}^{\infty} C_n \cos \frac{2\pi}{T} n\tau \tag{2.79}$$

where

$$C_0 = \langle R_{\hat{x}_a}(\tau) \rangle = \left(\frac{1}{T}\right)^2$$

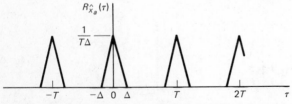

Figure 2.24   Autocorrelation of the periodic pulse sequence.

and

$$C_n = \frac{2}{T} \int_{-T/2}^{T/2} R_{\hat{x}_a}(\tau) \cos\left(\frac{2\pi}{T} n\tau\right) d\tau = \frac{2}{T^2} \operatorname{sinc}^2 \frac{n}{T} \Delta$$

Since the Fourier transforms of C_0 and $C_n \cos[(2\pi/T)n\tau]$ are $C_0\delta(f)$ and $\frac{1}{2}C_n[\delta(f - n/T) + \delta(f + n/T)]$, respectively, it follows that

$$S_{\hat{x}_a}(f) = \frac{1}{T^2} \sum_{n=-\infty}^{\infty} \operatorname{sinc}^2\left(\frac{n}{T}\Delta\right)\delta\left(f - \frac{n}{T}\right) \tag{2.80}$$

As $\operatorname{sinc}^2[(n/T)\Delta] \to 1$ when $\Delta \to 0$,

$$S_{x_a}(f) = \frac{1}{T^2} \sum_{n=-\infty}^{\infty} \delta\left(f - \frac{n}{T}\right) \tag{2.81}$$

Equations 2.78 and 2.81 are plotted in Table 2.2. The transform of an infinite sequence of impulses spaced T s apart is an infinite sequence of impulses spaced $1/T$ Hz apart.

Let us now define the *bipolar periodic impulse sequence* by

$$x_b(t) = \sum_{n=-\infty}^{\infty} (-1)^n \delta(t - nT) \tag{2.82}$$

Following the identical arguments for $\hat{x}_a(t)$, it is seen that $R_{\hat{x}_b}(\tau)$ is as shown in Figure 2.25. Taking limits as $\Delta \to 0$, we have

$$R_{x_b}(\tau) = \frac{1}{T} \sum_{n=-\infty}^{\infty} (-1)^n \delta(\tau - nT) \tag{2.83}$$

Expanding $R_{\hat{x}_b}(\tau)$ in a Fourier series (see Problem 2.15), we obtain

$$R_{\hat{x}_b}(\tau) = \sum_{n=1}^{\infty} C_n \cos\frac{2\pi}{2T} n\tau,$$

where

$$C_n = \frac{2}{2T} \int_{-T}^{T} R_{\hat{x}_b}(\tau) \cos\left(\frac{2\pi}{2T} n\tau\right) = \begin{cases} \dfrac{2}{T^2} \operatorname{sinc}^2\left(\dfrac{n}{2T}\Delta\right) & n \text{ odd} \\ 0 & n \text{ even} \end{cases}$$

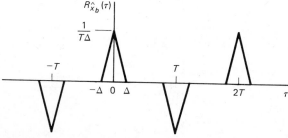

Figure 2.25 Autocorrelation of the bipolar periodic pulse sequence.

TABLE 2.2 IMPULSE SEQUENCES

Signal	Model	$R(\tau)$
Periodic impulse sequence	$x_a(t) = \Sigma\, \delta(t - nT)$	$R_{x_a}(\tau) = (1/T)\,\Sigma\,\delta(\tau - nT)$
Bipolar periodic impulse sequence	$x_b(t) = \Sigma\,(-1)^n\,\delta(t - nT)$	$R_{x_b}(\tau) = (1/T)\,\Sigma\,(-1)^n\,\delta(\tau - nT)$
Bipolar random impulse train	$x_c(t) = \Sigma\, a_n \delta(t - nT)$ $a_n = \pm 1$	$R_{x_c}(\tau) = (1/T)\delta(\tau)$
(Monopolar) random impulse train	$x_d(t) = \Sigma\, a_n \delta(t - nT)$ $a_n = +1$ or 0	$R_{x_d}(\tau) = \dfrac{1}{4T}\delta(\tau) + \dfrac{1}{4T}\,\Sigma\,\delta(\tau - nT)$
Ideal impulse noise	$x_e(t) = \Sigma\,\delta(t - t_n)$ d(average) $= 1/\gamma$ γ = average number of impulses per second	$R_{x_e}(\tau) = \gamma\delta(\tau) + \gamma^2$

It follows that

$$S_{\hat{x}_b}(f) = \frac{1}{T^2} \sum_{\substack{n=-\infty \\ n\text{ odd}}}^{\infty} \text{sinc}^2\!\left(\frac{n}{2T}\Delta\right)\delta\!\left(f - \frac{n}{2T}\right)$$

Since $\text{sinc}^2[(n/2T)\Delta] \to 1$ as $\Delta \to 0$,

$$S_{x_b}(f) = \frac{1}{T^2} \sum_{n\text{ odd}} \delta\!\left(f - \frac{n}{2T}\right) \tag{2.84}$$

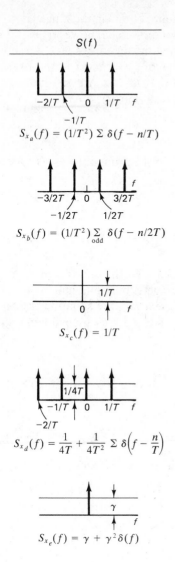

$$S(f)$$

$$S_{x_a}(f) = (1/T^2) \Sigma \, \delta(f - n/T)$$

$$S_{x_b}(f) = (1/T^2) \underset{\text{odd}}{\Sigma} \, \delta(f - n/2T)$$

$$S_{x_c}(f) = 1/T$$

$$S_{x_d}(f) = \frac{1}{4T} + \frac{1}{4T^2} \Sigma \, \delta\left(f - \frac{n}{T}\right)$$

$$S_{x_e}(f) = \gamma + \gamma^2 \delta(f)$$

Equations 2.83 and 2.84 are plotted in Table 2.2.

The next three examples concern random impulse sequences. The first of these, called the *bipolar random impulse train*, $x_c(t)$, is very similar to $x_b(t)$ except that the signs of the impulses are random. Half the time they are positive, and the other half they are negative, much as a flipped coin may come up heads or tails. Thus

$$x_c(t) = \sum_{n=-\infty}^{\infty} a_n \delta(t - nT) \tag{2.85}$$

where a_n is $+1$ or -1 (see Table 2.2). One calculates $R_{\hat{x}_c}(\tau)$ in the range $|\tau| < \Delta$ much as for the periodic signal $\hat{x}_b(t)$. There is a difference, however, in the vicinity of $\tau = nT$. For this random case, half the time the pulses of $\hat{x}_c(t - \tau)$ have the same polarity as the overlapping pulses of $\hat{x}_c(t)$, and half the time they have the opposite polarity. Hence the average $\langle \hat{x}_c(t)\hat{x}_c(t - \tau) \rangle$ remains zero for all $|\tau| > \Delta$. $R_{\hat{x}_c}(\tau)$ is shown in Figure 2.26. Letting $\Delta \to 0$,

$$R_{x_c}(\tau) = \frac{1}{T}\delta(\tau) \quad \text{and} \quad S_{x_c}(f) = \frac{1}{T} \tag{2.86}$$

Thus the bipolar random impulse train has a white spectrum of level $1/T$ (see Table 2.2).

Now we consider the *random impulse train* $x_d(t)$ (or the *monopolar random impulse train*)

$$x_d(t) = \sum_{n=-\infty}^{\infty} a_n \delta(t - nT) \tag{2.87}$$

where a_n is either $+1$ or 0. The autocorrelation $R_{\hat{x}_d}(\tau)$ is the same for $|\tau| < \Delta$, as for the previous signals, except that the average is only one half as large since there are half as many pulses. When τ is near nT, the situation is like $\hat{x}_a(t)$ (the periodic pulse sequence) except that positive pulses are overlapping with other positive pulses one fourth of the time. Hence

$$R_{\hat{x}_d}(nT) = \tfrac{1}{2}R_{\hat{x}_d}(0) = 1/4T\Delta$$

$R_{\hat{x}_d}(\tau)$ is as shown in Figure 2.27. Taking limits as $\Delta \to 0$, after separating half the $\tau = 0$ triangular portion, we obtain

$$R_{x_d}(\tau) = \frac{1}{4T}\delta(\tau) + \frac{1}{4T} \sum_{n=-\infty}^{\infty} \delta(\tau - nT) \tag{2.88}$$

We have seen [for $x_a(t)$] that the second term of Equation 2.88 transforms to

$$\frac{1}{4T^2} \sum_{n=-\infty}^{\infty} \delta\left(f - \frac{n}{T}\right)$$

It follows that

$$S_{x_d}(f) = \frac{1}{4T} + \frac{1}{4T^2} \sum_{n=-\infty}^{\infty} \delta\left(f - \frac{n}{T}\right) \tag{2.89}$$

as shown in Table 2.2. This spectrum is similar to that of the periodic impulse sequence with the addition of a white component.

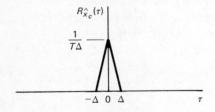

Figure 2.26 Autocorrelation of the bipolar random pulse train.

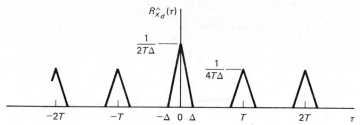

Figure 2.27 Autocorrelation of the random pulse train.

In the final example, called ideal impulse noise, all of the impulses are positive but their locations occur randomly rather than every T s apart. The definition of the *ideal impulse noise* $x_e(t)$ is

$$x_e(t) = \sum_{n=-\infty}^{\infty} \delta(t - t_n) \tag{2.90}$$

where the t_n (called the *random arrival times*) can be any time. We shall define an average number of impulses per second γ by

$$\gamma = \lim_{T \to \infty} \frac{1}{T} \int_0^T x_e(t) \, dt = \lim_{T \to \infty} \frac{1}{T} K(T) \tag{2.91}$$

where $K(T)$ is the number of impulses in the time slot $[0, T]$, and integrating $x_e(t)$ corresponds to counting the impulses since every impulse contributes unity to the integral. The average distance between adjacent impulses is $1/\gamma$ s.

Ideal impulse noise is similar (as seen from Table 2.2) to the random impulse train if $1/\gamma = 2T$ (the average distance between adjacent impulses for the random impulse train). We can argue that $R_{\hat{x}_e}(\tau) = R_{\hat{x}_d}(\tau)$ with $2T$ replaced by $1/\gamma$, but only in the range $|\tau| < \Delta$. Beyond this range, the pulses of $\hat{x}_e(t)$ do not appear at designated time slots as in $\hat{x}_d(t)$, and there is nothing special about the vicinity of $\tau = n/\gamma$. Thus $R_{\hat{x}_e}(\tau)$ appears as in Figure 2.28. By relating this figure to Equation 2.70 (for signals with a nonzero average value), we see that the pedestal has the value $\langle x_e(t) \rangle^2$, and from Equation 2.91,

$$\langle x_e(t) \rangle^2 = \gamma^2 \tag{2.92}$$

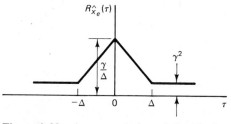

Figure 2.28 Autocorrelation of the ideal pulse noise.

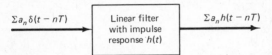

$\Sigma a_n \delta(t - nT)$ → Linear filter with impulse response $h(t)$ → $\Sigma a_n h(t - nT)$

Figure 2.29 Generation of binary random pulse trains.

Note that the triangular regions of Figure 2.27 (not including the one in the vicinity of $\tau = 0$) are spread uniformly in Figure 2.28 because there is nothing special about the points $\tau = nT$.

Letting $\Delta \rightarrow 0$ in Figure 2.28 gives

$$R_{x_e}(\tau) = \gamma\delta(\tau) + \gamma^2 \quad \text{and} \quad S_{x_e}(f) = \gamma + \gamma^2\delta(f) \tag{2.93}$$

These results are also included in Table 2.2.

2.12 BINARY RANDOM PULSE TRAINS

Signals that will be quite useful for the purpose of communicating information can be visualized as the response of a linear filter to a random impulse train (either bipolar or monopolar). These signals, which are called *binary random pulse trains*, are often generated this way (see Figure 2.29). If $h(t)$ is the generating impulse response, then the signals become

$$x(t) = \sum_{n=-\infty}^{\infty} a_n h(t - nT) \tag{2.94}$$

where a_n is ±1 for bipolar inputs, $+1$ or 0 for monopolar inputs. The spectrum of this signal follows from Equation 2.63:

$$S_x(f) = S_{\hat{x}}(f)|H(f)|^2 \tag{2.95}$$

where $H(f)$ is the Fourier transform of $h(t)$ and $\hat{x}(t)$ is either $x_c(t)$ or $x_d(t)$ (from Table 2.2), depending on whether the random impulse train that is the input is bipolar or monopolar.

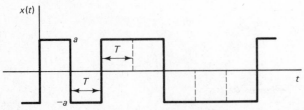

Figure 2.30 Binary random pulse train.

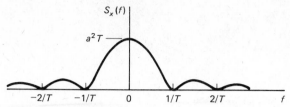

Figure 2.31 Spectrum of a binary random pulse train.

EXAMPLE 2.11 ■

Consider the binary random pulse train in Figure 2.30. This is generated by the system in Figure 2.29 if the input is the bipolar random impulse train $[x_c(t)]$ and the impulse response is given by

$$h(t) = \begin{cases} a & 0 \le t \le T \\ 0 & \text{elsewhere} \end{cases} \tag{2.96}$$

From Table 2.1, $|H(f)|^2 = a^2 T^2 \text{sinc}^2 Tf$ and from Table 2.2, $S_{x_c}(f) = 1/T$. Thus

$$S_x(f) = a^2 T \text{sinc}^2 Tf \tag{2.97}$$

which is plotted in Figure 2.31. This spectrum has the same shape as the energy density of a flat-topped pulse (Figure 2.8), which follows directly from Equation 2.95 because $S_{x_c}(f)$ is white. As before, the bandwidth of this signal train is $m(1/T)$, where $1 \le m \le 3$. It is convenient to formulate the bandwidth B in terms of the pulse rate $f_p = 1/T$:

$$B = mf_p \tag{2.98}$$

where m assumes a value between 1 and 3 depending on how the signal is to be processed. ■ ■

EXAMPLE 2.12 ■

Consider now the monopolar binary pulse train of Figure 2.32. This signal can be generated by passing the monopolar random impulse train $x_d(t)$ through the same filter, as characterized by Equation 2.96. Using the input spectrum from Table 2.2, we obtain

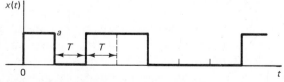

Figure 2.32 Monopolar binary random pulse train.

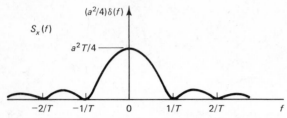

Figure 2.33 Spectrum of a monopolar binary random pulse train.

$$S_x(f) = \left[\frac{1}{4T} + \frac{1}{4T^2}\sum_{n=-\infty}^{\infty}\delta\left(f - \frac{n}{T}\right)\right]a^2 T^2\,\mathrm{sinc}^2\,Tf$$

$$= \frac{a^2 T}{4}\,\mathrm{sinc}^2\,Tf + \frac{a^2}{4}\delta(f) \tag{2.99}$$

since $\mathrm{sinc}^2\,Tf = 0$ when $f = n/T$ and n is an integer other than zero. This spectrum is plotted in Figure 2.33. ■ ■

We could have obtained this result in another way. Observe that the signal of Figure 2.32 is simply a version (scaled by a factor of $\frac{1}{2}$) of the signal in Figure 2.30 shifted up. The shifting introduces a dc component of $a/2$ which causes an impulse of magnitude $a^2/4$ at the origin of the spectrum. The rest of the spectrum is the same as before with a scale of $\frac{1}{4}$.

2.13 IMPULSE/SHOT NOISE

A wide class of noise processes that are encountered can be modeled as the response of a linear system to ideal impulse noise $[x_e(t)]$. These are defined by

$$n(t) = \sum_{m=-\infty}^{\infty} h(t - t_m) \tag{2.100}$$

where there are, on the average, γ pulses/s of identical shape $h(t)$ and where the arrival times t_m are random. From the model of Figure 2.29 and from Table 2.2, the spectrum of $n(t)$ is given by

$$N(f) = \gamma^2|H(0)|^2\delta(f) + \gamma|H(f)|^2 \tag{2.101}$$

where $H(f)$ is the Fourier transform of $h(t)$. Alternatively, this signal can be regarded as having a dc component of $\gamma|H(0)|$ and a zero-average-value random-noise component whose spectrum is $\gamma|H(f)|^2$.

If the width of the pulse $h(t)$ is large compared to $1/\gamma$ (the average distance between pulse arrivals), there is considerable overlapping of pulses. In this case, the noise bears little resemblance to the shape of the pulses, as in Figure 2.34b. This noise is called *shot noise*. When there is only a small amount of overlapping pulses, the noise is called *impulse noise* and appears as in Figure 2.34a.

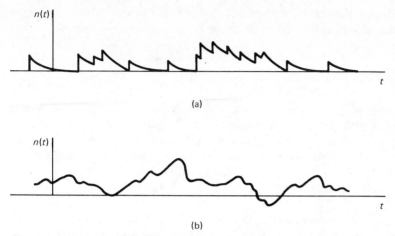

Figure 2.34 (a) Impulse and (b) shot noise signals.

Shot noise occurs, for example, in transistors when electrons pass through a narrow connecting region (called the base) and induce current pulses in the output, as in Equation 2.100. The pulse widths correspond to the transition times and are small in absolute terms. Nevertheless, the number of electrons per second is so enormous that the shot noise model is valid.

EXAMPLE 2.13 ■

Let us model $h(t)$ for shot noise in a transistor by

$$h(t) = \frac{q}{d} e^{-t/d} u(t)$$

where d is the transition time and q is the charge of one electron in coulombs (note that $\int_0^\infty h(t)\,dt = q$). Taking Fourier transforms (see Example 2.3), we obtain

$$H(f) = q/(1 + j2\pi fd)$$

From Equation 2.101,

$$N(f) = (\gamma q)^2 \delta(f) + \frac{\gamma q^2}{1 + (2\pi fd)^2} \tag{2.102}$$

Thus the current has a dc component of γq A plus a zero-average-value random-shot-noise component. If the transition time is small (in the order of nanoseconds) the spectrum is quite flat. Very often shot noise is considered white for this reason.   ■ ■

Impulse noise is a more appropriate model for signals generated from such things as lightning, low-level radiation, or distant sunspots. Whereas the spectral formulas and plots are identical to those of shot noise, it is clear from

Figure 2.34 that the signals are different in the time domain. Something other than power spectra is needed to characterize a random process fully.

2.14 CONCLUSIONS

The purpose of this chapter has been to develop the concept of power spectrum for visualizing random signals in the frequency domain. In the next two chapters we shall study many communication techniques and the resulting signals that are transmitted. We now have the tools to evaluate the power spectra of these signals and to determine the bandwidths that the channels must have in order for these signals to be transmitted without distortion.

The power-spectrum concept is not sufficient, however, to evaluate the performance of the communication techniques in a noisy environment. A time-domain model that utilizes concepts from probability will be presented for this purpose in Chapter 5.

PROBLEMS

2.1. Find the Fourier transforms for the two waveforms shown:

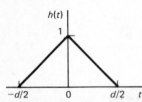 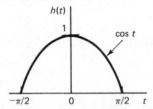

2.2. Determine the Fourier transform of the double pulse shown:

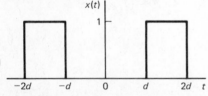

2.3. Reconsider the three signals in Problems 2.1 and 2.2.
   (a) For each case sketch the energy density.
   (b) Give an indication of the bandwidth of each signal.

2.4. Consider a linear system characterized by the following impulse response:

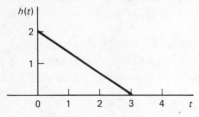

Determine the output for the input shown:

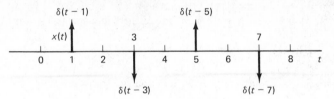

2.5. Consider a continuous, linear, causal system whose impulse response is given by

$$h(t) = \tfrac{1}{5}e^{-t/5}u(t)$$

(Note that the electric circuit shown has this response when $RC = 5$.)

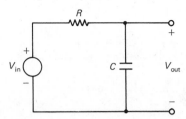

Determine the system response to the input $v_{in}(t) = u(t)$.

2.6. Suppose that the input to the previous system is the following square wave:

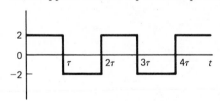

Sketch the output for $\tau \gg 5$ and $\tau \ll 5$. For which condition does the system act like an integrator? Explain.

2.7. A meter is an electromechanical device that provides a mechanical motion of an indicator in response to an applied electric signal. Often the inertia of the meter movement is such that the indicator reads the average value of the signal; that is,

$$D = \frac{1}{T}\int_{t-T}^{t} x(\xi)\,d\xi$$

where $x(t)$ is the electric signal and D is the indicator displacement.
(a) Is this meter a linear system? If so, what is its impulse response?
(b) Suppose the electric signal is the square wave of Problem 2.6, sketch the output (indicator displacement) when $\tau = T$.
(c) For what relationship between τ and T will this meter measure the average value of the signal?

2.8. Find the impulse response $h(t)$, of the system whose input is $x(t)$ and is described by the following differential equation:

$$\frac{d^2y(t)}{dt^2} + 9\frac{dy(t)}{dt} + 20y(t) = x(t)$$

Solve by finding the transfer function first. (*Hint*: Factor the transform first and then use Example 2.3.)

2.9. From the definition of the Fourier transform, prove the following transform pairs:

(a) $tx(t) \Leftrightarrow (j/2\pi)d/df\{X(f)\}$.

(b) $x(t)/t \Leftrightarrow 2\pi j\int_f^\infty X(f)\,df$.

2.10. An electric network has the following transfer function (where $\omega = 2\pi f$):

$$\frac{I(\omega)}{V(\omega)} = \frac{(j\omega)^2 + (j\omega) + 1}{(j\omega)^4 + 3(j\omega)^2 + j\omega + 5}$$

Calculate the steady sinusoidal output (i.e., the magnitude and phase shift) if the input is

(a) $v(t) = 10 \sin 2t$;

(b) $v(t) = 100 \sin 20t$.

2.11. Consider the ideal differentiator $y(t) = dx(t)/dt$ when the input is $20 \sin 3t$. Find the transfer function of the ideal differentiator and determine the output using $y(t) = 20 |H(3)| \sin(3t + \angle H(3))$.

2.12. Find the first three nonzero coefficients a_k and b_k of the Fourier series representation of the waveforms

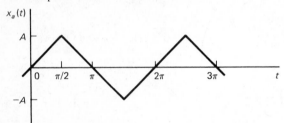

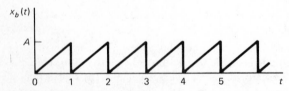

2.13. Prove that Equations 2.38 and 2.39 are valid by carrying out the substitution suggested in the text.

2.14. Consider the periodic signal in Figure 2.24. Carry out the Fourier series expansion and verify the answer in the text.

2.15. Repeat for the signal in Figure 2.25.

2.16. Find the Fourier series for the half-wave rectified sinusoid shown:

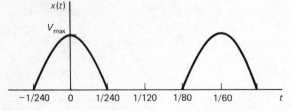

2.17. Suppose the signal in the previous problem were passed through a filter whose characteristics are shown. Determine an expression for the output.

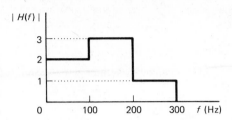 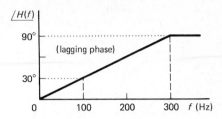

2.18. Reconsider the periodic signals in Problem 2.12.
   (a) Sketch the distribution of average power in the frequency domain for each signal.
   (b) Determine the percentage of average power that is contained in the average value and fundamental component of each signal.

2.19. Repeat Problem 2.18 for the signal in Problem 2.16.

2.20. The signal shown has the given Fourier series expansion:

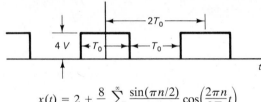

$$x(t) = 2 + \frac{8}{\pi} \sum_{n=1}^{\infty} \frac{\sin(\pi n/2)}{n} \cos\left(\frac{2\pi n}{2T_0}t\right)$$

Let $2T_0 = \frac{1}{60}$ s, and suppose this signal is passed through a linear filter as follows:

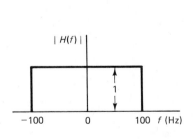 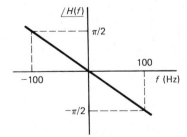

   (a) Write an expression for the output.
   (b) What is the average power of the output?
   (c) What percentage of the input power is this?

2.21. Show that if white noise (two-sided spectrum level $\eta_0/2$ V^2/Hz) is passed through a causal linear system whose impulse response is $h(t)$, the correlation function of the output is given by

$$R_y(\tau) = \frac{\eta_0}{2} \int_0^\infty h(\xi)h(\tau + \xi) \, d\xi$$

2.22. White noise with a two-sided power spectral density of $\eta_0/2$ V^2/Hz is passed through a first-order, low-pass filter with time constant A [i.e., $H(f) = 1/(1 + j2\pi fA)$] and thereafter through an ideal amplifier with a gain of 2.

(a) Write the expression for the power spectral density of the noise at the output of the amplifier.

(b) Write an expression for the autocorrelation of the output noise.

2.23. Repeat Problem 2.22 except that the filter is an ideal low-pass filter with transfer function

$$H(f) = \begin{cases} 1 & |f| \leq B \\ 0 & \text{otherwise} \end{cases}$$

2.24. Repeat Problem 2.22 for an input noise that is not white but rather has an autocorrelation function $R(\tau) = 50e^{-10^5|\tau|}$. Assume the time constant of the filter is 10^{-5} s.

2.25. Repeat Problem 2.22 for a filter that is an ideal differentiator [$y(t) = (1/20\pi) dn(t)/dt$] followed by an ideal low-pass filter whose cutoff frequency is 10^3 Hz.

2.26. Which of the following functions are suitable for autocorrelation functions? Indicate the reason for those that are not.

(a) $Ae^{-a\tau}u(\tau)$, $a > 0$.

(b) $Ae^{-2|\tau|}$.

(c) A for $|\tau| \leq \tau_0/2$ and zero otherwise.

(d) $A(1 - |\tau|/\tau_0)$ for $|\tau| \leq \tau_0$ and zero otherwise.

2.27. The input to a linear filter is $z(t) = x(t) + n(t)$, where the processes $x(t)$ and $n(t)$ are both zero-average-value processes with spectra $S_x(f) = 2/[1 + (2\pi f)^2]$ and $S_n(f) = \eta_0/2$ and where

$$R_{xn}(\tau) = \langle x(t)n(t - \tau) \rangle = 0 \qquad \text{for all } \tau$$

The transfer function of the linear filter is $H(f) = 1/(1 + j2\pi f)$.

(a) Find the power spectra of the input $z(t)$ and output $y(t)$ of the filter. [*Hint:* Show first that the autocorrelation function of $z(t)$ is the sum of the autocorrelation functions of $x(t)$ and $n(t)$.]

(b) What is the average value and average power of the processes $x(t)$, $n(t)$, $z(t)$, and $y(t)$?

2.28. Consider a discrete random signal . . . $x_{-2}, x_{-1}, x_0, x_1, x_2, \ldots$ (which can be thought of as samples of a continuous signal Δt s apart) that can be modeled as a wide-sense-stationary sequence of zero-average-value variables with correlation sequence $R_x(m) = \langle x_j x_{j-m} \rangle$. Show that the correlation sequence of the process defined by $y_i = \sum_{r=0}^{L} h_r x_{i-r}$ is given by

$$R_y(n) = \sum_{r=0}^{L} \sum_{s=0}^{L} h_r h_s R_x(s - r + n)$$

2.29. Show that if the bipolar pulse train (Equation 2.85) is extended from binary pulses to multilevel pulses (i.e., a_n has one of K positive values or their negatives), then Equation 2.86 becomes

$$S(f) = \frac{C}{T} \quad \text{where} \quad C = \frac{1}{K} \sum_{n=1}^{K} a_n^2$$

2.30. Show that if the monopolar pulse trains (Equation 2.87) is extended from binary pulses to multilevel pulses [i.e., a_n has one of K-positive (or zero) values], then Equation 2.89 becomes

$$S(f) = \frac{C_1}{T} + \frac{C_2}{T^2} \sum_{n=-\infty}^{\infty} \delta\left(f - \frac{n}{T}\right)$$

where $C = (1/K) \sum_{n=1}^{K} a_n^2$, $C_2 = 4[(1/K) \sum_{n=1}^{K} a_n]^2$, and $C_1 = C - C_2$

2.31. Find the autocorrelation function of the random signal whose spectrum is shown.

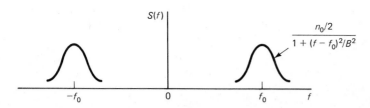

2.32. Suppose the pulse shape in the shot noise model is $(q/d)[u(t) - u(t - d)]$ and the number of pulses per second $\gamma = 20/d$.
   (a) What is the average value or dc component of the signal?
   (b) Compute and sketch the power spectrum of the random component. What is the bandwidth?

2.33. Let $x(t)$ be a WSS input and $y(t)$ be the output of a linear system whose impulse response is $h(t)$ and transfer function $H(f)$.
   (a) Determine an expression for the cross correlation $R_{yx}(\tau) = \langle y(t)x(t - \tau)\rangle$ of the output and input in terms of the autocorrelation function of the input $R_x(\tau)$.
   (b) By taking Fourier transforms of this expression show that if $S_{yx}(f)$ is the Fourier transform of $R_{yx}(\tau)$, $S_{yx}(f) = H(f)S_x(f)$.
   (c) Argue that this represents a convenient way to measure the transfer function of a linear system—particularly if the input is white.

REFERENCES

There are many excellent textbooks that discuss spectral analysis and measurement and noise mechanisms in detail. A few of them are listed here.

1. J. S. Bendat and A. G. Piersol, *Random Data: Analysis and Measurement Procedures*, Wiley, New York, 1971.
2. W. R. Bennet, *Electrical Noise*, McGraw-Hill, New York, 1960.
3. W. R. Bennet and J. R. Davey, *Data Transmission*, McGraw-Hill, New York, 1965.
4. W. B. Davenport, Jr., and W. L. Root, *Random Signals and Noise*, McGraw-Hill, New York, 1958.

5. P. F. Panter, *Modulation, Noise and Spectral Analysis*, McGraw-Hill, New York, 1965.
6. A. Papoulis, *Probability, Random Variables and Stochastic Processes*, McGraw-Hill, New York, 1965.
7. J. B. Thomas, *An Introduction to Statistical Communication Theory*, Wiley, New York, 1969.
8. A. Van der Zeil, *Noise*, Prentice-Hall, Englewood Cliffs, NJ, 1954.

Chapter 3

Overview of Discrete Communications

3.1 INTRODUCTION

Many communication tasks involve transmitting a series of symbols or characters that can take on one of M *values*. Examples are

1. computer-to-computer communications where each symbol takes on one of two values (i.e., a 1 or a 0)
2. man-to-computer communications via a teletype where each symbol takes on one of 64 values (i.e., 32 keys both small and capitals)
3. man-to-computer communications via a "touchtone" headset where $M = 12$ (i.e., 12 buttons)

When a series of symbols is transmitted and the task of the receiver is to decide repeatedly which symbol was transmitted, then we are using discrete communications.

In this chapter we discuss many of the most important techniques of discrete communications. We shall concern ourselves with the implementation of these techniques and the required bandwidth of the channels through which the communication signals are transmitted.

A principal goal of this text (although not of this chapter) is to show how well these techniques perform in the presence of noise. This result suggests using discrete communications for continuous signals, such as those arising from speech or from acoustic or visual measurements. A brief discussion of the process of digitizing continuous signals, which is necessary if we are to employ the techniques of discrete communications, is included in this chapter.

3.2 BINARY BASEBAND TECHNIQUES

We first consider discrete signals whose average power is concentrated in the low or *baseband* frequencies and are applicable for transmission through cables. All of these signals consist of a sequence of pulses of T-s duration and are characterized by the pulse rate f_p pulses/s, where $f_p = 1/T$. There might be M different kinds of pulses where each type of pulse represents one of the M characters that is being transmitted.

If $M = 2$ (i.e., binary pulses), this signal is useful for transmitting the 1s and 0s of computer-to-computer communications. This binary pulse train might look like one of the signals in Figure 3.1.

A possible receiver for any of these signals is shown in Figure 3.2. The output of the comparator is either a positive or negative voltage depending on whether the input signal exceeds some fixed preset threshold voltage (V_t).

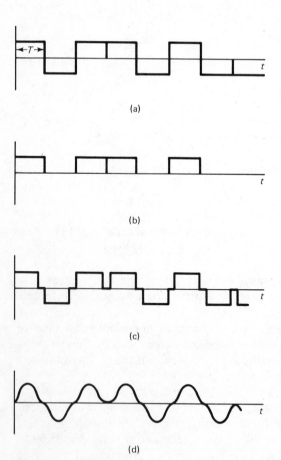

(a)

(b)

(c)

(d)

Figure 3.1 Various binary signals: (a) bipolar or non-return-to-zero pulses, (b) monopolar pulses, (c) bipolar, return-to-zero pulses, and, (d) raised-cosine-shaped pulses.

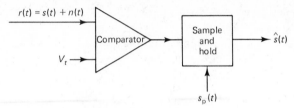

Figure 3.2 Simple binary receiver.

The sample and hold circuit samples the input when there is a narrow pulse present in the pulse-sampling signal $s_p(t)$ and maintains that voltage as its output until the next pulse arrives. If the input to this receiver is a noisy version of one of the signals in Figure 3.1, the output will be a clean non-return-to-zero (NRZ) pulse train which indicates the estimates of the 1s and 0s [$\hat{s}(t)$], as shown in Figure 3.3.

A study of this simple receiver yields a number of important concepts. First the receiver is a decision making mechanism and largely insensitive to noise. If the noise power is small, there will be no errors in the output estimate as seen in Figure 3.3. Because the receiver is deciding between a 1 or a 0, based on a single sample, it is not necessary that the bandwidth of the channel (or cable) be large enough to preserve the signal waveshape accurately. Finally, generating a pulse-sampling signal whose pulse rate corresponds precisely to the incoming signal pulse rate is a difficult task. If the pulse rates are slightly different, the receiver could not continue to sample the incoming pulses once per pulse and in the middle of the pulse as in Figure 3.3. The generation of an appropriate sampling signal that remains synchronized

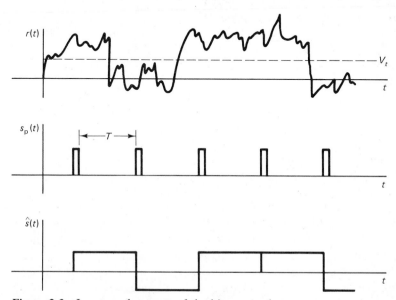

Figure 3.3 Inputs and outputs of the binary receiver.

with the input pulses is called *pulse synchronization*. Pulse synchronization techniques will be discussed in Section 3.6.

The signals of Figure 3.1 are the binary random pulse trains whose power spectra were determined in Chapter 2. Specifically, for the bipolar pulse trains (see Equations 2.95 and 2.86)

$$S(f) = \frac{1}{T}|H(f)|^2 \tag{3.1}$$

where $H(f)$ is the Fourier transform of one of the pulses. If the pulses are flat topped (of height A and width W, where $W \leq T$), then

$$|H(f)|^2 = A^2W^2 \operatorname{sinc}^2 Wf \quad \text{and} \quad S(f) = \frac{1}{T}A^2W^2 \operatorname{sinc}^2 Wf \tag{3.2}$$

The bandwidth of such a signal is not uniquely defined, but, using the multiplying factor m, we see that

$$B = m(1/W) \geq m(1/T) = mf_p \tag{3.3}$$

As we have noted, the binary receiver does not require that the pulse shapes be preserved accurately, and $m = 1$ (or maybe smaller) is adequate.

The relationship between the signal and the acceptable channel bandwidth can be quite complicated. A restricted bandwidth not only distorts the signal but also causes the pulses to spread into each other. This phenomenon, called *intersymbol interference*, has a major influence on the ability of a receiver to decide between symbols. For example, the return-to-zero (RZ) signal has a slightly wider spectrum than the NRZ signal ($W < T$), but, because of the interpulse zero-level interval, the intersymbol interference is reduced. As a result, the required channel bandwidth (or factor m) is slightly smaller for the RZ signal despite its wider spectrum. It does make sense to consider the effect of pulse shape on the required channel bandwidth (or factor m) in an effort to minimize this bandwidth. This is not, however, the same as minimizing the spectral width, and intersymbol interference must be considered.

3.3 PULSE-SHAPING TECHNIQUES

Suppose the pulses were generated by passing impulse functions (or approximations to impulse functions) through an ideal low-pass filter. The resultant sinc-shaped pulses are shown in Figure 3.4 along with the magnitude of their Fourier transform. If the filter bandwidth is $1/2T = f_p/2$, the pulse "width" is larger than $2T$, and adjacent pulses spread into each other. If, however, the pulses are sampled precisely at their peak values ($t = nT$), all adjacent pulses will have a zero value, and there will be no intersymbol interference. A binary pulse train using these sinc-shaped pulses is shown in Figure 3.5. For this signal, samples at $t = nT$ have a value of $\pm A$ V, independent of the values of

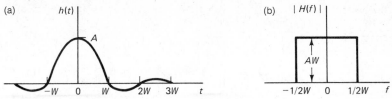

Figure 3.4 (a) sinc$(1/W)t$ pulse and (b) its transform.

adjacent pulses. This pulse train has a compact spectrum and a well-defined bandwidth of $\frac{1}{2}f_p$ Hz $\left(\text{i.e., } m = \frac{1}{2}\right)$.

We have just argued that the channel bandwidth for a binary pulse train required in order to avoid intersymbol interference is mf_p, where $m \geq \frac{1}{2}$. The lower bound can be achieved with sinc-shaped pulses, but precise pulse synchronization is required in addition to accurate pulse-shaping filters. Actually, the pulse zero crossings must be at the correct locations at the receiver, not the transmitter. Thus the pulse-shaping filters must compensate for the transfer function of the channel as well.

In order to achieve the lower bound, the overall transfer function must be that of an ideal low-pass filter with a linear phase (which causes a delay that must be accounted for by the pulse synchronization). We have seen in Chapter 2 that while such filters are noncausal, good causal approximations are possible if the delay is large enough (see Example 2.5). We must concern ourselves with the question of whether the filter approximations affect the location of the pulse zero crossings.

Suppose the pulse-shaping filter (or, more correctly, the overall transfer function) is an approximation to an ideal low-pass filter as in Figure 3.6a. The

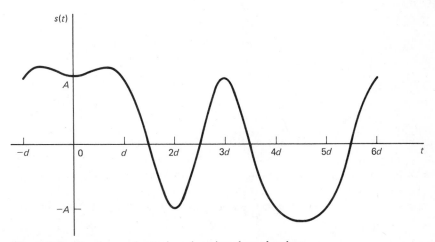

Figure 3.5 Random pulse train using sinc-shaped pulses.

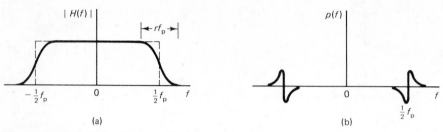

Figure 3.6 (b) Roll-off approximation to (a) an ideal low-pass filter.

bandwidth of the resulting signal will be slightly increased to $\frac{1}{2}(1 + r)f_p$, or $m = \frac{1}{2}(1 + r)$, where r is called the *rolloff factor*. If r is small ($r \ll 1$), we have achieved a good approximation to a low-pass filter, and the required channel bandwidth is near the minimum.

Let us decompose the filter into two parts, an ideal low-pass filter [with inverse transform $h_1(t)$] and the perturbation $p(f)$ as shown in Figure 3.6b [with inverse transform $h_p(t)$]. Since transforms are linear, we can apply superposition or

$$h(t) = h_1(t) + h_p(t) \tag{3.4}$$

Because $p(f)$ has symmetry about $f = 0$, the inverse Fourier transform can be written

$$h_p(t) = 2\int_{(1-r)f_p/2}^{(1+r)f_p/2} p(f) \cos(2\pi ft) \, df \tag{3.5}$$

Replacing f by $f' + \frac{1}{2}f_p$ or, equivalently, $f' + 1/2T$, in Equation 3.5 and evaluating it at the appropriate sampling times ($t = nT$) produces

$$\begin{aligned}
h_p(nT) &= 2\int_{-rf_p/2}^{rf_p/2} p(f' + \tfrac{1}{2}f_p) \cos\left[2\pi\left(f' + \frac{1}{2T}\right)nT\right] df' \\
&= \left[2\int_{-rf_p/2}^{rf_p/2} p(f' + \tfrac{1}{2}f_p) \cos(2\pi nTf') \, df'\right] \cos \pi n \\
&\quad - \left[2\int_{-rf_p/2}^{rf_p/2} p(f' + \tfrac{1}{2}f_p) \sin 2\pi nTf' \, df'\right] \sin \pi n \\
&= 2(-1)^n \int_{-rf_p/2}^{rf_p/2} p(f' + \tfrac{1}{2}f_p) \cos(2\pi nTf') \, df' \tag{3.6}
\end{aligned}$$

since $\cos \pi n = (-1)^n$ and $\sin \pi n = 0$ for all n. It is seen from Equation 3.6 that if $p(f)$ has odd symmetry about $\frac{1}{2}f_p$, then $h_p(nT) = 0$ for all n. Thus if the rolloff of the transfer function has odd symmetry about $\frac{1}{2}f_p$, the pulses (while not having the precise sinc shape) will still cross through zero at the correct sampling times and therefore not cause any intersymbol interference.[1]

---

[1]This problem was first described by Nyquist [1] and is discussed fully by Bennet and Davey [2].

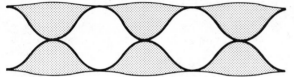

Figure 3.7 Eye pattern: noise free.

There is a slight increase in the bandwidth $[m = (1 + r)/2]$ owing to the rolloff approximations.

Finally, we must be concerned with the precision of the design of the pulse-shaping filters. The design problem is particularly important because in the communications systems where these techniques are employed, such as telephony, the channel transfer function (which must be compensated for) is often changing. In this environment the pulse-shaping filters, which are distributed through the channel, need constant adjustment. The design of self-adjusting, or adaptive, filters has received considerable attention [2–4] in order to deal with this problem. Whether the filters are adaptive or manually adjustable, it is desirable to observe the signals on an oscilloscope in such a way that the intersymbol interference, as well as the effect of noise, is clearly highlighted. These displays will help us to understand the interactions of noise and intersymbol interference.

When the pulse train is displayed on an oscilloscope with the sweep triggered at a fraction of the pulse rate by the pulse-sampling signal $[s_p(t)]$, the resultant pattern is called an *eye pattern*. Such patterns are shown in Figures 3.7 and 3.8, where the sweep is triggered at every third pulse. When there is no noise and the filters are adjusted perfectly, the eye pattern appears as in Figure 3.7. In this case the magnitude of the values of the pulse train at the sampling times is always the same, and there is no intersymbol interference. When there is some noise or the filters are not precisely adjusted, the "eye" tends to close, as in Figure 3.8. As long as the eye is partially open, the receiver in Figure 3.2 will work without error. Filter adjustments are made with the goal of maximizing the "eye opening."

The eye patterns also give us some insight to the pulse synchronization problem. The narrow pulses of the pulse-sampling signal (see Figure 3.3) must be aligned with the maximum "opening of the eye." As we shall discuss later, in most cases some self-adjusting means of pulse synchronization is used that utilizes the "information" that is visually presented to us with the eye pattern.

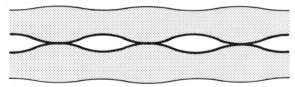

Figure 3.8 Noisy eye pattern.

3.4 MULTILEVEL BASEBAND TECHNIQUES

The most common low-frequency or baseband technique employed to transmit a series of characters that take on one of M values is a binary encoded scheme that employs the binary pulse trains previously discussed. This scheme, called *pulse code modulation* (PCM), is the technique used for computer-to-computer communications where the actual messages or numerical values are multilevel but encoded into a binary code. We can represent one of M messages by a unique sequence of N 1s and 0s, where

$$N = \text{smallest integer greater than or equal to } \log_2 M \quad \text{bits/symbol} \quad (3.7)$$

The term *bits* comes from the first two and last two letters of "binary digits." Thus, for example, any integer that lies in the range between plus and minus 32,000 can be encoded into a 16-bit block of 1s and 0s called a *word*. If the message comes from an external device, such as a teletype, the signal can be generated by the hardware in Figure 3.9. It is presumed that one of the M inputs to the encoder is enabled by the external device. The pulse-shaping filter can be used to generate any of the binary pulse trains shown in Figure 3.1 or any other, such as the sinc-shaped pulse train. For computer-to-computer communications, the computer output is the N-bit word in parallel form.

    If the word or symbol rate is denoted f_s, where

$$f_p = Nf_s \tag{3.8}$$

it follows from Equation 3.3 that the required channel bandwidth is

$$B_{PCM} = mNf_s \tag{3.9}$$

As before, $m \geq \frac{1}{2}$, where $m = \frac{1}{2}$ can be achieved with careful pulse-shaping techniques and $m = 1$ is adequate for the signals of Figure 3.1. The first part, or front end, of a PCM receiver is identical to a binary receiver such as that shown in Figure 3.2. The entire PCM receiver is shown in Figure 3.10, where S&H stands for sample and hold and P/S and S/P for parallel-to-series and

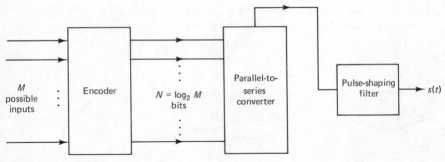

Figure 3.9 PCM generator.

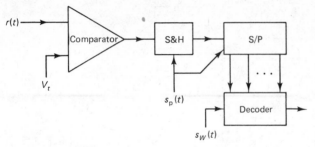

Figure 3.10 PCM receiver.

series-to-parallel converters, respectively. The series-to-parallel converter is a memory or storage device, sometimes called a *shift register*. It is necessary to store all N decisions prior to decoding. The input to the decoder [$s_W(t)$] is a word synchronizing or sampling signal that is necessary to enable the decoder to divide the pulses into the correct blocks of N bits. PCM requires this second timing function, called *word synchronization*, which will be discussed in Section 3.6. When one is communicating to a teletype, the decoder output can be used to enable an appropriate character generator for a visual display of the message. When one is communicating to a computer, the decoder is not used and the binary word in parallel form is the input to the computer. Word synchronization is still needed to "tell" the computer when to read.

EXAMPLE 3.1 ■

Let us determine the channel bandwidth needed for a computer-to-computer communications link. Assume that the computers can read and write 16-bit words at the rate of 50,000 words/s. Furthermore, assume that pulse-shaping filters are used with a rolloff factor $r = 0.2$.

The channel must handle a pulse rate f_p of

$$f_p = 16f_s = 800 \text{ kHz}$$

The bandwidth required is mf_p, where

$$m = \tfrac{1}{2}(1.2) = 0.6$$

Thus

$$B_{PCM} = 480 \text{ kHz} \qquad\qquad ■ ■$$

Let us now consider an alternative technique in which each symbol or character is encoded into a single pulse that can have one of M different amplitude levels. This technique is called *pulse amplitude modulation* (PAM), and such a signal can be generated as in Figure 3.11. The device labeled D/A is a *digital-to-analog converter* whose output is one of 2^N voltages (held constant between enabling impulses of the pulse synchronization signal). The voltages are set by the N-bit input word. As before, when one is communicat-

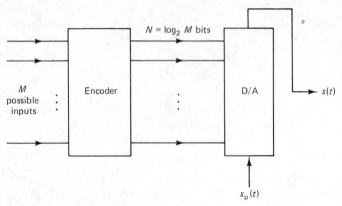

Figure 3.11 PAM generator.

ing from a computer, the input is the N-bit binary word in parallel form. For PAM there is no distinction between the pulse rate and the word rate, and word synchronization is unnecessary.

A possible receiver could use a series of comparators with different threshold voltages, as in Figure 3.12, to determine the pulse amplitudes. The array of resistors in Figure 3.12 sets the different threshold voltages. The task performed by the dashed box, the *analog-to-digital converter* (labeled A/D), can be accomplished more efficiently than indicated in Figure 3.12 with a single integrated component.

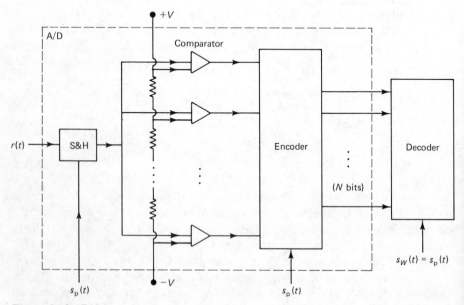

Figure 3.12 PAM receiver.

The spectrum of a multilevel random pulse train can be determined (see Problems 2.29 and 2.30). The result is much the same as that of the binary random pulse train. Thus Equation 3.3 ($B = mf_p$) still holds. The only significant difference is that the decision mechanism of the receiver is much more sensitive to signal distortions and noise. As a result, the factor m is usually larger. Because a small amount of intersymbol interference is damaging, pulse shaping is rarely used, and a bandwidth factor of 2 or more is needed. With the reduced pulse rate, the required bandwidth is normally less than PCM.

EXAMPLE 3.2 ■

Let us consider the use of PAM for the computer-to-computer problem of Example 3.1. Assume a bandwidth factor $m = 2$. Disregard the unreasonable expectation of transmitting a pulse with one of 2^{16} (65,536) different amplitudes. (One significant problem with such an input to the receiver of Figure 3.12 is that a slight amount of additive noise will make many errors.) Now the pulse rate is equal to the word or symbol rate, or

$$f_p = f_s = 50 \text{ kHz}$$

The bandwidth required is mf_p, where for $m = 2$

$$B_{PAM} = 100 \text{ kHz}$$

■ ■

In order to use PAM it is not necessary that the number of different amplitude levels (L) be equal to the number of different symbols (M). In some cases, as in Example 3.2, that would be an unreasonable requirement. In this case the PAM generator is as shown in Figure 3.13. Notice that there is once again a distinction between pulse and word synchronization. The receiver is the same as that shown in Figure 3.12 except that the final decoder

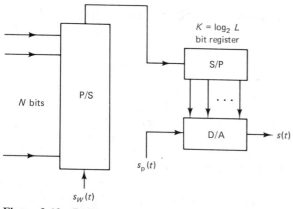

Figure 3.13 PAM generator when $L < M$.

must have memory to store the decision of each pulse and must have a word synchronizing signal as an input.

EXAMPLE 3.3 ■

Let us reconsider Example 3.2, using a 16-level pulse ($L = 16$) PAM technique. Now

$$f_p = (\log_L M)f_s = \frac{\log_2 M}{\log_2 L}f_s$$

$$= \frac{16}{4} \times 50,000 = 200 \text{ kHz} \tag{3.10}$$

As before, the bandwidth is $2f_p$ or

$$B_{\text{PAM}(L=16)} = 400 \text{ kHz}$$

Notice that the required channel bandwidth is still less than that of PCM with pulse-shaping filters.
                                                                              ■ ■

For the rest of this section we shall consider two methods of transmitting L different kinds of pulses when the "information" is associated with pulse timing rather than with the pulse amplitude. These techniques, called *pulse position modulation* (PPM) and *pulse duration modulation* (PDM), are both generated by the hardware shown in Figure 3.14. The signals themselves are also shown in Figure 3.14, where it is assumed that $L = 8$. The clock generates a sequence of narrow pulses at a pulse rate that is L times larger than the actual pulse rate specified by $s_p(t)$. The clock output is synchronized (using techniques that will be discussed shortly) with the pulse synchronizing signal, as shown in Figure 3.14. The PPM scheme transmits pulses whose width is always a single clock interval and whose leading edge is determined by the K input bits. The PDM pulses have their leading edge determined by $s_p(t)$ and their trailing edge determined by the K input bits. The "information" is related to the leading and trailing edges of the pulses (i.e., the timing of the pulses).

The bandwidth of these signals cannot be related only to the pulse rate, but rather

$$B_{\text{PPM}} = B_{\text{PDM}} = mLf_p \tag{3.11}$$

where

$$f_p = (\log_2 M/\log_2 L)f_s \tag{3.12}$$

We can argue the validity of Equation 3.11 a number of ways. The PPM signal looks like impulse noise, whose spectrum was examined in Chapter 2. We saw that the spectral width is related not to the pulse rate but rather inversely to the pulse width ($1/Lf_p$). Alternatively, the channel bandwidth must be large enough to permit pulses to be transmitted relatively undistorted at the clock

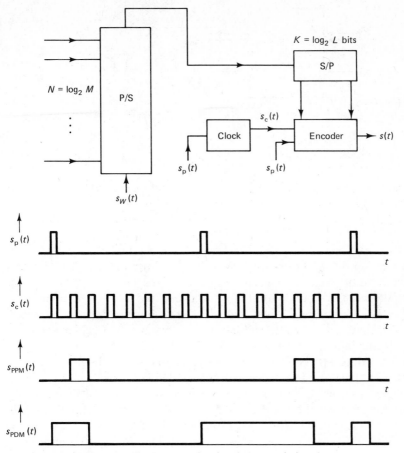

Figure 3.14   PPM and PDM generating hardware and signals.

rate Lf_p, otherwise the edges would spread to the point where timing information is lost. For similar reasons we can argue that the factor m cannot be small and values of 2 or more are needed.

EXAMPLE 3.4 ■

Let us reconsider Example 3.3. Instead of a PAM scheme ($L = 16$), we shall now consider a PPM ($L = 16$) scheme. Assume that $m = 2$. As in the PAM case (or for any 16-level scheme),

$$f_p = \tfrac{16}{4}f_s = 200 \text{ kHz}$$

Now, however,

$$B_{PPM} = 2 \times 16 \times 200 \times 10^3 = 6.4 \text{ MHz}$$

There are advantages for this large bandwidth. In the use of optical

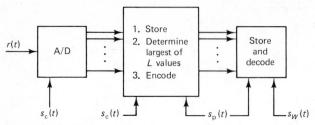

Figure 3.15   PPM receiver.

channels employing lasers the amplitude of an occasional narrow pulse can be made quite large, and PPM can perform exceptionally well (few decision errors) in the presence of noise. This will be shown in Chapter 7.   ■ ■

A receiver that will work for PPM signals is shown in Figure 3.15. If $L = M$, there is no distinction between pulse and word synchronization, and the last decode stage is not required. In any event, the receiver is somewhat complex. An alternative receiver that is easier to realize is shown in Figure 3.16. This receiver relies on the pulse edges; unfortunately, it will make more decision errors than the receiver of Figure 3.15.[2] The counter in this receiver measures the time between the impulses of $s_p(t)$ and the leading (or positive) edge of the signal. The last decoder is not needed if $L = M$.

A modification of the receiver shown in Figure 3.16 results in an easily realizable receiver for the PDM technique, shown in Figure 3.17, where it is assumed that $L = M$. Word synchronization is not needed, and pulse synchronization is not required. These features make it easy to realize.

3.5 TIME DIVISION MULTIPLEXING

If the bandwidth of a channel is sufficiently large, a number of messages of lesser bandwidth can be transmitted simultaneously. *Time division multiplexing* (TDM), which separates the messages in the time domain, is primarily used in connection with PCM. A commutator, or switch, is employed that sequentially samples the pulses in each of the PCM encoded messages, as seen in Figure 3.18. The relationship between the individual messages and the time multiplexed signal is seen in Figure 3.19 for the case of two messages. If A signals, each having a pulse rate f_p, are time multiplexed, the combined signal has a pulse rate of Af_p. Thus the bandwidth of the combined signal is the sum of the bandwidths.

[2]See Schwartz, Bennet, and Stein [5] for an analysis of the performance of PPM.

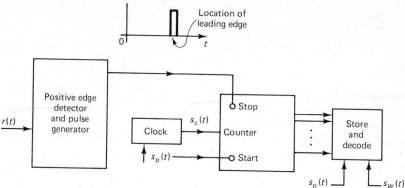

Figure 3.16 Another PPM receiver.

EXAMPLE 3.5 ■

Present machine printing and reading speeds are on the order of 20,000 bits/s. These machines have the capability of printing or reading characters at a rate of 2000 char/s encoded into 10 bits/char. These printers are often the limiting factor in a computer-to-computer link. A typical telephone circuit that has been in use since 1940 has a bandwidth of 1 MHz and can handle a pulse rate of 10^6 bits/s. The number of computer-to-computer links that this channel can handle using TDM is

$$A \le 10^6/(2 \times 10^4) = 50$$

This number can be increased with good pulse-shaping filters.   ■ ■

In time multiplexing it is normally. assumed that all messages are transmitted at the same pulse rate. It is possible, however, to transmit messages having different pulse rates. Consider as an example the problem of multiplexing three messages two of which (m_1 and m_2) require 10-kHz pulse rates while the third (m_3) requires a pulse rate of 20 kHz. If the commutator in Figure 3.20 is revolving at a rate of 5000 revolutions/s or 40,000 taps/s, the output has a 40-kHz pulse rate, where $f_{P1} = f_{P2} = 10$ kHz and $f_{P3} = 20$ kHz.

Often in TDM a few words from each message are lumped together into a frame of a few hundred bits. If we know where the frame begins (this

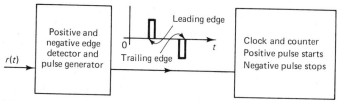

Figure 3.17 PDM receiver.

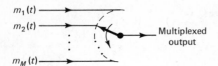

Figure 3.18   TDM commutator.

involves another type of synchronization, called *frame synchronization*), we can readily achieve word synchronization from our knowledge of the prearranged ordering. As we shall discuss in Section 3.6, frame synchronization is usually achieved with the help of a synchronizing word between frames. The extra synchronizing word reduces the number of messages that can be multiplexed through a given channel. Thus, if f_{p_I} is the "information" pulse rate or bit rate and f_b is the total pulse rate, usually called the *baud rate*, the relationship is

$$f_b = \frac{b_i + b_s}{b_i} f_{PI} \tag{3.13}$$

where b_i is the number of *information bits* per frame and b_s is the number of synchronization bits between frames.

In many communication problems the baud rate is slightly higher than the bit rate and the distinction is not significant. In some cases, particularly computer network applications, the extra bits perform functions other than synchronization, and there can then be a significant difference between the baud rate and bit rate. These functions include such things as identification of the source and location of the receiver.

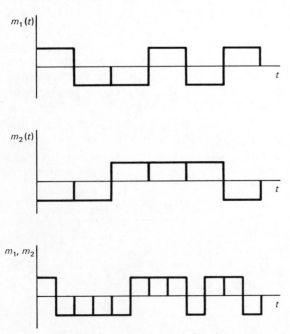

Figure 3.19   Two time multiplexed signals.

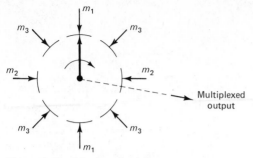

Figure 3.20   Nonstandard commutator.

EXAMPLE 3.6 ■

Let us reconsider the communication problem of Example 3.5. Assume that a frame consists of one 10-bit word from each message and that between frames there is a 20-bit synchronization word. Let us recalculate the number of messages that can be multiplexed through the telephone line. Because $b_i = 10A$, it follows that

$$f_b = \frac{10A + 20}{10A} 20{,}000A \leq 10^6$$

Solving for A yields

$$A \leq 48 \hspace{4cm} ■ ■$$

Although the commutation (and decommutation) of TDM is implemented with PCM signals, this does not mean that TDM cannot be used with the other L-level techniques. The generator, or transmitter, can use PCM hardware except for an L-level encoder at the last stage. Similarly, the receiver can use an L-level decision device followed by a binary encoder, and the rest of the hardware can be PCM. When this procedure is used, as in Figure 3.21, the final stage of the transmitter and first stage of the receiver are called *modems*. The channel pulse rate is $1/K$th that of the binary pulse rate, where $K = \log_2 L$. The channel pulse synchronizing signal [$s_{p_2}(t)$ in Figure 3.21] is obtained by dividing the frequency of the binary pulse synchronizing signal $s_{p_1}(t)$ by K.

3.6 PULSE, WORD, AND FRAME SYNCHRONIZATION

Not all communication problems involve transmitting data over a large distance. For example, when microprocessors are operating in parallel to solve different facets of a complex problem, there is a need to communicate back and forth between processors. In such situations it is possible to have an

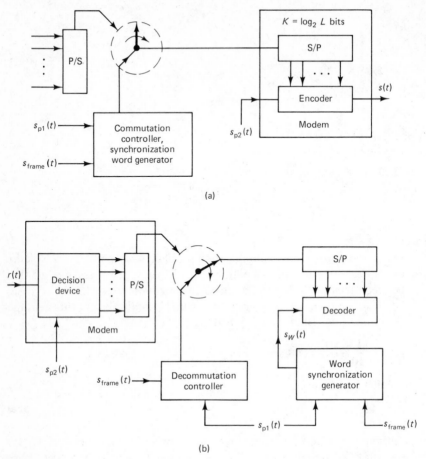

(a)

(b)

Figure 3.21   General transmitter/receiver structure: (a) transmitter; (b) receiver.

externally controlled clock. One microprocessor, or a clock circuit, can generate word- and pulse-sampling signals that are transmitted directly to each processor. This simple technique, which guarantees synchronization, is not available for communicating over a long distance.

A common technique is to transmit a separate external signal for synchronization purposes only. This signal can be transmitted on a separate wire or modulated on a carrier sinusoid. Typically these signals are low-frequency square waves or sinusoids that are used as an input to a *phased-locked-loop* (PLL) circuit. We shall briefly discuss PLLs because of their importance in communication hardware.[3]

The key component of a PLL is the *voltage-controlled oscillator* (VCO) whose frequency can be increased or decreased according to an input signal

---

[3]For a more complete discussion see Blanchard [6], Viterbi [7], Gardner [8], or Stiffler [9].

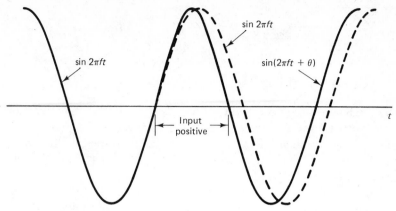

Figure 3.22 Increasing the phase of a VCO output.

that is positive or negative. The frequency changes are usually small and are made for fine tuning. The phase of the VCO output can also be increased (decreased) if the input signal is positive (negative) for only a short time. (See Figure 3.22, where it is assumed that the VCO output is a sinusoid.) The other necessary component is a *phase comparator*. The hardware of Figure 3.23 is one way to implement a phase comparator for either sinusoids or square waves. If the inputs have the same frequency, the output is a sequence of pulses that are positive or negative, depending on whether the top input precedes (the case of positive relative phase angle) or follows the bottom input. The width of the pulse corresponds to the relative delay between the inputs.

The system shown in Figure 3.24 is the phase-locked loop. The output of the VCO is locked in both phase and frequency to the input. This is true even if the phase or frequency of the input is changing, provided it is changing slowly. If the nominal (uncontrolled) frequency of the VCO is the same as

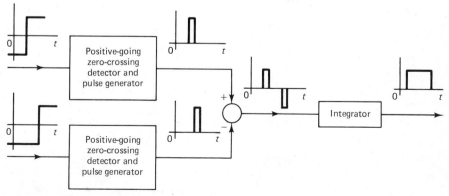

Figure 3.23 Phase comparator implementation.

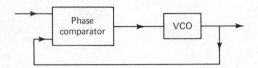

Figure 3.24   Phase-locked loop.

that of the input, the phase comparator output will have occasional narrow positive and negative pulses to keep the phases aligned. The average value will be zero. If the nominal VCO frequency is slightly different, the phase comparator output will have a nonzero average value that will lock the frequency of the VCO to that of the PLL input.

Sometimes a sinusoid or square wave is available, or can be generated, that is harmonically or subharmonically related to the desired oscillator output. Under these circumstances a modified PLL, such as those shown in Figure 3.25, can be used. The frequency dividers are easy to implement for square waves. The modified PLL of Figure 3.25b has more than one stable operating point for the phase which is a problem. If the frequency of the input is twice the desired frequency, the phase will lock in at either 0° or 180°. The modified PLL of Figure 3.25a does not have this problem.

Phase-lock loops are used for synchronization purposes. Suppose both the transmitter and receiver employ the circuit of Figure 3.26 to generate the pulse and word synchronizing signals. If the low-frequency input of the transmitter is sent as a separate signal to the receiver, which uses it as the input to the phase comparator, then the receiver timing signals will be synchronized with those of the transmitter. It is necessary that the delays introduced by the channel be the same for the communication signal, which

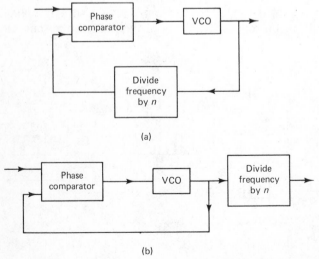

Figure 3.25   Modified PPLs based on (a) subharmonic and (b) harmonic input.

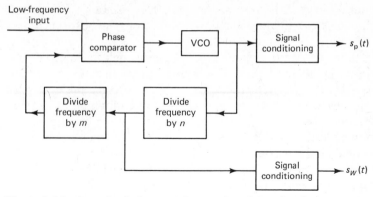

Figure 3.26 A method of generating synchronizing signals.

conveys the message information, and for the low-frequency tone, which conveys the timing information.

A different kind of external signal is transmitted between words. In this method one channel is time shared between the information signal and the timing or synchronizing signal. It is commonly used with time division multiplexing, as we have indicated. An easily identifiable word or sequence of bits is transmitted between every frame. If, for example, the receiver misidentifies some information bits as the synchronizing word, it will decode a frame of bits erroneously. It will not find the synchronizing word immediately afterward, but it will seek another group of bits with the same pattern as that of the synchronizing word. In this way the receiver will soon find the correct synchronizing word and remain in frame synch thereafter. Because the receiver "knows" the number and order of the messages that have been multiplexed, as well as the number of bits per word, word synchronizing signals are easily generated once the beginning of the frame is identified. This discussion has assumed that pulse synchronization has been achieved, and this can be accomplished in a number of ways.

If the bandwidth of the channel is sufficiently large ($m \geq 2$) to preserve the edges of the pulses, then the synchronizing word can be used to achieve pulse synchronization. Suppose, for example, the synchronizing word consists of a series of positive pulses followed by a single low pulse, as shown in Figure 3.27. When the receiver "recognizes" the synchronizing word (i.e., the signal goes "high"), the clock that generates the pulse-sampling signal is turned off. The leading, negative-going edge of the low pulse turns on the clock after a

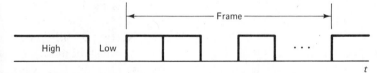

Figure 3.27 Synchronizing word between frames.

delay of 1.5 pulse widths. It is presumed that the frame is not long enough for the pulse-sampling clock to lose synchronization before the next synchronizing word.

If the bandwidth of the channel is not large enough to preserve pulse edges $\left(\frac{1}{2} < m \leq 1\right)$, some method of self-pulse synchronization is needed. These methods involve generating a signal that points to the previous pulse peaks through some means of waveshaping or signal conditioning. This signal is then the input to a PLL that generates the pulse-sampling signal. For example, if signals such as the RZ or raised-cosine-shaped pulses (see Figure 3.1) are full-wave rectified, the result is a periodic pulse train that indicates the pulse locations. This signal can be used to drive the PLL. More often, the zero crossings, which occur halfway between the pulse peaks, are used to generate the signal that drives the PLL. The circuit in Figure 3.47 (at the end of this chapter) uses this method of self-pulse synchronization. Sometimes wave-shaping techniques can be used for detection. For example, the time between a positive-going zero crossing and the next negative-going zero crossing is an indication of the number of consecutive positive pulses. A receiver that uses this detection method may require a pulse synchronizing signal for the decommutators and decoders.

3.7 DIGITIZING ANALOG SIGNALS

If a continuous signal, such as speech, music, and environmental or industrial measurements, is the input to an analog-to-digital converter (see Figure 3.12), the output is a sequence of binary words that can take on one of M values. This procedure is called *digitizing*. If the A/D output sequence is communicated, using any of the techniques that we have considered, to a receiver whose output is put through a D/A converter, the resulting continuous signal is an approximation to the original continuous signal. Such a system, using PCM, is shown in Figure 3.28. The continuous signal $m(t)$ is said to be sampled at a rate set by $s_W(t)$ (the word-sampling signal) and quantized into M levels by the A/D converter. Presumably, if the sampling rate (f_s) is fast enough and the number M of quantization intervals is large enough, the output $m_q(t)$ is a close approximation to the input (in the absence of receiver decision errors).

EXAMPLE 3.7 ■

We wish to communicate a speech signal (whose bandwidth is 3.6 kHz) over a telephone wire using PCM. Suppose that it has been determined subjectively that if the signal is quantized into 32 levels and sampled at a 9-kHz rate, then human observers will not discern the difference between the original signal and the discrete approximation of the output. Let us determine the required channel bandwidth. The pulse rate is

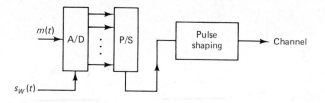

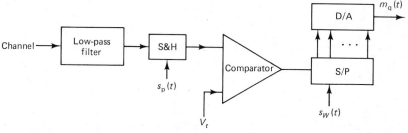

Figure 3.28 Communicating a continuous signal by PCM.

$$f_p = Nf_s = (\log_2 32)f_s = 5f_s = 45 \text{ kHz}$$

Assume $m = 1$ is adequate; then

$$B_{PCM} = mf_p = 45 \text{ kHz.}$$ ■ ■

We can visualize the digitizing process with the help of Figure 3.29. The signal is sampled at a rate $f_s = 1/T$ samples/s and quantized into M levels. The quantization intervals are all of equal value (Δ, where $\Delta = V_{pp}/M$ and where V_{pp} is the peak-to-peak value of the signal). This type of quantization is called

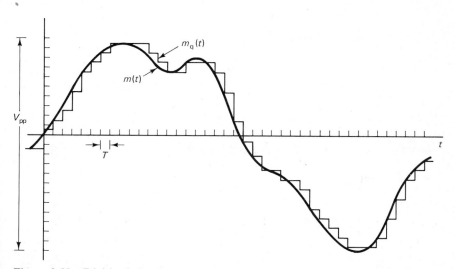

Figure 3.29 Digitized signal.

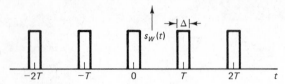

Figure 3.30 Word sampling signal.

uniform quantization. The A/D converter determines the quantization interval where a sample of the signal is placed, and the D/A converter output corresponds to the midpoint of this interval. Thus the quantization error ($|m(t) - m_q(t)|$) has a maximum value of $\Delta/2$ or $V_{pp}/2M$ for uniform quantization.

Very often the number of quantization levels is made large; it can therefore be argued that there is no real discernible error. For speech, music, or pictorial signals, one can determine the number of levels subjectively, that is, by the number above which human observers cannot distinguish between the original signal and the quantized signal. Depending on the signal, this usually occurs somewhere between 32 and 128 levels. While subjectivity does not apply to measured data, there is always a measurement inaccuracy (or measurement error) involved in obtaining the signal. The quantization error is inconsequential if it is smaller than the measurement error. More specifically, we can ignore the quantization error if the peak quantization error is less than the measurement error or if the measurement error is $k\%$ of V_{pp},

$$\frac{V_{pp}}{2M} < \frac{k}{100}V_{pp} \quad \text{or} \quad M > \frac{50}{k} \tag{3.14}$$

3.8 SAMPLING ANALYSIS

Suppose the word-sampling signal $s_W(t)$ is as shown in Figure 3.30, where the width Δ of the pulses is small [i.e., the signal $m(t)$ can be assumed to be constant over a time interval Δ]. The sampling is accomplished if $m(t)$ is the input to a transistor that is turned "on" by the pulses of $s_W(t)$ (i.e., the output corresponds to the input during the pulses and is zero otherwise). An example is shown in Figure 3.31, where the sampled signal $m_s(t)$ is given by

$$m_s(t) = m(t)s_W(t) \tag{3.15}$$

The average value of $s_W(t)$ [$\langle s_W(t) \rangle$] is Δ/T, and its autocorrelation function was computed in Chapter 2 (see Figure 2.24) and shown again in Figure 3.32.[4] The average value of the sampled signal, $\langle m_s(t) \rangle$, is given by

[4]In Chapter 2, the pulse heights were assumed to be $1/\Delta$ rather than 1. Because of this difference, there is a difference in the scale factor of Δ^2.

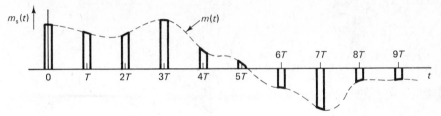

Figure 3.31 Sampled signal.

$$\langle m_s(t) \rangle = \langle s_W(t)m(t) \rangle \approx \frac{\Delta}{T} \left[\lim_{N \to \infty} \frac{1}{2N} \sum_{n=-N}^{N} m(nT) \right] \qquad (3.16)$$

assuming a small pulse width Δ. If $m(t)$ is random, rather than periodic,[5] we can assume that

$$\lim_{N \to \infty} \frac{1}{2N} \sum_{n=-N}^{N} m(nT) = \lim_{T' \to \infty} \frac{1}{T'} \int_{0}^{T'} m(t) \, dt = \langle m(t) \rangle \qquad (3.17)$$

and therefore

$$\langle m_s(t) \rangle = \frac{\Delta}{T} \langle m(t) \rangle \qquad (3.18)$$

[Note that if $m(t)$ is a deterministic, periodic signal, Equation 3.17 holds only for small sampling intervals T. If T corresponds to the period of the signal or a multiple of the period, then Equation 3.17 does not hold. We assume, however, that for random signals Equation 3.17 holds for all values of T.] For simplicity we assume that $\langle m_s(t) \rangle = \langle m(t) \rangle = 0$.

Let us now calculate the autocorrelation function of the sampled signal $m_s(t)$:

$$R_{m_s}(\tau) = \langle s_W(t)s_W(t - \tau)m(t)m(t - \tau) \rangle \qquad (3.19)$$

We recognize that if $\tau = nT + \tau'$, where n is any integer and $|\tau'| < \Delta$, then the product $s_W(t)s_W(t - \tau)$ has the same shape as the sampling signal $s_W(t)$ except that the pulse widths are $\Delta - |\tau'|$. (See Section 2.11 and Figure 2.23 for a more detailed explanation.) For other values of τ the product $s_W(t)s_W(t - \tau)$ is equal to zero. From Equations 3.18 and 3.19,

$$R_{m_s}(\tau) = \begin{cases} \dfrac{\Delta - |\tau - nT|}{T} \langle m(t)m(t - \tau) \rangle & |\tau - nT| < \Delta \\ 0 & \text{elsewhere} \end{cases}$$

$$\approx \begin{cases} \dfrac{\Delta}{T} \left(1 - \dfrac{|\tau - nT|}{\Delta} \right) R_m(nT) & |\tau - nT| < \Delta \\ 0 & \text{elsewhere} \end{cases} \qquad (3.20)$$

[5]The mathematical notion of ergodicity or strong mixing, which will be defined in Chapter 5, is needed to guarantee the validity of Equation 3.17.

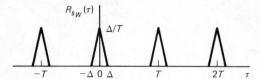

Figure 3.32   Autocorrelation function of $s_W(t)$.

Equation 3.20 is plotted in Figure 3.33, and comparing this plot with that of Figure 3.32, we recognize that

$$R_{m_s}(\tau) = R_m(\tau)R_{s_W}(\tau) \tag{3.21}$$

$R_{s_W}(\tau)$ is an even periodic signal and can be expanded in the Fourier series

$$R_{s_W}(\tau) = \left(\frac{\Delta}{T}\right)^2 \left[1 + \sum_{n=1}^{\infty} C_n \cos\left(\frac{2\pi n}{T}\tau\right)\right]$$

where $C_n = 2 \operatorname{sinc}^2(n\Delta/T)$ (see Equation 2.79 or Problem 2.14). Equation 3.21 can now be written

$$R_{m_s}(\tau) = \left(\frac{\Delta}{T}\right)^2 \left[R_m(\tau) + \sum_{n=1}^{\infty} C_n R_m(\tau) \cos\frac{2\pi n}{T}\tau\right] \tag{3.22}$$

Since the Fourier transform of $R_m(\tau)$ is $S_m(f)$ and the transform of $R_m(\tau)$ $\cos[(2\pi n/T)\tau]$ is $\frac{1}{2}[S_m(f + n/T) + S_m(f - n/T)]$ (see Equations 2.72 and 2.73), it follows that

$$\left(\frac{T}{\Delta}\right)^2 S_{m_s}(f) = S_m(f) + \sum_{n=1}^{\infty} \frac{C_n}{2}\left[S_m\left(f - \frac{n}{T}\right) + S_m\left(f + \frac{n}{T}\right)\right] \tag{3.23}$$

This relationship between the spectrum of the message and that of the sampled message is shown in Figure 3.34. It is not surprising that the sampled signal has a much wider spectrum than the original signal because of the sharp edges (see Figure 3.31). It is assumed in Figure 3.34 that the message $m(t)$ is strictly bandlimited [i.e., $S_m(f) = 0$ for $f > B$].

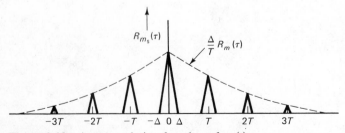

Figure 3.33   Autocorrelation function of $m_s(t)$.

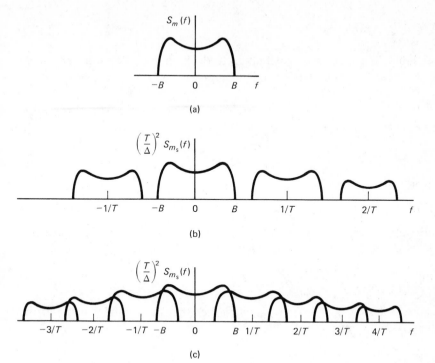

Figure 3.34 (a) Unsampled signal spectrum; (b) sampled spectrum when $1/T>2B$; (c) sampled spectrum when $1/T<2B$.

It is seen from Figure 3.34 that the spectral zones of the sampled signal do not overlap if and only if

$$f_s = 1/T \geq 2B \tag{3.24}$$

where B is the bandwidth of the continuous signal. A low-pass filter will reconstruct a bandlimited message with no distortion provided $f_s \geq 2B$. This remarkable fact is *Nyquist's theorem* and the $2B$ sampling rate is called the *Nyquist sampling rate*. If the sampling rate is less than $2B$, the original message cannot be reconstructed without distortion. This distortion, whether due to an insufficient sampling rate or a message that is not strictly bandlimited, is called *aliasing*. Although we can reconstruct a message that was sampled at its Nyquist rate with an ideal low-pass filter, we normally sample faster, making the reconstruction or filtering easier to accomplish.

Let us examine the reconstruction process more closely. Assume that the sampled message can be approximated by

$$m_s(t) = T \sum_{n=-\infty}^{\infty} m(nT)\delta(t - nT) \tag{3.25}$$

The response of an arbitrary linear filter with impulse response $h(t)$ is given by

$$y(t) = T \sum_{n=-\infty}^{\infty} m(nT)h(t - nT) \tag{3.26}$$

Equation 3.26 is a close approximation to the response of an arbitrary linear filter to the actual sampled signal. The impulse response of an ideal low-pass filter,

$$|H(f)| = \begin{cases} 1 & \text{if } f \leq B \\ 0 & \text{elsewhere} \end{cases}$$

$$\angle H(f) = 0$$

is given by

$$h(t) = 2B \operatorname{sinc} 2Bt, \tag{3.27}$$

(see Example 2.5). It follows that, if the signal is sampled at the Nyquist rate $(T = 1/2B)$, the response to the ideal low-pass reconstruction filter is

$$y(t) = \sum_{n=-\infty}^{\infty} m\left(\frac{n}{2B}\right) \operatorname{sinc}\left[2B\left(t - \frac{n}{2B}\right)\right] \tag{3.28}$$

Because sinc $k = 0$, if k is an integer other than zero (sinc $0 = 1$), at the Nyquist sample times $(t = k/2B)$

$$y(k/2B) = m(k/2B) \tag{3.29}$$

The output at a Nyquist sample time is the sample itself. Equation 3.28 also indicates the way in which the ideal low-pass filter interpolates the message between the samples. Thus, if $t = (k + p)/2B$, where $|p| < \frac{1}{2}$ (i.e., $k/2B$ is the "nearest" Nyquist sample time),

$$y\left(\frac{k + p}{2B}\right) = \sum_{q=-\infty}^{\infty} m\left(\frac{k + q}{2B}\right) \operatorname{sinc}(p - q) \tag{3.30}$$

The magnitude of the sinc$(p - q)$ weights fall off as q increases. Specifically,

$$|\operatorname{sinc}(p - q)| = \left|\frac{\sin \pi(p - q)}{\pi(p - q)}\right| = \left|\frac{\sin \pi p}{\pi(p - q)}\right| < \frac{1}{|q - \frac{1}{2}|\pi}$$

Therefore, to a good approximation,

$$y(t) \approx \sum_{q=-Q}^{Q} m\left(\frac{k + q}{2B}\right) \operatorname{sinc}\left[2B\left(t - \frac{k + q}{2B}\right)\right] \tag{3.31}$$

where $k/2B$ is the nearest Nyquist sample time and where Q can be as small as 3 or 4. We have just argued that a finite segment of $y(t)$ (of duration T' s) can be reconstructed accurately from its Nyquist samples provided there are a large number of such samples $(2BT' \gg 1)$. The resultant truncation errors will be limited to small segments (roughly three samples) at each end.

The digital-to-analog converter method of interpolation (see Figure 3.29) does not reconstruct the continuous message as acurately as the ideal low-pass filter. The D/A behaves like a linear filter whose impulse response is given by

$$h(t) = \begin{cases} 1 & 0 \le t < T \\ 0 & \text{elsewhere} \end{cases} \tag{3.32}$$

The interpolation consists of "holding" the value of a Nyquist sample until the next sample arrives.

We can get some insight to the hold method of interpolation by looking at the transfer function of this linear filter, $|H(f)| = T\,|\text{sinc } f/f_s|$, together with the sampled signal spectrum (Figure 3.35). Although it is not an ideal low-pass filter, the hold technique does behave like a low-pass filter. There is some distortion of the desired spectral zone because $|H(f)|$ is not constant over the band B Hz. There is an additional error contributed by the other spectral zones that are not completely filtered out. If the sampling rate is sufficiently larger than the Nyquist rate, the distortion of the desired spectral zone is minimal, but the other spectral zones are still present and appear as a staircase approximation, as in Figure 3.29. This effect can be largely eliminated with additional low-pass filtering.

We conclude that both the quantization error and the sampling distortion can be made insignificant and that discrete communication techniques can be applied to bandlimited continuous signals as well as discrete signals.

EXAMPLE 3.8 ■

We wish to multiplex, using PCM and TDM, A acoustical measurements (each having a bandwidth of 10 kHz) through a cable that can handle a pulse rate of 1.5 MHz. The accuracy of the measurements is 1% of the peak-to-peak value (or $\frac{1}{2}$% of the peak value). After each frame, consisting of one sample from each measurement, there is a 12-bit synchronizing word. How big can A be?

From Nyquist's theorem $f_s > 20$ kHz. Set $f_s = 25$ kHz. From Equation 3.14, $M > 50$. Set $M = 64$ (or $\log_2 M = 6$). It follows that the information pulse rate is $A \times 150$ kHz. Since the baud rate must be no larger than 1.5 MHz,

$$A \times 150 \times 10^3 (6A + 12)/6A \le 1.5 \times 10^6$$

Solving, we get $A \le 8$.   ■ ■

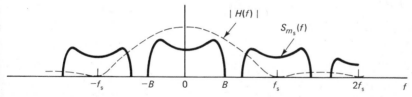

Figure 3.35 Reconstruction by the hold method.

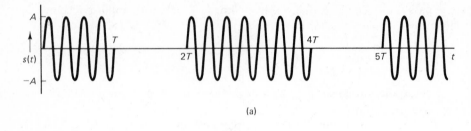

(a)

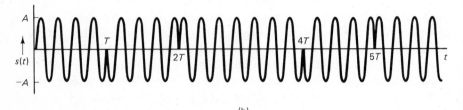

(b)

Figure 3.36   Modulated binary signals.

3.9 BINARY MODULATION TECHNIQUES

In order to communicate using electromagnetic waves it is necessary to employ a high-frequency (carrier) sinusoid (see Section 2.10 and Figure 2.21). If the pulses used in PCM are sinusoidal bursts, the signals would be applicable for these high-frequency, bandpass channels as well as cables. For example, if the monopolar and bipolar pulses of Figure 3.1 were replaced with sinusoidal bursts, they would appear as in Figure 3.36. The signal in Figure 3.36a is generated by turning a sinusoid on or off and is called *amplitude shift keying* (ASK). The signal in Figure 3.36b is generated by switching the phase of the sinusoid between 0° and 180° and is called *phase shift keying* (PSK). Both schemes can be implemented by the modulator or modem of Figure 3.37. The gate, or analog switch, passes the sinusoid when the controlling signal is positive. When the controlling signal is negative, the gate either inverts the sinusoid or turns it off.[6]

We saw in Chapter 2 that the spectrum for any bipolar random impulse train (see Equation 2.95 and Table 2.2) is

$$S_{bip}(f) = \frac{1}{T}|H(f)|^2 \tag{3.33}$$

and for any monopolar random impulse train

$$S_{mon}(f) = \left[\frac{1}{4T} + \frac{1}{4T^2}\sum_{n=-\infty}^{\infty}\delta\left(f - \frac{n}{T}\right)\right]|H(f)|^2 \tag{3.34}$$

---

[6]Other multiplying circuits, such as the chopper circuit of Problem 3.26, can also be used.

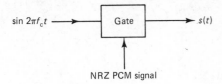

Figure 3.37 Modem for modulating binary signals.

For ASK and PSK, $H(f)$ is the Fourier transform of a sinusoidal burst. If we let

$$h(t) = Af(t) \cos 2\pi f_c t \tag{3.35}$$

where $f(t)$ is the constant 1 for a T-s interval and 0 elsewhere, it follows from Table 2.1 that

$$|H(f)| = \frac{A}{2}|F(f - f_c) + F(f + f_c)|$$

$$= \frac{A}{2}|T \operatorname{sinc} T(f - f_c) + T \operatorname{sinc} T(f + f_c)|$$

For large enough f_c, the positive and negative zones do not overlap, and

$$|H(f)|^2 = \frac{A^2 T^2}{4}[\operatorname{sinc}^2 T(f - f_c) + \operatorname{sinc}^2 T(f + f_c)] \tag{3.36}$$

Combining Equations 3.33 and 3.36 yields

$$S_{PSK}(f) = \frac{A^2 T}{4}[\operatorname{sinc}^2 T(f - f_c) + \operatorname{sinc}^2 T(f + f_c)] \tag{3.37}$$

Let us assume that $f_c = k/T$, where k is an integer (i.e., there is an integral number of cycles per pulse). Then

$$\operatorname{sinc}^2 T\left(\frac{n}{T} - f_c\right) = \operatorname{sinc}^2(n - k) = \begin{cases} 1 & n = k \\ 0 & n \neq k \end{cases} \tag{3.38}$$

Combining Equations 3.34, 3.36, and 3.38 yields

$$S_{ASK}(f) = \frac{A^2 T}{16}[\operatorname{sinc}^2 T(f - f_c) + \operatorname{sinc}^2 T(f + f_c)]$$

$$+ \frac{A^2}{16}[\delta(f - f_c) + \delta(f + f_c)] \tag{3.39}$$

These spectra are shown plotted in Figure 3.38. As in the amplitude modulated signal (see Section 2.10), the spectra of ASK and PSK are centered about f_c and the bandwidth of these signals is twice that of baseband PCM:

$$B_{ASK} = B_{PSK} = 2mf_p \tag{3.40}$$

where $f_p = 1/T$ pulses/s. Pulse shaping to reduce the factor m to slightly more

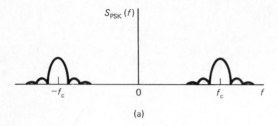

(a)

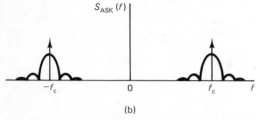

(b)

Figure 3.38 Spectra of modulated binary signals: (a) a PSK signal; (b) an ASK signal.

than $\frac{1}{2}$ can be accomplished in two ways. The signals of Figure 3.36 can be passed through a bandpass, pulse-shaping filter that is centered at the carrier frequency. Alternatively, the pulses of the baseband PCM signal can be pulse shaped and the gate in Figure 3.37 replaced with a multiplier circuit. Pulse shaping is used less frequently, however, for the modulated signals since the bandwidth of electromagnetic channels is usually less restricted than the bandwidth of cables.

The front end of the receiver for these modulated binary signals could be a decision device, as in Figure 3.21. More often, it is a demodulator or modem that converts the signal into a noisy baseband PCM signal, and the rest of the receiver is a PCM receiver as in Figure 3.10. For example, a demodulating modem for ASK is shown in Figure 3.39. (This envelope detector was discussed and analyzed in Example 2.7.) Demodulating PSK is more difficult. Since the "information" is in the phase of the carrier sinusoid, a phase reference is essential. Thus the receiver must generate a sinusoid that is locked in phase with the carrier. This represents *phase synchronization*.

If a separate subharmonic sinusoid is transmitted (via amplitude modulation on a different carrier) for synchronizing purposes, the PLL of either

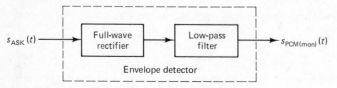

Figure 3.39 ASK demodulating modem.

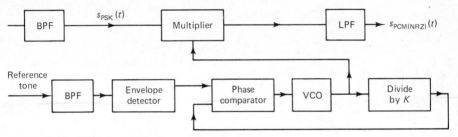

Figure 3.40 PSK demodulating modem.

Figure 3.24 or 3.25b can be used to generate a phase reference. Using this phase reference, the modem of Figure 3.40 will demodulate the PSK signal. If the signal pulses are modeled by $\pm A \sin(2\pi f_c t + \theta)$, the output of the VCO is presumed to be $\sin(2\pi f_c t + \theta)$ (i.e., locked in frequency and phase). The output of the multiplier becomes

$$\pm A \sin^2(2\pi f_c t + \theta) = \pm \frac{A}{2} \mp \frac{A}{2} \cos(4\pi f_c t + 2\theta)$$

The low-pass filter removes the high-frequency sinusoidal component, and its output becomes an NRZ, PCM pulse train.

Other methods of achieving phase synchronization are possible [3]. For example, self-synchronization can be achieved if the PSK signal is full-wave-rectified, and the fundamental component can then be used to drive the PLL of Figure 3.25a. As indicated earlier, however, the VCO output will be locked into either 0° or 180°, and some method of resolving this uncertainty must be found. If TDM is being used, the synchronizing word can be utilized to resolve this problem.

A signal that is similar to PSK but avoids phase synchronization is called *differential phase shift keying* (DPSK). The modulating and demodulating circuits for this system are shown in Figure 3.41. The phase of the carrier going into the gate is locked to that of the previous pulse. Thus the information is carried not by the value of the phase but by the difference

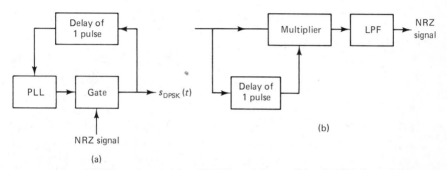

Figure 3.41 DPSK modems: (a) a DPSK modulator; (b) a DPSK demodulator.

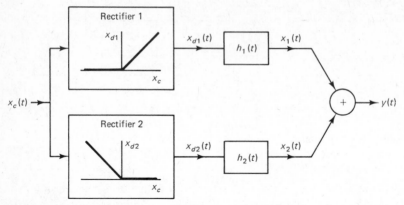

Figure 3.42   Model for a two-pulse binary pulse train.

between phases of adjacent pulses. The receiver demodulates by "comparing" adjacent pulses, rather than by "comparing" the input pulse with a phase reference. Unfortunately, this system does not work as well as PSK in the presence of noise [3, 5].

Let us now consider binary pulse trains that consist of two different kinds of pulses. These pulses can be generated from the bipolar random impulse train $[x_c(t)]$ with the hardware of Figure 3.42. The rectifier (half-wave) circuits, however, are not really needed and $x_{d1}(t)$, indicating the location of the 1s, and $x_{d2}(t)$, indicating the location of the 0s, could have been generated directly. If the two impulse responses are sinusoidal bursts of different frequencies, so that for $i = 1, 2$

$$h_i(t) = \begin{cases} A\, \sin\!\left(2\pi \dfrac{k_i}{T}t\right) & 0 \le t \le T \\ 0 & \text{elsewhere} \end{cases} \tag{3.41}$$

the signal is called *frequency shift keying* (FSK). The filters are normally replaced by gate circuits, and the FSK signal is generated by switching between two different oscillators operating at different frequencies (k_1/T and k_2/T).

Since $x_{d1}(t)$ and $x_{d2}(t)$ are (monopolar) random impulse trains, the spectra of $x_1(t)$ and $x_2(t)$ are identical to that of ASK; for $i = 1, 2$, then,

$$S_{x_i}(f) = \frac{TA^2}{16}\left[\operatorname{sinc}^2 T\!\left(f - \frac{k_i}{T}\right) + \operatorname{sinc}^2 T\!\left(f + \frac{k_i}{T}\right)\right]$$
$$+ \frac{A^2}{16}\left[\delta\!\left(f - \frac{k_i}{T}\right) + \delta\!\left(f + \frac{k_i}{T}\right)\right] \tag{3.42}$$

It might be assumed that the spectrum of FSK would be the sum of these two spectra. Unfortunately, this is not true, although the FSK spectrum of Figure 3.43 does look similar to that sum. The autocorrelation of the combined signal is given by

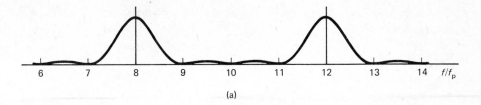

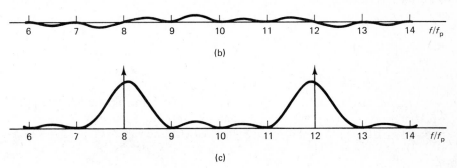

Figure 3.43 Spectrum of an FSK signal.

$$R_y(\tau) = \langle [x_1(t) + x_2(t)][x_1(t - \tau) + x_2(t - \tau)] \rangle$$
$$= R_{x_1}(\tau) + R_{x_2}(\tau) + \underbrace{R_{x_1 x_2}(\tau) + R_{x_2 x_1}(\tau)}_{R_{\dot{x}}(\tau)} \qquad (3.43)$$

The speculation would be correct if and only if $R_{\dot{x}}(\tau) = 0$ for all τ. It is true that $R_{\dot{x}}(0) = 0$, since $x_1(t)$ and $x_2(t)$ do not overlap. Furthermore, $R_{\dot{x}}(\tau) = 0$ for $|\tau| > T$, since for these large delays $x_1(t)$ and $x_2(t - \tau)$ are unrelated.[7] In the range $0 < |\tau| < T$, $R_{\dot{x}}(\tau)$ is not zero (the signals are random sinusoidal bursts, not sinusoids). $R_{\dot{x}}(\tau)$ is small, however, in this range provided k_1 and k_2 are much greater than 1 (i.e., there are many cycles of each frequency in each pulse).

In the range $0 < \tau < T$, half the sinusoidal pulses of $x_1(t)$ overlap a delayed sinusoidal pulse of $x_2(t)$ for T s. Because $x_1(t) = 0$ half of the time, there is on the average one overlap for every $4T$ s. Hence,

$$R_{\dot{x}}(\tau) = R_{x_1 x_2}(\tau) + R_{x_2 x_1}(\tau)$$
$$= \frac{A^2}{4T} \int_0^\tau \sin\left(2\pi \frac{k_1}{T} t\right) \sin\left[2\pi \frac{k_2}{T}(t - \tau)\right] dt$$
$$+ \frac{A^2}{4T} \int_0^\tau \sin\left(2\pi \frac{k_2}{T} t\right) \sin\left[2\pi \frac{k_1}{T}(t - \tau)\right] dt$$

[7]This undefined notion will be clarified in Chapter 5 along with the proof that the value of $R_{\dot{x}}(\tau)$ is zero for unrelated signals.

Evaluating these integrals and recognizing the symmetry for $\tau < 0$, one obtains for $|\tau| < T$

$$R_{\dot{x}}(\tau) = \frac{k_2 A^2}{4\pi(k_1^2 - k_2^2)} \sin 2\pi \frac{k_1}{T} |\tau|$$

$$- \frac{k_1 A^2}{4\pi(k_1^2 - k_2^2)} \sin 2\pi \frac{k_2}{T} |\tau| \tag{3.44}$$

where the tedious but straightforward calculations have been omitted. Taking Fourier transforms gives

$$S_{\dot{x}}(f) = \frac{-k_2 T A^2}{4\pi^2(k_1^2 - k_2^2)} \frac{[\sin \pi T(f - k_1/T)]^2}{(fT - k_1)}$$

$$+ \frac{k_1 T A^2}{4\pi^2(k_1^2 - k_2^2)} \frac{[\sin \pi T(f - k_2/T)]^2}{(fT - k_2)} \tag{3.45}$$

where there are two similar terms for the negative frequency range. This cross spectrum, the spectral terms of Equation 3.42, and their sum $S_{FSK}(f)$ are plotted for positive frequencies in Figure 3.43 for the values $k_1 = 12$ and $k_2 = 8$.

By inspection of Figure 3.43,

$$B_{FSK} = (f_1 - f_2) + 2mf_p \tag{3.46}$$

where $f_p = 1/T$ is the pulse-sampling rate. If the spectral zones do not overlap $(f_1 - f_2 \geq 2mf_p)$, the demodulator of Figure 3.44 will work. The outputs of the bandpass filters are not precisely the same as $x_1(t)$ and $x_2(t)$, but if k_1, $k_2 \gg 1$, the outputs will be close enough to $x_1(t)$ and $x_2(t)$ for this receiver to work. In this event the bandwidth must satisfy

$$B_{FSK} \geq 4mf_p \tag{3.47}$$

or be at least twice that of either ASK or PSK.

If the sinusoidal frequencies are much closer together (i.e., the receiver of Figure 3.44 will not work), demodulation is still possible with the demodulators used for PSK (Figure 3.40). In both branches the bandpass filter and envelope detector are replaced by these demodulators. Separate phase synchronization is needed for each sinusoid. This method is rarely if ever used, however, since there are no advantages relative to PSK.

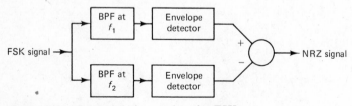

Figure 3.44 Demodulating modem for FSK.

EXAMPLE 3.9 ■

Let us reconsider the problem of multiplexing the acoustic measurement of Example 3.8. We have a microwave link that has a bandwidth of 6 MHz. As before the baud rate, if A messages are multiplexed, is

$$f_b = A \times 150 \times 10^3 \frac{6A + 12}{6A}$$

If we use ASK or PSK and if $m = 1$,

$$B = 2f_b = 2A \times 150 \times 10^3 \frac{6A + 12}{6A} \leq 6\,\text{MHz}$$

Solving, we get $A \leq 18$. If we use FSK with the receiver of Figure 3.44, then

$$B \geq 4f_b = 4A \times 150 \times 10^3 \frac{6A + 12}{6A} \leq 6\,\text{MHz}$$

and $A \leq 8$.   ■ ■

3.10 MULTILEVEL MODULATION TECHNIQUES

Most of the common multilevel modulation techniques are natural extensions of the binary modulation techniques. We can extend ASK, for example, into an L-level ASK technique. Just as PAM has the same spectrum as the binary random pulse train, L-level ASK has the same spectrum as ASK, and Equation 3.40 applies to L-level ASK. Similarly, PSK can be extended to L-level PSK where the bandwidth is also given by Equation 3.40. The same ASK demodulators can be used for L-level ASK (Figure 3.39). This demodulator would be followed by a PAM receiver, however, rather than a PCM receiver. The demodulator decision mechanism for L-level PSK is slightly different than that of PSK (see Problem 3.25).

The four-level PSK scheme, called *quadrature phase shift keying* (QPSK), has received some attention. Like all of the L-level amplitude or phase schemes, its principal advantage is bandwidth reduction. Since the pulse rate is always $f_p = f_s \log_L M$, the larger the number of levels, the smaller the pulse rate, and hence the smaller the bandwidth. However, we shall see that, unlike the other L-level techniques, QPSK performs as well as PSK in the presence of noise. The QPSK receiver is more complicated, and phase synchronization is more critical. All the phase modulated schemes have the problem of phase synchronization, but they perform better than the amplitude modulated schemes. It is possible to use amplitude and phase modulation simultaneously, and such schemes have been employed where bandwidth reduction is essential and performance in the presence of noise can be sacrificed.

A very efficient scheme is the natural extension of FSK to L different

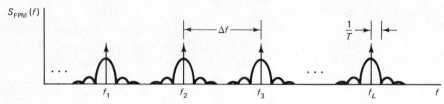

Figure 3.45   Spectrum of an FPM signal.

frequencies, which is called *frequency position modulation* (FPM). The power spectrum of FPM and a workable receiver are natural extensions of those shown in Figures 3.43 and 3.44 for FSK. The bandwidth of FPM can be quite large; in Figure 3.45 it is seen to be

$$B_{\text{FPM}} = (f_L - f_1) + 2mf_p \tag{3.48}$$

where $f_p = 1/T$ is the pulse rate and $f_p = (\log_2 M/\log_2 L)f_s$. If each of the frequency differences $\Delta f = f_K - f_{K-1}$ are equal (as in Figure 3.45), then Equation 3.48 becomes

$$B_{\text{FPM}} = (L - 1)\Delta f + 2mf_p \tag{3.49}$$

While the demodulator for FPM is similar to that of FSK, it must be combined with the decision mechanism, as in Figure 3.46. For this receiver to work it is necessary that the frequency zones do not overlap, or $\Delta f > 2mf_p$;

$$B_{\text{FPM}} \geq 2mLf_p \tag{3.50}$$

Theoretically, FPM can be demodulated even when the zones overlap, but this is rarely done.

It is of interest to compare the bandwidth formulas of Equations 3.11 and 3.50. Can we argue that FPM is a modulated form of PPM? Surely not!

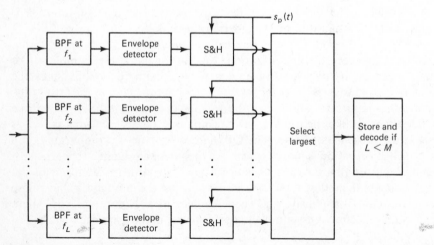

Figure 3.46   FPM receiver.

The signals and the receivers look very different. Nevertheless, we shall see in Chapter 7 that there is a correspondence between PPM and FPM.

EXAMPLE 3.10 ■

Let us summarize some of the key bandwidth formulas by way of an example. Consider the use of a teletype that has 64 characters with a maximum typewriter speed of 15 char/s. We calculate the required channel bandwidth for a variety of communication schemes. The factor m is chosen arbitrarily. The results are given in Table 3.1. ■ ■

3.11 INTRODUCTION TO REPEATERS

It is possible to introduce hardware along the communication channel that can detect the signal pulses using the decision mechanisms of a receiver and subsequently generate and transmit new noise-free pulses. The receiver decision errors caused by the channel noise are greatly reduced. This effect will be analyzed in Chapter 6. These devices, called *repeaters*, are used regularly when the channel is a cable such as a telephone line or a submarine cable. In the use of communications satellites, signals are not simply reflected: There are repeaters on the satellite.

We have introduced the concept of repeaters for the primary purpose of combining many of the hardware techniques and circuits that have been discussed in this chapter. For example, a possible repeater for a PSK communications scheme is shown in Figure 3.47. The demodulation and phase synchronization techniques shown in Figure 3.40 are used. Because pulse-shaping filters are employed to reduce the required channel bandwidth, the gate in the modulator of Figure 3.37 is replaced with a multiplier. The method of self-pulse synchronization, based on the zero crossings (as discussed in Section 3.6), is used. Although a frame synchronizing word may be transmitted, the repeater treats all pulses alike and is not concerned with

TABLE 3.1

Communication scheme	Bandwidth formula	m	N (pulses/ character)	Pulse rate f_p	B (Hz)
PAM ($L = 64$)	mf_p	2	1	15	30
PCM	mf_p	1	6	90	90
PPM ($L = 64$)	mLf_p	2	1	15	1920
PPM ($L = 8$)	mLf_p	2	2	30	480
PSK	$2mf_p$	1	6	90	180
FSK	$\geq 4mf_p$	1	6	90	≥ 360
FPM ($L = 64$)	$\geq 2mLf_p$	2	1	15	≥ 3840
FPM ($L = 8$)	$\geq 2mLf_p$	2	2	30	≥ 960

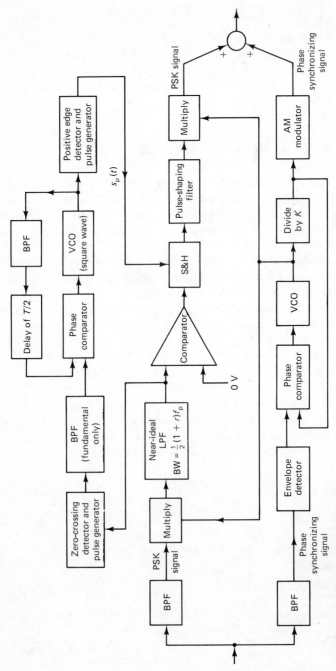

Figure 3.47 A possible PSK repeater.

word or frame synchronization. Decommutation and decoding (which require word and frame synchronization) are performed at the receiver and not at the repeaters.

3.12 CONCLUSIONS

This chapter presented the basic techniques of discrete communications. We concentrated on the hardware realizations and the required channel bandwidths. Chapters 5–8 are dedicated to analyzing the performance of these techniques in the presence of noise. The final choice of which scheme to use is, of course, an engineering compromise between performance in the presence of noise, receiver complexity or cost, and the channel bandwidth required.

Very often the communication task encountered has unique problems that require modification of these schemes. Such modifications require a great deal of insight. A good example of this can be seen by studying the communication signal for a push-button telephone. After studying the various communication techniques, we would probably conclude that FPM is natural for this task. Not only is FPM very efficient, but no significant hardware cost lies in the transmitter or head set, since it is an easy task to energize one of 12 oscillators by depressing a button. Also, the required bandwidth is quite modest. Using a maximum symbol rate of 15 char/s, we can determine that a bandwidth of 720 Hz (assuming $m = 2$) is adequate, significantly less than the 3.3 kHz bandwidth of a typical telephone channel.

The receiver avoids the pulse synchronization problem by fixing the length of each sinusoidal burst (let us say 60 ms) and then having the receiver sample the signal once every 60 ms. As a consequence of this procedure, the receiver must decide which of the 12 tones and a noise-only message is most likely. Unfortunately, the major source of error is something other than the usual additive noise. A sudden connection change at the switching station or a sudden lifting or hanging up of an extension phone introduces a ringing in the line at the resonant frequency of the system. This ringing causes the receiver to mistake the noise-only message for a sinusoidal burst. The solution to this problem is to modify the FPM signal set so that each message corresponds to a pair of sinusoidal tones determined by the location (row and column) of the button depressed.

Continuous signals, such as speech, can utilize many special methods of communications, both discrete and continuous, that we have not considered. We shall discuss some of these techniques in the next chapter prior to beginning the analysis of communicating in the presence of noise.

PROBLEMS

3.1. We wish to communicate the output of a teletype that generates 5000 char/s (with 32 possible characters) by a PCM system. Determine
(a) the number of bits per character,

    (b) the number of pulses per second,

    (c) the bandwidth required if $m = r = 1$.

3.2. Consider the system

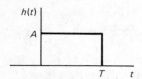

$$x(t) = \sum_{n=-\infty}^{\infty} a_n \delta(t - nT) \qquad \boxed{h(t)} \qquad y(t) = \sum_{n=-\infty}^{\infty} a_n h(t - nT)$$

    where $a_n = \pm 1$ as determined by the flip of a coin.

    (a) Sketch the output $y(t)$ if $h(t)$ is given by

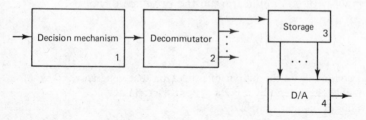

    (b) Determine and sketch the power spectrum of the output.

3.3. Repeat Problem 3.2 if

$$h(t) = \begin{cases} \dfrac{A}{2}(1 + \cos \pi t/T) & |t| < T \\ 0 & |t| > T \end{cases}$$

    (*Hint*: Use Table 2.1.)

3.4. We are given the problem of communicating a series of symbols, each of which can take on 512 values, at a rate of $f_s = 10^5$ symbols/s.

    (a) What is the required bandwidth if PCM is used (with $m = 1$)?

    (b) Repeat for PAM (with $m = 2$) if each pulse can take on eight amplitudes.

    (c) Repeat for PPM (with $m = 2$) if each pulse can take on eight different time slots.

3.5. Ten analog measurements whose bandwidth is 4 kHz are to be transmitted simultaneously through a cable by time division multiplexing using PCM. Each message is sampled at a 10-kHz rate and quantized into 128 levels. The pulses are shaped with a near-ideal low-pass filter with a rolloff factor r of 0.2. At the beginning of each frame (consisting of one word from each of the ten signals) there is a seven-bit synchronizing word.

    (a) Why is each measurement sampled at a 10-kHz rate?

    (b) What is the multiplexed pulse rate f_{p_I} (information bits only)?

    (c) What is the baud rate f_b?

    (d) What is the required bandwidth of the cable?

       Assume the following simplified block diagram for the receiver:

(e) Indicate which (if any) types of synchronization are needed for each of the four boxes.

3.6. The continuous signal whose spectrum is shown below is quantized, sampled, and transmitted using a baseband PCM system:

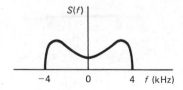

(a) If each data sample must be known to within ±1% of the peak-to-peak full-scale value, how many binary symbols must each transmitted digital word contain?

(b) Assuming the signal is sampled at the Nyquist rate, estimate the bandwidth of the resulting PCM signal.

3.7. Suppose the signal of Problem 3.6 is to be transmitted via a PPM modulation system. Estimate the required bandwidth assuming $m \approx 3$.

3.8. You wish to transmit an analog signal (bandwidth $B = 15$ kHz) via a baseband PCM code employing seven bipolar pulses per sample.

(a) Sketch a reasonable receiver up to the analog output.

(b) Which types of synchronization are needed to implement the receiver?

3.9. Suppose the signal of Problem 3.1 was transmitted via FSK using the two frequencies 1 and 1.1 MHz.

(a) What is the bandwidth necessary?

(b) Sketch the form of a workable receiver.

(c) Which types of synchronization are needed to implement this receiver?

(d) Would this receiver work if the two frequencies were 1 and 1.02 MHz? Why?

3.10. Your problem is to design the communication link for transmitting radiation measurements from a satellite to a ground station. The analog signal has a bandwidth of 20 kHz, and the measurement error is approximately 1% of the maximum peak-to-peak signal. You wish to convert this signal to a discrete signal.

(a) Indicate the sampling rate and number of quantization levels per sample that you would use.

(b) You wish to consider some binary modulation technique. Indicate which modulation scheme you would use and why.

(c) Determine the bandwidth of your signal.

(d) Sketch the receiver you would use and indicate which types of synchronization signals you would need.

3.11. Now consider transmitting the signal of Problem 3.10 using an encoded FPM scheme where $L = 12$ (12 frequencies per pulse).

(a) Indicate the frequencies you might use.

(b) Repeat parts (c) and (d) of Problem 3.10.

3.12. Calculate the bandwidth required for the transmission of five messages via a cable using a time multiplexed PCM system. Each message occupies a 20-kHz bandwidth, is sampled at the minimum (Nyquist) rate, and is quantized into 16 levels. It is assumed that the transmission bandwidth must be 1.5 times the pulse rate in order for the intersymbol interference to be acceptable.

3.13. Suppose the multiplexed signal of Problem 3.12 is converted into an FSK signal, the two frequencies being 4 and 7 MHz.
   (a) What is the required channel bandwidth?
   (b) Show that the receiver of Figure 3.44 can be used for this signal.

3.14. Consider a source that transmits symbols or messages at a rate f_s symbols/s. Each symbol must represent one out of a possible 64 messages. We wish to consider two methods of transmission. The first is a binary PSK code and the second is a four-level FPM code (i.e., each pulse is one of four frequencies). For each scheme
   (a) Indicate the number of pulses/message,
   (b) calculate the signal bandwidth (assume $m = 1$ for both schemes), and
   (c) determine which method requires phase synchronization.

3.15. Consider again the signal of Problem 3.14 where the symbol rate is 8 kHz. Estimate the bandwidth for each of the following modulation schemes (assume $m = 2$ for all cases):
   (a) FSK with $f_1 = 4$ MHz and $f_2 = 4.5$ MHz,
   (b) PPM,
   (c) FPM where the tones are spaced 100 kHz apart, and
   (d) a PPM code using eight different positions for each pulse.

3.16. We desire to communicate simultaneously (using time division multiplexing) ten similar series of messages, each of which can take on 1024 values, through a microwave link. Each source is transmitted at a rate of $f_s = 10^5$ messages/s.
   (a) If PSK is used (with $m = 2$), what is the required bandwidth?
   (b) If FSK is used (with $m = 2$), what is the minimum bandwidth which permits the receiver to avoid the need for phase synchronization?
   (c) If FPM (each pulse can be one of ten frequencies; i.e., $L = 10$) is used ($m = 2$) with the same hardware constraint of part (b), what is the minimum required bandwidth?

3.17. A single information channel carries voice frequencies in the range 50–3300 Hz. The channel is sampled at an 8-kHz rate and quantized into 32 levels.
   (a) Assume the signal is transmitted by baseband PCM. Find the required transmission bandwidth.
   (b) Assume now the signal is transmitted by FSK by switching between two sine waves of frequencies 100 and 104 MHz. Find the required bandwidth (assuming $m = 2$).

3.18. Consider three ways of communicating one of 60 messages, each of which can be regarded as an L-level phase modulation technique. The first method is to encode in a binary code where each sinusoidal pulse has a phase of 0° or 180° (PSK). The second method is to encode in a tertiary or three-level code, where each pulse has a phase of 0°, 120°, or −120°. The final method is to encode in a four-level code, where each pulse has a phase of 0°, 90°, 180°, or 270° (QPSK).
   (a) How many pulses per message are needed for each scheme?
   (b) Calculate the necessary bandwidth for each scheme. Which scheme would be useful when bandwidth is at a premium?
   (c) How would you store the decisions on each pulse?
   (d) Discuss the practical hardware difficulties for each scheme.

(e) Could you use differential techniques for the three- and four-level codes as well as DPSK?

3.19. Twelve messages, each having a word rate of 8 kHz, where each word can have 64 values, are encoded using a baseband PCM technique and time multiplexed over a single channel.

(a) Ignoring any extra synchronizing pulses, and assuming pulse shaping with a rolloff factor $r = 0.5$, find the bandwidth required.

(b) Assume now that a single frame will accommodate three entire words from all 12 messages and that 34-frame synchronization bits are inserted between every frame. What is the bandwidth required?

3.20. Suppose the multiplexed signal of Problem 3.19b is next modulated into an FSK signal, the two frequencies being 4 and 7 MHz.

(a) What is the required bandwidth?

(b) Show that the two frequencies are workable for simple detection as in Figure 3.44.

(c) Sketch in block diagram form such a receiver, including the commutator.

3.21. Reconsider Problem 3.20 but assume now that we wish to multiplex more than 12 messages. The available bandwidth is 6 MHz (i.e., 2.5–8.5 MHz channel).

(a) Determine the two frequencies that you would use in order to maximize the number of messages that can be multiplexed while still guaranteeing that the receiver of Figure 3.44 will work.

(b) What is the maximum number of messages that can be communicated using your design?

3.22. Now reconsider Problem 3.21 with the change that, instead of using FSK, we wish to consider PSK and QPSK. Determine the maximum number of messages we can communicate over the 6-MHz channel using both these modulation schemes.

3.23. Sketch a complete repeater that can be used for FSK which does not employ phase synchronization or pulse shaping.

3.24. Repeat Problem 3.23 for an FPM signal.

3.25. The PSK demodulator of Figure 3.40 must be modified for QPSK as shown on page 104.

(a) What are the four possible inputs to the decision mechanism (assuming no noise and a pulse amplitude of A V)?

(b) Indicate a possible circuit for the decision mechanism/encoder.

3.26. A gate that is turned on and off rapidly with a high-frequency square wave is called a chopper since the output appears as shown below when the input is an NRZ PCM signal.

If $m(t)$ is the input, the output can be modeled as $m(t)x(t)$ where $x(t) = \frac{1}{2} + (2/\pi)(\cos 2\pi f_c t + \frac{1}{3} \cos 6\pi f_c t + \cdots)$, where f_c is the chopping frequency (see Problem 2.20).

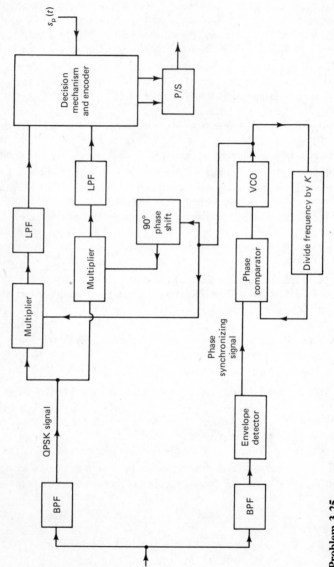

$s_p(t)$

Decision mechanism and encoder

P/S

LPF

LPF

Multiplier

90° phase shift

Multiplier

QPSK signal

Phase synchronizing signal

VCO

Phase comparator

Divide frequency by K

BPF

Envelope detector

BPF

Problem 3.25

(a) Show that if the chopper is followed by an appropriate filter and the bandwidth of the input signal $m(t)$ is less than f_c, then the output is $m(t) \cos 2\pi f_c t$.

(b) Show that this technique can be used as the multiplier in the demodulation circuit of Figure 3.40.

REFERENCES

1. H. Nyquist, "Certain Topics in Telegraph Transmission Theory," *Trans. AIEE* 47:617–644 (April 1928).
2. W. R. Bennet and J. R. Davey, *Data Transmission*, McGraw-Hill, New York, 1965.
3. J. G. Proakis, *Digital Communications*, McGraw-Hill, New York, 1983.
4. W. C. Lindsey and M. K. Simon, *Telecommunications Systems Engineering*, Prentice-Hall, Englewood Cliffs, NJ, 1973.
5. M. Schwartz, W. R. Bennet, and S. Stein, *Communication Systems and Techniques*, McGraw-Hill, New York, 1966.
6. A. Blanchard, *Phase Locked Loops: Application to Coherent Receiver Design*, Wiley, New York, 1976.
7. A. J. Viterbi, *Principles of Coherent Communications*, McGraw-Hill, New York, 1966.
8. F. M. Gardner, *Phaselock Techniques*, Wiley, New York, 1966.
9. J. J. Stiffler, *Theory of Synchronous Communications*, Prentice-Hall, Englewood Cliffs, NJ, 1971.
10. M. Schwartz, *Information, Transmission, Modulation, and Noise*, McGraw-Hill, New York, 1980.
11. W. R. Bennet and S. O. Rice, "Spectral Density and Autocorrelation Functions Associated with Binary Frequency Shift Keying," *Bell Syst. Tech. J.* 42:2355–2385 (September 1963).

Chapter 4

Communicating Analog Signals

4.1 INTRODUCTION

Continuous signals (also called analog signals) can be communicated using the techniques of discrete communications. Many analog signals, however, have unique problems that are not shared by discrete signals. The average power or bandwidth of some analog signals changes by significant amounts over a period of time. Such signals are said to be nonstationary. Let us consider some examples.

Music has occasional quiet passages as well as quite loud passages. In other words, the average power and peak-to-peak voltage swings (dynamic range) are changing. Subjectively, during either the quiet or the loud passages, it is adequate to quantize the signal into 64 levels (or six bits). If the occasional loud passages have values 64 times larger than the quiet ones and uniform quantization is used, then the maximum dynamic range must be quantized into 4096 levels (or 12 bits), even though for any particular passage six bits is adequate. This large number of quantization levels not only causes hardware problems, but the signal bandwidth is twice the necessary bandwidth (12 bits instead of six per sample).

Another example involves the task of monitoring the radiation level in an experimental satellite. Most of the time the radiation levels are small, yet because of the possibility of the satellite passing through a radioactive belt, a larger number of quantization levels (and hence a large bandwidth) is needed.

As an example of a nonstationary bandwidth, consider the problem of monitoring the temperature levels in a communication satellite. Although the temperatures may change significantly, depending on whether the Earth is blocking the sun, they will not change rapidly; and slow sampling rates are normally acceptable. When the satellite enters the atmosphere, however, or undergoes a sustained rocket thrust, the temperature may change rapidly. In order to monitor these important temperature changes, relatively large sampling rates are needed. Thus during normal operations, the sampling rate, and hence the bandwidth, is larger than necessary.

In some cases, and the example of radiation measurements could be such a case, both nonstationary effects occur simultaneously. In the next two sections we shall discuss these effects and ways of reducing the unnecessarily large bandwidth.

Analog signals offer other unique possibilities. In telephony, for example, analog speech signals can be transmitted directly rather than by discrete communications after digitizing. While this signal is more susceptible to distortion by additive noise, its bandwidth is that of the original message, that is, the smallest possible bandwidth. If analog signals are modulated directly, the resultant continuous waveform (cw) signals tend to have significantly less bandwidth than the discrete signals considered in Chapter 3. Because of the restricted frequency range allocated to the radio bands, the citizens' band, mobile-land communications, etc., and the demand to use these bands, cw techniques are attractive. The individual channel bandwidths allocated for these applications permit low-baud-rate digital communications, but cw techniques are essential for real-time transmission of analog signals such as speech or music. Much of this chapter will be devoted to cw communications.

4.2 COMPANDING

If $m(t)$ is an analog signal that has been digitized and $m_q(t)$ is the quantized signal, then the quantization noise power N_q is defined by

$$N_q = \lim_{N \to \infty} \frac{1}{2N} \sum_{n=-N}^{N} [m(nT) - m_q(nT)]^2 \tag{4.1}$$

where T is the time between samples. The effect of the quantization error, whether it is subjective or relative to the measurement inaccuracies, depends on the signal power S, defined as

$$S = \langle m^2(t) \rangle = \lim_{N \to \infty} \frac{1}{2N} \sum_{n=-N}^{N} m^2(nT) \tag{4.2}$$

The signal-to-noise ratio S/N_q is generally regarded as a reasonable measure of the quantization error. For example, it has been determined that the quantization error when music is digitized is negligible if the signal-to-noise ratio is greater than or equal to 10^4, or 40 dB.[1]

---

[1]Power ratios P_1/P_2 are often described in terms of decibels (dB), defined as $10 \log_{10}(P_1/P_2)$.

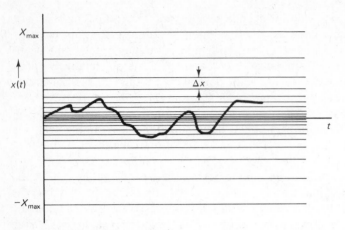

Figure 4.1 Unequal quantization.

As we shall see in Chapter 6, for most signals uniform quantization (as implemented by standard A/Ds) achieves a near maximum S/N_q for a given number of quantization levels. For nonstationary signals and certain kinds of impulse noise, however, unequal quantization is more efficient. In such cases, a more efficient quantization is carried out as in Figure 4.1. In this way, if the signal is weak relative to the largest anticipated dynamic range $(2x_{max})$, as in Figure 4.1, there is an adequate number of quantization levels. Furthermore, the number of quantization levels for strong signals is adequate and far less than with uniform quantization techniques. Hence, a signal quantized in this manner would require less bandwidth than one quantized in a uniform manner.

The typical analog-to-digital converter carries out uniform quantization. Similarly, the typical digital-to-analog converter at the receiver presumes uniform quantization. Rather than design A/D and D/A devices that employ nonuniform quantization, it is more convenient to distort the signal with a nonlinearity prior to the A/D and subsequently to remove the distortion with an inverse nonlinearity after the D/A. The characteristic of a component with such a nonlinearity, called a *compander*, is shown in Figure 4.2. We observe from the figure that the output of the compander, $y(t)$, does not vary over as wide a range as the input signal. Equal width quantization on the output is equivalent to the kind of unequal quantization that we desire for the signal $x(t)$. More explicitly, if we can model the compander by $y(t) = g[x(t)]$ and if we assume that the number of levels M is large enough that Δy and Δx are small, then

$$\frac{\Delta y}{\Delta x} \simeq \frac{dg(x)}{dx}$$

or

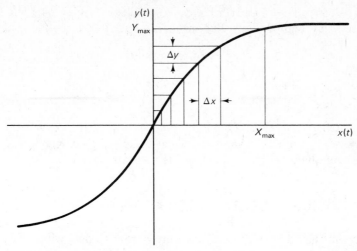

Figure 4.2   Nonlinear companding characteristic.

$$\Delta x \cong \frac{\Delta y}{g'(x)} = \frac{2y_{max}/M}{g'(x)} \tag{4.3}$$

where $g'(x) = dg(x)/dx$ and $y(t)$ is quantized uniformly.

A commonly used compander, known as the μ-law compander, whose response has the shape of Figure 4.2 is modeled by

$$g(x) = \begin{cases} \dfrac{\ln(1 + \mu x/x_{max})}{\ln(1 + \mu)} & x \geq 0 \\[3mm] \dfrac{-\ln(1 - \mu x/x_{max})}{\ln(1 + \mu)} & x < 0 \end{cases} \tag{4.4}$$

This equation has been normalized; that is, $y_{max} = g(x_{max}) = 1$. Since $d[\ln(1 + ax)]/dx = a/(1 + ax)$, it follows that

$$g'(x) = \frac{\mu/x_{max}}{(1 + \mu|x|/x_{max})\ln(1 + \mu)}$$

and, substituting into Equation 4.3, we get

$$\Delta x \cong \frac{2x_{max}}{M}\gamma(x) \tag{4.5}$$

where

$$\gamma(x) = \frac{(1 + \mu|x|/x_{max})\ln(1 + \mu)}{\mu} \tag{4.6}$$

and μ is generally a large number on the order of 200. Since $\Delta x = 2x_{max}/M$ for uniform sampling, γ represents the variation from uniform sampling.

Observe that for small x $\gamma \ll 1$ and for large x $\gamma \gg 1$ (see Problems 4.1 and 4.2). The inverse μ-law companding relationship is found from Equation 4.4:

$$x = \begin{cases} x_{max}\left[\dfrac{(1 + \mu)^y - 1}{\mu}\right] & y \geq 0 \\[4mm] -x_{max}\left[\dfrac{(1 + \mu)^{-y} - 1}{\mu}\right] & y < 0 \end{cases} \tag{4.7}$$

We shall see in Chapter 6 that comparable signal-to-noise ratios can be achieved with fewer quantization intervals when using μ-law companding, rather than uniform quantization, if the signal is nonstationary.

4.3 SOURCE CODING 1: DELTA MODULATION

Nonstationarities in the bandwidth of a signal are harder to deal with using digital techniques. Reducing the bandwidth of a signal that is sampled too rapidly involves specialized coding techniques. Coding for this purpose is referred to as *source coding*.

Let us consider the signal of Figure 4.3, which is intended to represent the type of nonstationarity to which we refer. The sampling rate (indicated by the marks on the time scale) is chosen so that during brief periods of high activity ($a < t < b$) it is slightly higher than the Nyquist rate. Let us assume for this signal that eight-bit words are adequate. During low activity, because the sampling rate is "too fast," we do not need eight bits to identify each sample. Thus, if we relate two adjacent samples by (see Figure 4.3)

$$x(kT) = x[(k - 1)T] + \epsilon(kT) \tag{4.8}$$

the difference $\epsilon(t)$ is normally quite small except during periods of high activity. If, instead of transmitting quantized values of the samples of $x(t)$, we transmit quantized values of $\epsilon(t)$ and use Equation 4.8 to reconstruct the signal $x(t)$, we then require significantly less than eight bits per sample except

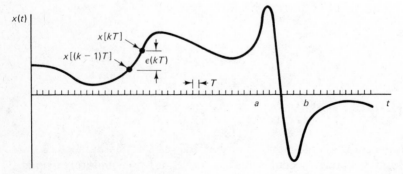

Figure 4.3 Signal whose activity is nonstationary.

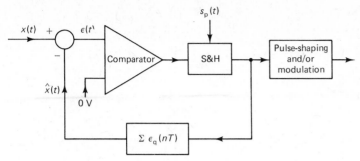

Figure 4.4 Delta modulator.

during the active periods. If we assume that $x(0) = 0$ (which will be justified shortly), this reconstruction can be carried out from

$$x(kT) = \sum_{n=1}^{k} \epsilon(nT) \tag{4.9}$$

If we presume the difference signal $\epsilon(t)$ is very small, comparable to the measurement error, it can be quantized into two levels to denote whether the difference is positive or negative:

$$\epsilon_q(kT) = \pm\epsilon \tag{4.10}$$

This leads to simple hardware, and the technique is called *delta modulation* (DM). The delta modulator is shown in Figure 4.4. The return or feedback path, which adds the appropriately scaled binary pulses to the previous total, is the basis of the receiver hardware (see Figure 4.5). The output is denoted $\hat{x}(t)$ because it is not the same as $x(t)$ because of the quantization of $\epsilon(t)$. Note that there is no need for A/D and D/A quantization and interpolation, nor is there need for word synchronization. There is, however, a price to pay as we would expect.

Let us reexamine the signal in Figure 4.3 (reproduced in Figure 4.6), use a delta modulation scheme, and observe the output of the receiver. Much of the time the error between the received signal $\hat{x}(t)$ and the transmitted signal $x(t)$ is comparable to the measurement error. When the signal changes rapidly, as in the interval $a < t < b$, there is a significant error. This error,

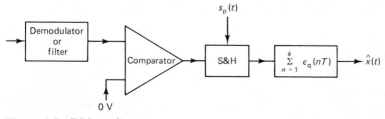

Figure 4.5 DM receiver.

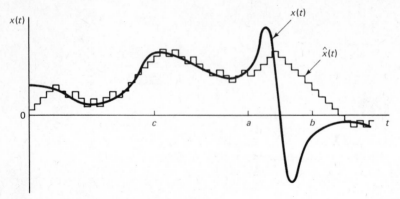

Figure 4.6 Output of a delta modulator.

called the *slope overload error*, also occurs as an initial transient because of the assumption that $x(0) = 0$, which is rarely true. We also note that there are occasional small errors (such as at time $t = c$) that are larger than the measurement error. Because highly active periods are often quite important, it could be a major drawback that during this time the signal is not being transmitted accurately. This, of course, must be traded against the hardware simplicity and the enormous bandwidth reduction (one instead of eight bits per sample).

There is still another problem with delta modulation. All receivers will occasionally, if rarely, make decision errors. The analysis of these errors will be carried out in detail in Chapters 6–8. For the standard discrete techniques, such as PCM, this results in a single sample or word error. We can see from Equation 4.9, however, that if one sample of the difference signal is received as an error, it will affect the entire reconstructed signal $\hat{x}(t)$ (i.e., the errors are said to propagate). More specifically, if a negative pulse is mistaken as a positive one, $\hat{x}(t)$ will increase one quantization level rather than decrease and subsequently remain two quantization levels higher than intended. Usually when DM is used the pulse train is divided up into large frames (note the reintroduction of word or frame synchronization) and restarted at the beginning of every frame. While this procedure introduces overload transients at the beginning of every frame, the effect of errors due to noise will not propagate from one frame to the next.

4.4 SOURCE CODING 2: DPCM AND IP CODING

There are two ways of reducing the overload error of delta modulation. One method involves quantizing the difference signal $\epsilon(t)$ into more than two levels (see Problem 4.3 for an example). This procedure will reintroduce the need for A/D and D/A quantization and interpolation, as well as the need for word synchronization. This procedure will also require more bandwidth than DM and is a compromise between DM and straight PCM.

The other method uses a more complicated relationship between the samples of $x(t)$ than that of Equation 4.8 and subsequently a more complicated reconstruction algorithm than that of Equation 4.9. Let us first view the block in the feedback path of the delta modulator as an estimator, where for DM the prediction is trivial: $x_{pred}(k) = \hat{x}(k - 1)$, and the estimation becomes

$$\hat{x}(k) = \hat{x}(k - 1) \pm \epsilon \qquad (4.11)$$

The parameter ϵ is the magnitude of the quantization interval and the T has been dropped from the argument of the estimated values for the sake of convenience. The difference signal is now viewed as the error between the actual signal $x(t)$ and the estimated signal $\hat{x}(t)$. We can improve our prediction, and hence reduce the error $\epsilon(t)$, by using a predictor based on additional previous estimations:

$$\hat{x}(k) = \sum_{j=1}^{N} a_j \hat{x}(k - j) \pm n\epsilon \qquad (4.12)$$

where it is now presumed that the error is quantized into $2K$ levels and $n \leq K$. Thus, for example, if we wish to use the slope of the signal for prediction $[x_{pred}(k) = x(k - 1) + T \, dx/dt|_{(k-1)T}]$, this can be approximated in terms of previous estimations by

$$\hat{x}(k) = \hat{x}(k - 1) + \frac{\hat{x}(k - 1) - \hat{x}(k - 2)}{T} T \pm n\epsilon$$

$$= 2\hat{x}(k - 1) - \hat{x}(k - 2) \pm n\epsilon \qquad (4.13)$$

Determining the best choice of N and the coefficients a_j to minimize the error $\epsilon(t)$ is beyond the scope of this treatment.[2]

A binary encoding scheme that employs one or both of these improvements is called *differential pulse code modulation* (DPCM); a typical encoder is shown in Figure 4.7. As in DM the feedback path is the receiver hardware after demodulation and detection of the binary pulses. The number of quantization levels that one uses is a trade-off between the desire to reduce the bandwidth and the desire to reduce the slope overload error. Observe the need for a small amount of memory in both the modulator and demodulator.

By using a computer to generate a special, more complex, variable length code, instead of the standard binary codes generated by A/D converters, it is possible to eliminate the overload error completely. One must pay a heavy price, however, for such codes, which are called *information-preserving* (IP) *codes*. These codes rely on the fact that the quantized error tends to be small most of the time. They assign short code words to the small errors and long code words to the large errors that seldom occur. In this way the average number of bits per sample ($\bar{b}$) can be kept low. Specifically, if p_j is the percentage of time the jth level occurs and b_j is the number of bits used for the jth level code word, then

---

[2]See Rabiner and Schafer [1] for a development of the optimal predictor.

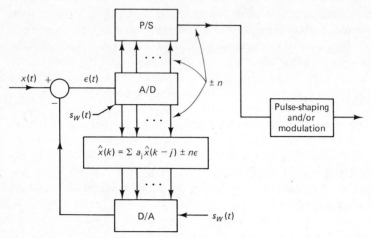

Figure 4.7   DPCM modulator.

$$\bar{b} = \sum \frac{p_j}{100} b_j \tag{4.14}$$

An example of such a code is the Huffman code [2], shown in Table 4.1. A significant problem with this code is the need to know the percentage of time the various levels of the quantized error occur. In the following example, the percentages are assumed to be known and are given in Table 4.1 along with the Huffman code.

EXAMPLE 4.1 ■

Let us assume that after prediction the quantized error lies between $\pm 6\epsilon$, where ϵ is comparable to the measurement inaccuracy. A variety

TABLE 4.1   VARIOUS SOURCE CODES

Quantized error level	Percent of occurrence time	One-bit or DM code	Two-bit DPCM code	Three-bit DPCM code	Huffman code
+6	1.5625	0	01	011	111110
+5	1.5625	0	01	011	111100
+4	3.125	0	01	011	11100
+3	6.25	0	01	010	1100
+2	12.5	0	01	001	100
+1	25	0	00	000	00
−1	25	1	10	100	01
−2	12.5	1	11	101	101
−3	6.25	1	11	110	1101
−4	3.125	1	11	111	11101
−5	1.5625	1	11	111	111101
−6	1.5625	1	11	111	111111

of possible source codes are listed in Table 4.1. For DM or two-bit DPCM we must either increase the quantization interval ϵ and accept a quantization noise that is larger than the measurement inaccuracies or, alternatively, accept a significant slope overload (50% of the time for DM and 25% of the time for two-bit DPCM). For three-bit DPCM slope overload occurs when $|\epsilon_q(kT)| \geq 5\epsilon$, which happens 6.25% of the time. This might be acceptable. If the Huffman code, listed in Table 4.1, is used, it follows from Equation 4.14 that

$$\begin{aligned} \bar{b} &= 2[0.25(2) + 0.125(3) + 0.0625(4) + 0.03125(5) + 0.015625(6) \\ &\quad + 0.015625(6)] \\ &= 2.9375 \text{ bits/sample} \end{aligned}$$ ■ ■

This code provides a trivial improvement in compression when compared to the three-bit DPCM scheme, but it has no slope overload.

A major problem with variable length codes, however, is the need for a sizable memory buffer that accepts the code words asynchronously and then transmits them synchronously at the average bit rate. Word synchronization, however, is not a problem since the Huffman code is self-synchronizing. If in Example 4.1 the error sequence were $3\epsilon, -\epsilon, 2\epsilon, -3\epsilon, \epsilon, -\epsilon$, the transmitted code would be 11000110011010001 (see Table 4.1). If the receiver erroneously began with the fourth pulse (i.e., out of synchronization), the output would decode as $\epsilon, 3\epsilon, -3\epsilon, \epsilon, -\epsilon$. Thus after an initial error, synchronization is regained.

4.5 CONTINUOUS WAVEFORM COMMUNICATION TECHNIQUES

If a continuous or analog message $m(t)$ is either transmitted directly (as in telephony) or modulated directly (as in radio), the resulting signal is called a *continuous waveform* (cw) *signal*. While these signals are more susceptible to distortions due to additive noise, their bandwidths tend to be less than those of signals used for discrete communications. As a consequence of this, they are quite useful for applications where bandwidth is at a premium, such as broadcast communications.

Modulated cw signals are most often characterized by the form

$$s(t) = E(t) \sin[2\pi f_c t + \phi(t)] \tag{4.15}$$

where $E(t)$ and $\phi(t)$ are directly related to the message $m(t)$ or are constant. In Section 2.9 we considered the case

$$E(t) = A[1 + m(t)] \qquad \phi(t) = \theta_0 \tag{4.16}$$

which is called *amplitude modulation* (AM); it will be discussed in detail in Section 4.6. We saw, however, that the power spectrum of the AM signal was a bandpass spectrum centered about the carrier frequency f_c. This is true for all forms of cw signals.

As another example, instead of modulating the amplitude (or envelope) of the signal with the message, it is possible to vary the instantaneous frequency $[f_i(t)]$ with the message:

$$f_i(t) = f_c + m(t) \tag{4.17}$$

where we define the instantaneous frequency of any signal of the form $E(t)$ $\cos \psi(t)$ by

$$f_i(t) = \frac{1}{2\pi} \frac{d\psi(t)}{dt} \tag{4.18}$$

This definition is consistent with our notions of frequency since, for a sinusoid $[A \cos (2\pi ft + \theta_0)]$, the term $\psi(t)$ is $2\pi ft + \theta_0$ and, from Equation 4.18, $f_i(t) = f$. Thus, for sinusoids, $f_i(t)$ is a constant that corresponds to the traditional definition of frequency. For the instantaneous frequency to vary as in Equation 4.17, we require that

$$\frac{1}{2\pi} \frac{d}{dt} [2\pi f_c t + \phi(t)] = f_c + m(t)$$

or

$$\frac{1}{2\pi} \frac{d}{dt} \phi(t) = m(t)$$

or

$$\phi(t) = 2\pi \int m(t)\, dt + \theta_0 \tag{4.19}$$

where θ_0 can be regarded as the constant of integration. We conclude that if

$$s(t) = A \sin\left\{ 2\pi \left[f_c t + \int m(t)\, dt \right] + \theta_0 \right\} \tag{4.20}$$

then the instantaneous frequency varies linearly with the message $m(t)$ as in Equation 4.17. This signal is called a *frequency modulated* (FM) *signal* and will be discussed in detail in Section 4.7. As in AM, the spectrum will be seen to be bandpass and centered about the carrier frequency f_c.

A major advantage of cw signals, in addition to their relatively small bandwidths, is the ease with which they can be shifted in the frequency domain. Consider the system in Figure 4.8, where the input is any cw signal. Using the identity $2 \sin A \sin B = \cos(A - B) - \cos(A + B)$, we see that

$$x(t) = E(t) \sin[2\pi(f_c - f_b)t + \phi(t) - \theta + \pi/2]$$

$$+ E(t) \sin[2\pi(f_c + f_b)t + \phi(t) + \theta - \pi/2]$$

The bandpass filter passes one of the components of $x(t)$ and removes the other. Thus the output $y(t)$ is the same cw signal as the input except that the

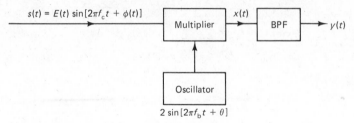

Figure 4.8   Frequency shifting technique.

carrier frequency has changed from f_c to either $f_c - f_b$ or $f_c + f_b$. The change in the constant phase angle is unimportant.

Frequency shifting makes it relatively easy to build a receiver that can select, or tune in, any one of many cw signals that are being transmitted. For commercial broadcasting to be economically viable, it is essential that a number of transmitters, using different carrier frequencies, be operating simultaneously and that it be easy to build a single receiver that can filter out (or tune in) these signals selectively. Tuning is accomplished for all broadcast cw signals (AM and FM radio, television, etc.) with the *superheterodyne receiver* that is shown in Figure 4.9. The principal filtering and amplifying is carried out by a fixed bandpass amplifier in the intermediate frequency (i.f.) stage. Changing the frequency of the oscillator moves a different signal into and through the i.f. filter.

We have seen the need for a multiplying circuit (see, for example, the PSK demodulator of Figure 3.40). A true multiplying circuit is difficult to implement. For our applications, however, multiplication can be accomplished with a simple circuit called a mixer.

A *mixer* is an adder with one of a wide class of nonlinearities both of which can be implemented with a single transistor. Suppose we replace the multiplier of Figure 4.9 with a mixer modeled as in Figure 4.10. The output can be written

$$y(t) = as(t) + ac(t) + bs^2(t) + bc^2(t) + 2bs(t)c(t) \qquad (4.21)$$

where the last term is the multiplication that we desire. This multiplication term can be expressed as

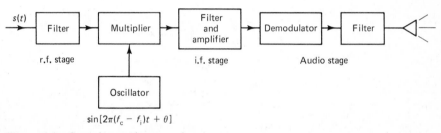

Figure 4.9   Superheterodyne receiver.

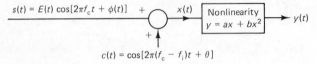

Figure 4.10 Mixer model.

$$2bs(t)c(t) = bE(t)\cos[2\pi f_i t + \phi(t) - \theta] - bE(t)\cos[2\pi(2f_c - f_i)t + \phi(t) + \theta]$$

where one of these components is passed by the i.f. filter. None of the other terms of Equation 4.21 is passed by this filter, as can be seen in the spectral plot in Figure 4.11. Only the spectrum of the first two terms of Equation 4.21 (which is proportional to the mixer input) and the last product term are labeled. The spectrum of the $bs^2(t) + bc^2(t)$ terms are shown but unlabeled (see Problem 4.7). The dashed spectra shown in Figure 4.11 represent the output of the mixer for other signals using different carrier frequencies. We see that they are not passed by the i.f. filter, but if the frequency of the oscillator is changed, they will be moved into the i.f. filter passband.

4.6 AMPLITUDE MODULATION

Let us assume that an AM signal is generated by first adding a constant or dc component of A V to the analog message $m(t)$ and then multiplying this sum by a sinusoid whose frequency (called the carrier frequency) is large relative to the bandwidth B of $m(t)$. Thus

$$s(t) = [A + m(t)]\cos(2\pi f_c t + \theta_0) \tag{4.22}$$

The multiplication can be implemented with a simple mixer followed by a bandpass filter (see Problem 4.8). If the message is bounded, that is, if there exists a voltage V such that $|m(t)| \leq V$ for all t, then we can define the normalized message $m_n(t)$ by

$$m_n(t) = m(t)/V \tag{4.23}$$

where $|m_n(t)| \leq 1$ for all t. Equation 4.22 can now be expressed

$$s(t) = A[1 + km_n(t)]\cos(2\pi f_c t + \theta_0) \tag{4.24}$$

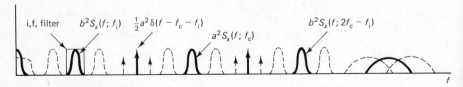

Figure 4.11 Spectrum of a mixer output.

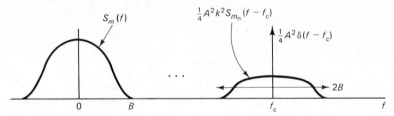

Figure 4.12 AM spectrum.

where $k = V/A$ is called the *modulation index*. The spectrum of this signal was determined in Section 2.9, where it was found that

$$S_s(f) = \tfrac{1}{4}A^2\delta(f - f_c) + \tfrac{1}{4}A^2k^2S_{m_n}(f - f_c) + \tfrac{1}{4}A^2\delta(f + f_c) + \tfrac{1}{4}A^2k^2S_{m_n}(f + f_c)$$
(4.25)

and where

$$\tfrac{1}{4}A^2k^2S_{m_n}(f - f_c) = \tfrac{1}{4}V^2S_{m_n}(f - f_c) = \tfrac{1}{4}S_m(f - f_c)$$
(4.26)

In Figure 4.12 the positive frequency portion of the spectrum (the first two terms of Equation 4.25) is plotted,[3] along with the spectrum of $m(t)$ $[S_m(f)]$. As we saw before, the bandwidth of AM is given by

$$BW_{AM} = 2B$$
(4.27)

where B is the bandwidth of the message $m(t)$. The portion of the spectrum to the right of the carrier frequency is called the *upper sideband*, and the symmetrical portion to the left of the carrier frequency is called the *lower sideband*. Sometimes this spectrum is referred to as a *double sideband AM* (DSB-AM) spectrum.

The AM signal (Equation 4.24) is plotted in the time domain in Figure 4.13 for different values of the modulation index k. Observe that, for $k < 1$, $A[1 + km_n(t)]$ is always greater than zero, and the envelope of the signal $[E(t)]$ is a replica of the message. Broadcast AM restricts k to be ≤ 1 because the simple demodulator (full-wave rectifier and low-pass filter) called an envelope detector that was used for ASK (or L-level ASK) works for DSB-AM if $k < 1$. If $k > 1$ is allowed (as in Figure 4.13c), this receiver will cause distortion called *overmodulation distortion*.

Restricting to $k \leq 1$ is costly. The sidebands $Akm_n(t)\cos(2\pi f_c t + \theta)$ contain the "information" about the message $m(t)$, whereas the carrier $A\cos(2\pi f_c t + \theta)$ does not and is needed only to make the simple envelope detector work. At least two-thirds of the total power is wasted on the carrier (see Problem 4.10). The total power is a transmitter constraint, and for the same transmitter the power in the sidebands can be increased approximately

---

[3]This practice will be followed throughout this chapter.

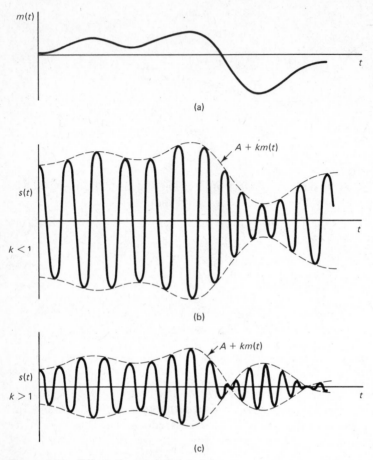

Figure 4.13 Effect of modulation index.

by a factor of 3 (or 4.77 dB) if the carrier can be suppressed. *Suppressed carrier AM* (SC-AM) is defined by

$$s(t) = Am_n(t) \cos(2\pi f_c t + \theta) \tag{4.28}$$

where the multiplication can be carried out by either a chopper (Problem 3.26) or a balanced modulator (see Problem 4.26). This signal can be demodulated with the same hardware as PSK (see Figure 3.40), where the multiplier can be implemented with a chopper or balanced modulator. The output of this demodulator, called a *coherent demodulator*, is $\frac{1}{2}Am_n(t) \cos(\Delta\theta)$, where $\Delta\theta$ is the phase difference between the oscillator and the signal carrier. With phase synchronization $\Delta\theta = 0$. Without phase synchronization, because the AM transmission path may vary with atmospheric conditions, $\Delta\theta$ may vary with time. This time variation, called *fading*, can be severe and is undesirable. As indicated in Section 3.9, there are methods of self-phase synchronization that can be used which do not require transmitting a separate synchronizing signal. A full-wave rectified SC signal can be used to drive the modified PLL of Figure 3.25b. Recall that this circuit has the drawback that the phase can lock into 180° as well as 0°. Fortunately, this is not a problem as human listeners cannot discern this phase difference. Other kinds of modified PLLs, such as the Costa loop [3, 4], can also be used.

In another sense, transmitting both sidebands may be wasteful since one is just a mirror image of the other. If only one sideband is transmitted, the process called *single-sideband suppressed carrier AM* (SSB-SC-AM), the coherent demodulator will work (see Problem 4.12). This not only reduces the bandwidth to half,

$$BW_{SSB-AM} = B \tag{4.29}$$

but as a result only half as much noise power will pass through the i.f. stage, whose bandwidth is now B. Because cutting the noise power in half has the same effect as increasing the signal power by 2, SSB-SC is an additional 3 dB better than SC-AM, a total of 7.8 dB better than standard broadcast AM. The receiver, which requires some PLL for phase synchronization, is more costly than the simple envelope detector.

We distinguish between two kinds of SSB-AM depending on how the signals are generated. If the signal is generated by passing a DSB-AM signal through a near ideal bandpass filter, it is called *vestigial single sideband* (VSSB). If these filters have a rolloff with the same symmetry as the pulse-shaping filters, the coherent detector will demodulate with no distortion (see Problem 4.13). Because of the filter rolloff factor r, the bandwidth of the signal is slightly higher:

$$BW_{VSSB} = (1 + r/2)B \tag{4.30}$$

It is possible to generate a true SSB-SC signal directly. A number of references [4–6] discuss the techniques and circuits to generate these signals.

4.7 FREQUENCY AND ANGLE MODULATION

Let us rewrite Equation 4.17

$$f_i(t) = f_c + \Delta f m_n(t) \tag{4.31}$$

where the message has been normalized ($|m_n(t)| \le 1$), and Δf is the *frequency deviation*. The instantaneous frequency varies between $f_c + \Delta f$ and $f_c - \Delta f$, but this does not mean that the bandwidth is $2\Delta f$. We should compute the power spectrum of the signal,

$$s(t) = A \sin\left\{2\pi\left[f_c t + \Delta f \int m_n(t)\, dt\right] + \theta_0\right\} \tag{4.32}$$

in order to determine its bandwidth. Unfortunately, Equation 4.32 is too complex to permit computation of the spectrum $S_s(f)$ in terms of the spectrum of the message $S_m(f)$. We are forced to consider special cases and to generalize from them.

In Chapter 3 we determined the spectrum and bandwidth of FSK, which can be thought of as an FM signal where $m(t)$ is a baseband binary random pulse train. The bandwidth of FSK was determined to be $(f_2 - f_1) + 2mf_p$. If we interpret the frequencies as $f_2 = f_c + \Delta f$ and $f_1 = f_c - \Delta f$, then $f_2 - f_1 = 2\Delta f$ (the instantaneous frequency swing). Since mf_p is the bandwidth of this $m(t)$, it follows that for the special case of FSK (or FPM for that matter)

$$BW_{FM} = 2\Delta f + 2B \tag{4.33}$$

where B is the bandwidth of $m(t)$. As we shall see, this formula is a good approximation for other messages as well and is regarded as a general formula for the FM bandwidth.

Let us now assume that the message is periodic and can be expanded in a Fourier series. Even this is complicated because of the nonlinear relationship of Equation 4.32. Consider first the case $m_n(t) = \cos 2\pi f_m t$. (Note that superposition is not valid; the results for two sinusoids cannot be deduced from the results of one sinusoid.) For this sinusoidal message

$$\psi(t) = 2\pi \left(f_c t + \Delta f \int \cos 2\pi f_m t \, dt \right)$$

$$= 2\pi f_c t + \frac{\Delta f}{f_m} \sin 2\pi f_m t + \theta_0$$

We define the *modulation index* (β) by $\beta = \Delta f / f_m$ and set $\theta_0 = \pi/2$ for analytic convenience. In the general case β is defined as $\Delta f / B$ where B is the bandwidth of the message $m(t)$. It follows that

$$s(t) = A \cos(2\pi f_c t + \beta \sin 2\pi f_m t)$$

$$= A \cos(\beta \sin 2\pi f_m t) \cos 2\pi f_c t - A \sin(\beta \sin 2\pi f_m t) \sin 2\pi f_c t \tag{4.34}$$

The function $\cos(\beta \sin 2\pi f_m t)$ is an even periodic function and can be expanded in a Fourier series. Let us define

$$\theta_m = 2\pi f_m t \tag{4.35}$$

in order to simplify the calculations. Then

$$\cos(\beta \sin \theta_m) = a_0 + 2 \sum_{k=1}^{\infty} a_k \cos k\theta_m \tag{4.36}$$

where

$$a_k = \frac{1}{2\pi} \int_{-\pi}^{\pi} \cos(\beta \sin \theta_m) \cos k\theta_m \, d\theta_m \tag{4.37}$$

Similarly,

$$\sin(\beta \sin \theta_m) = 2 \sum_{k=1}^{\infty} b_k \sin k\theta_m \tag{4.38}$$

where

$$b_k = \frac{1}{2\pi} \int_{-\pi}^{\pi} \sin(\beta \sin \theta_m) \sin k\theta_m \, d\theta_m \tag{4.39}$$

Fortunately the integrals of Equations 4.37 and 4.39 have been tabulated, and

$$a_k = \begin{cases} J_k(\beta) & k \text{ even} \\ 0 & k \text{ odd} \end{cases} \qquad b_k = \begin{cases} 0 & k \text{ even} \\ J_k(\beta) & k \text{ odd} \end{cases} \tag{4.40}$$

where $J_k(\beta)$ is the kth-order Bessel function of the first kind [7]. Equation 4.34 becomes

$$s(t) = A \cos 2\pi f_c t \left[J_0(\beta) + 2 \sum_{k \text{ even}} J_k(\beta) \cos k 2\pi f_m t \right]$$

$$- A \sin 2\pi f_c t \left[2 \sum_{k \text{ odd}} J_k(\beta) \sin k 2\pi f_m t \right]$$

It is more convenient to manipulate this equation into the form

$$s(t) = A J_0(\beta) \cos 2\pi f_c t + A \sum_{k=1}^{\infty} J_k(\beta) \cos 2\pi (f_c + kf_m)t$$

$$+ A \sum_{k=1}^{\infty} (-1)^k J_k(\beta) \cos 2\pi (f_c - kf_m)t \tag{4.41}$$

We recognize the first term of Equation 4.41 as the carrier, the second term as defining the upper sideband components, and the third term as defining the lower sideband components. Since the sidebands contain frequency terms that are harmonically related to the frequency of $m(t)$, the bandwidth is greater than $2f_m$, which is the DSB-AM bandwidth for a sinusoidal message of frequency f_m.

An examination of the Bessel functions indicate that for a fixed β the $J_k(\beta)$ coefficients decrease as k increases. This means that from a practical point of view the bandwidth is finite. A standard rule of thumb is that $J_k(\beta)$ is negligibly small if $k > \beta + 1$. Based on this rule of thumb,

$$\text{BW}_{\text{FM}} \cong 2(\beta + 1)f_m = 2\Delta f + 2f_m \tag{4.42}$$

which agrees with Equation 4.33. If the message included two sinusoids $[m_n(t) = a \cos 2\pi f_1 t + (1 - a) \cos 2\pi f_2 t]$ the sidebands would be double sums containing terms like $J_{k_1}(\beta_1) J_{k_2}(\beta_2) \cos 2\pi (f_c + k_1 f_1 + k_2 f_2)t$. Thus not only are multiples of the message frequencies present, but there are also beats between them. Nevertheless, Equations 4.33 or 4.42 still apply:

$$\text{BW}_{\text{FM}} = 2(\beta + 1)B = 2\Delta f + 2B \tag{4.43}$$

where B is the bandwidth (or largest frequency) of $m(t)$ and $\beta = \Delta f / B$.

The coefficient of the carrier term can be small or zero as in FSK. Unlike in AM, there is no need to suppress the carrier even if it is possible. The beat frequencies combine in such a way that the sidebands are not

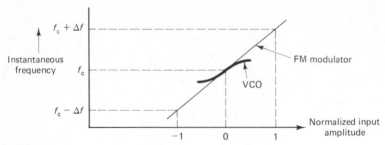

Figure 4.14 FM modulators and VCOs.

symmetric (see the FSK spectrum in Figure 3.43), and distortion would result
if SSB operation were possible. There is, however, little incentive for SSB
operation since, unlike AM, the performance in noise would be degraded.

We shall see in Chapter 8 that it is desirable to make the modulation
index β large, even though this means a large bandwidth. The bandwidth
allocated for standard broadcast FM is 200 kHz, and the message bandwidth
B is 15 kHz. It is possible for β to be as large as 5.7. If β were larger, the
bandwidth would still be limited to 200 kHz by bandpass filters (because of
strict FCC regulations) with the resultant distortion called *overmodulation
distortion*.

A frequency modulator is the same as a voltage-controlled oscillator
(VCO) whose frequency varies linearly with the amplitude of the input over a
wide range of frequencies. In Figure 4.14 such an FM modulator is compared
with a typical VCO that is used in PLLs. As one might expect, the
requirements of linearity and large dynamic range make the realization of FM
modulators more difficult than the realization of VCOs. A number of circuits
are available, however, for FM modulators [8].

If one can implement an FM modulator that is linear over a modest
range of frequencies ($\Delta f \leq B$), the resulting signal is called a *narrow-band
FM signal*. These signals are useful when the bandwidth is limited. It is
possible to start with a narrow-band FM modulator and then increase Δf (and
the bandwidth) with frequency multiplying techniques. Consider the circuit of
Figure 4.15. Since $2 \cos^2 \theta = 1 + \cos 2\theta$, both the instantaneous frequency
and Δf are doubled. Using multiple stages, the bandwidth can be increased
from a narrow band to a much broader band. This is a commonly used
procedure to generate FM signals.

An efficient demodulator can be realized if an ideal differentiator (or
close approximation) can be implemented. Let us assume a signal $s(t)$ given

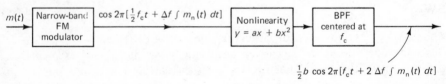

Figure 4.15 FM modulator and frequency doubler.

by $s(t) = A(t) \sin[2\pi f_c t + \phi(t)]$, where $\phi(t) = 2\pi \, \Delta f \int m_n(t) \, dt + \theta_0$ and where the amplitude is presumed to be varying with time because of the phenomenon of fading and the presence of noise in the channel. Passing the signal through an ideal differentiator yields

$$y(t) = \frac{ds(t)}{dt}$$

$$= \frac{dA(t)}{dt} \sin[2\pi f_c t + \phi(t)] + A(t) \cos[2\pi f_c t + \phi(t)] \frac{d}{dt}[2\pi f_c t + \phi(t)]$$

$$= A(t)2\pi [f_c + \Delta f m_n(t)] \cos[2\pi f_c t + \phi(t)] + \frac{dA(t)}{dt} \sin[2\pi f_c t + \phi(t)]$$

$$= \left\{ A^2(t)4\pi^2[f_c + \Delta f m_n(t)]^2 + \left[\frac{dA(t)}{dt}\right]^2 \right\}^{1/2} \cos[2\pi f_c t + \phi(t) + \psi(t)]$$

$$(4.44)$$

If the amplitude were constant $[A(t) = A, \, dA(t)/dt = 0]$, this equation would simplify to

$$y(t) = 2\pi A f_c \left[1 + \frac{\Delta f}{f_c} m_n(t)\right] \cos[2\pi f_c t + \phi(t)] \qquad (4.45)$$

The output of a standard AM demodulator, or envelope detector, would be proportional to the message $m(t)$ (the fact that the instantaneous frequency is varying would not affect the output of an envelope detector). A time-varying amplitude will cause distortion, however, as seen from Equation 4.44. Fortunately, a series of amplifying stages that limit the maximum instantaneous amplitude and bandpass filtering stages can remove the amplitude fluctuations if they are not severe. An FM demodulator that works on this principle is shown in Figure 4.16.

Another type of demodulator, which makes use of the PLL is shown in Figure 4.17. If the VCO output is written $\sin[2\pi f_c t + \theta(t)]$ and the phase comparator is just a multiplier followed by a low-pass filter, the output of the phase comparator is $\frac{1}{2}A \sin[\phi(t) - \theta(t)]$. If the loop is operating in a near locked condition $[\phi(t) \approx \theta(t)]$, the output can be replaced by $\frac{1}{2}A[\phi(t) - \theta(t)]$ since $\sin \epsilon \approx \epsilon$ for small values of ϵ. Thus, if at any instant of time they are locked and $\phi(t)$ changes by $2\pi \, \Delta f \int m_n(t) \, dt$, the phase comparator output will become $2\pi \, \Delta f \int m_n(t) \, dt$, which tends to lock the instantaneous frequency of the VCO into the new value. For the output to remain proportional to the integral of $m_n(t)$, the instantaneous frequency of the FM signal must vary slowly relative to the time it takes for the PLL to reach equilibrium.

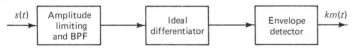

Figure 4.16 Typical FM demodulator.

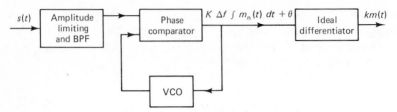

Figure 4.17 FM demodulator employing a PLL.

We shall see (in Chapter 8) that there can be advantages to prefiltering the message before modulation and subsequently postfiltering the output of the demodulator to remove the distortion of the first filter. Such filters are used in standard FM transmitters and receivers where they are called *preemphasis* and *deemphasis filters*, because the transmitter overemphasizes the high-frequency components of $m(t)$ and the receivers must compensate for this preemphasis. We can generalize FM to include such filters where the general form is called *angle modulation*. As an example, suppose the prefilter were an ideal differentiator and the postfilter an ideal integrator. The signal would become

$$s(t) = A \sin\{2\pi[f_c t + \Delta f\, m_n(t)] + \theta_0\} \tag{4.46}$$

and can be called a *phase modulated signal* (PM). For this example, the PLL receiver of Figure 4.17 would be convenient since the ideal differentiator and ideal integrator "cancel" each other out and neither would have to be implemented.

4.8 FREQUENCY DIVISION MULTIPLEXING

A number of messages can be transmitted simultaneously, if the channel bandwidth is large enough, by using cw modulation techniques to separate them in the frequency domain. This method of multiplexing, called *frequency division multiplexing* (FDM), is convenient because of the ease with which the messages can be separated at the receiver. For example, one can transmit a number of telephone conversations (each requiring a bandwidth of 3.3 kHz) through a single cable by using DSB-AM modulation techniques on 8 kHz, 16 kHz, etc., sinusoids (called *subcarriers* when used for this multiplexing purpose). The combined spectrum is shown in Figure 4.18 for the simultane-

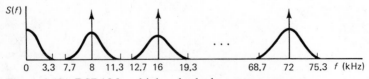

Figure 4.18 DSBAM multiplexed telephone messages.

ous transmission of ten conversations. Any of these telephone conversations can be heard by filtering and demodulation, precisely as in AM radio. It is a question of bookkeeping to determine the overall bandwidth required. Often *guardbands* are included, as in Figure 4.18, to make the filtering easier. These must be included in the bookkeeping. Thus, if a regular telephone line has a bandwidth of 1 MHz and the modulation scheme of Figure 4.18 is used, it follows that up to 125 phone conversations can be multiplexed. This number could be doubled by the use of single-sideband modulation.

The multiplexed signals could be modulated again on a carrier for the purpose of transmission via electromagnetic waves. This would be preferable to using a different carrier for each message since only one transmitting amplifier and antenna would be required. When two levels of modulation are used, the subcarrier modulation is referred to first. Thus, for example, DSB-SC-FM is used for stereo FM transmission. The second channel is separated from the standard monaural channel by DSB-SC modulation. The combined signal is then frequency modulated on a high-frequency carrier.

EXAMPLE 4.2 ■

Suppose we wished to transmit 125 multiplexed phone conversations via some communication satellite at microwave frequencies. If we use FM (that is, we are employing DSB-AM-FM multiplexing), the bandwidth is $2(\beta + 1)B$, where B is the bandwidth of the 1-MHz frequency multiplexed signal. If we set $\beta = 3$, this requires a bandwidth of 8 MHz.

Let us compare this with a discrete system that multiplexes the messages, using TDM, after they have been encoded by PCM and then uses an FSK signal. Working backward, since the bandwidth of FSK is $(f_2 - f_1) + 2B$ and since it is convenient to set $f_2 - f_1 \geq 2B$, with an 8-MHz channel, we can set $B = 2$ MHz and still set the two frequencies so that easy detection is possible. We saw in Example 3.7 that speech can be encoded into PCM with a pulse rate of 45 kHz. Ignoring the effects of possible frame synchronizing words, if we multiplex A messages using time division multiplexing, the pulse rate is $A \times 45$ kHz. Thus $mA \times 45 \times 10^3 < 2 \times 10^6$. With $m = 1$ we can transmit only 44 conversations using this technique.

Presumably, the discrete technique performs better in the presence of noise. On the other hand, speech is highly redundant; that is, it is intelligible even when noisy. For this reason FDM is used by communication satellites.   ■ ■

EXAMPLE 4.3 ■

Let us examine stereo FM more closely. Since we do not want a monaural receiver to hear one "side" of a stereo broadcast, the two signals we transmit are $x_1(t) = L(t) + R(t)$ and $x_2(t) = L(t) - R(t)$, where $L(t)$ stands for the left signal and $R(t)$ the right. If we use a full bandwidth of 15 kHz for both $x_1(t)$ and $x_2(t)$, there are two possible

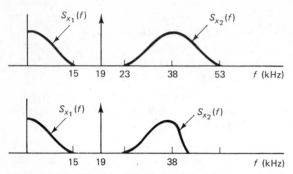

Figure 4.19 Possible FDM techniques for stereo FM.

ways we can use AM multiplexing to separate them as seen in Figure 4.19.

The first case uses DSB-SC modulation to separate the signals and the second SSB. The 19-kHz tone is needed to achieve the phase synchronization necessary for coherent demodulation. The first system is the one used in broadcast stereo FM. A third SSB message for separate, privately sold applications is permitted with the total base bandwidth restricted to 75 kHz. The modulated bandwidth is restricted to 200 kHz, which indicates a maximum Δf of only 25 kHz. We shall show in Chapter 8 that FM works better in noise for larger values of β, and thus there is a cost, in terms of noise performance, in order to transmit stereo. Presumably the subjective improvement caused by the stereo effect compensates for some of this cost, at least if the signal strength is large.   ■ ■

TABLE 4.2   TYPICAL SIGNALS

Signal	Analog BW or equivalent	Bits per sample	Digital bit rate (bits/s)
Very slow transmission[a]			
Typewriter speeds	15 char/s	6	90
Human reading speeds	300 char/s	6	2,000[b]
Machine printing and reading speeds			20,000
Telephone voice	3.3 kHz	5	45,000
Hi fi music	20 kHz	10[c]	400,000[d]
Picturephone	1 MHz	3	6,000,000
Color TV	4.6 MHz	6	60,000,000

[a] Remote burglar alarms, vehicle detectors, telephone handset on-off signals, and so on.

[b] Perhaps double for fast scanning.

[c] Using companding.

[d] Double for stereo.

TABLE 4.3 AVAILABLE CHANNELS

	Approximate date first in use	Bandwidth
Early telegraphy	1850	10 Hz
Telephone lines	1900	10 kHz
Regular (TD1) telephone lines	1940	1 MHz
Newer (TD2) telephone lines; early microwave links	1950	4 MHz
Today's coaxial cable (CATV) microwave highways communication satellites optical fiber cables	1960–1970	100–200 MHz
New communication satellites planned helical waveguide	1980s	1000 MHz

4.9 TYPICAL SIGNAL AND CHANNEL BANDWIDTHS

Let us complete this chapter by considering some of the typical signals encountered in practice and channels that are available to us. The signals are listed in Table 4.2. Available channels and their bandwidths are listed in Table 4.3. The number of signals that can be multiplexed in the available channels is a matter of bookkeeping. Thus the regular TD1 telephone line can accommodate one analog picturephone channel, or up to 250 analog or 20–30 digital phone conversations, or 75 computer-to-computer links, or 600 man- (via teletype) to-computer channels, etc. On the other hand, cable TV channels with 100–200 times the bandwidth can accommodate 20–40 color TV channels.

PROBLEMS

4.1. We wish to quantize a music signal into 64 levels during the quiet passages (defined by $|x| \le 10^{-4} x_{max}$, where $2x_{max}$ is the maximum anticipated dynamic range).
  (a) How many bits of quantization would be required to cover the maximum dynamic range if uniform quantization were used?
  (b) Repeat for a μ-law compander with $\mu = 255$.

4.2. Consider an 11-bit quantizer.
  (a) Prove that with the μ-law compander ($\mu = 255$) the largest quantization interval is 5.567 times larger than would be achieved with uniform quantization.
  (b) Show that the smallest quantization interval using μ-law companding is 46.083 times smaller than would be achieved with uniform quantization.

4.3. Trace the signal in Figure 4.3 or 4.6 and sketch the output of a DPCM system that employs the simple predictor of Equation 4.8 but employs four levels (two bits) of quantization of the error signal. Use the same ϵ as in Figure 4.6.

4.4. Consider once again the signal in Figure 4.3. Sketch the output of a source code that uses only two levels (one bit) of quantization but employs the following prediction algorithms:
(a) $\hat{x}(k) = 0.8\hat{x}(k - 1) \pm \epsilon$;
(b) $\hat{x}(k) = 2\hat{x}(k - 1) - \hat{x}(k - 2) \pm \epsilon$.

4.5. The prediction algorithm of Equation 4.13 (or Problem 4.4b) was an approximation to $\hat{x}(k) = x(k - 1) + T \, dx/dt|_{(k-1)T}$. Suppose we wished to generalize this to approximate

$$\hat{x}(k) = ax(k - 1) + bT \left.\frac{dx}{dt}\right|_{(k-1)T} + cT^2 \left.\frac{d^2x}{dt^2}\right|_{(k-1)T}$$

Determine the appropriate prediction algorithm and evaluate for $a = 0.7$, $b = 0.2$, and $c = 0.1$.

4.6. Argue that if the receiver makes a decision error on one of the binary pulses of the Huffman code shown in Table 4.1, word synchronization is temporarily lost but almost immediately regained. (Prove by example as in the last paragraph of Section 4.4.)

4.7. Determine the output of the mixer modeled by Figure 4.10 (i.e., expand Equation 4.21) and verify the spectrum plot of Figure 4.11. What are the unlabeled spectral zones?

4.8. The circuit shown below can be used for the generation of AM signals. Assume that the power spectra of $m(t)$ and $m^2(t)$ are both zero for frequencies greater than 20 kHz, $f_c = 2$ MHz, and $m_n(t)$ represents a normalized version of $m(t)$ (i.e., $|m(t)| < \alpha$).

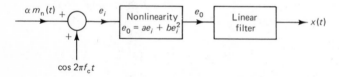

(a) Using arguments similar to those of Section 2.10, sketch the spectrum of the signal $e_0(t)$.
(b) Show that, with an appropriate linear filter, $x(t)$ can be regarded as an amplitude modulated version of $m(t)$.
(c) Determine the modulation index as a function of the parameters α, a, and b.

4.9. We have modulated (DSB-AM) an acoustic signal whose bandwidth is 10 kHz on a 2-MHz carrier. The receiver is a superheterodyne receiver which presumes all the modulated messages are limited to 10 kHz.
(a) What is the bandwidth of the AM signal?
(b) What is the bandwidth of the i.f. filter?
(c) What is the oscillator frequency if the i.f. center frequency is 455 kHz?
(d) What is the bandwidth of the final audio stage?

4.10. Assume a DSB-AM signal where $\langle m_n^2(t)\rangle = 0.5$. Compute the average power (in terms of A and k) in

(a) both sidebands (include negative frequencies),

(b) the upper sideband, and

(c) the carrier term.

(d) What percentage of the total power is in both sidebands?

(e) Evaluate part (d) for $k = 0.8$ and for $k = 1.0$.

4.11. For some reason we are communicating a 10-kHz tone $[m(t) = \cos(2\pi f_0 t + \theta_0)]$, where $f_0 = 10$ kHz, using some form of amplitude modulation with a carrier frequency $f_c = 1.1$ MHz.

(a) Determine the form of the modulated signal $s(t)$ for each of the following AM techniques:

    (1) DSB-SC,

    (2) DSB with $k = 0.7$, and

    (3) SSB-SC using the upper sidebands.

(b) Prove that the following demodulator works for the SSB-SC signal of part (3).

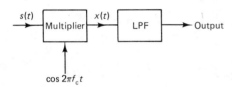

4.12. Suppose the message in an AM scheme can be approximated by

$$m(t) = \sum_{n=1}^{20} C_n \cos(n2\pi f_0 t + \theta_n) \qquad \text{where} \quad f_0 = 100 \text{ Hz}$$

(a) Expand a standard AM signal $s(t) = A[1 + km(t)] \cos 2\pi f_c t$ into a Fourier series by a simple trig identity. Identify the carrier, upper sideband, and lower sideband components. What is the bandwidth of $s(t)$?

(b) Show that the coherent demodulator works for SSB-SC by considering one sideband only as the input to the demodulator.

4.13. Consider again the message of Problem 4.13. Suppose it is passed by the filter shown here, where $|H(f_c + f')| + |H(f_c - f')| = 1$ for $0 < |f'| < (r/2)B$. Show that the coherent demodulator works for the resultant VSSB signal.

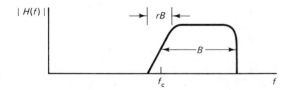

4.14. You are transmitting a 15-kHz signal using FM and have a 150-kHz allocated bandwidth.

(a) What is the largest value of β that can be used?

(b) What frequency deviation does this correspond to?

4.15. An AM signal of the form $s(t) = A[1 + km_n(t)] \cos 2\pi f_0 t$, where $f_0 = 1.1$ MHz and the bandwidth of $m_n(t)$ is 10 kHz is the input to the demodulator shown on page 132. Determine the desired values for f_c, f_m, B_1, B_2, θ_1, and θ_2.

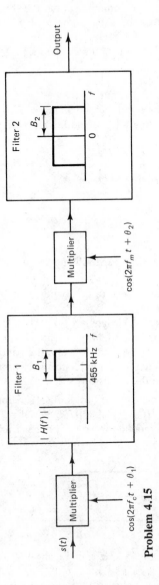

Problem 4.15

4.16. Assume that the maximum frequency deviation of your transmitter is 70 kHz and your allocated bandwidth is 160 kHz.
   (a) If $\Delta f = 70$ kHz, what message bandwidth (B) would achieve the maximum allocated bandwidth?
   (b) If $B = 20$ kHz, what is the maximum frequency deviation that you can use?

4.17. An audio signal that is bandlimited to 10 kHz is to be transmitted by either AM or FM. The amplitude is such as to provide 100% carrier modulation in the AM case and a frequency deviation $\Delta f = 75$ kHz in the FM case. Calculate the transmission bandwidth required in both cases.

4.18. Consider an analog signal whose bandwidth is 12.5 kHz which is to frequency modulate a 1-MHz carrier with a frequency deviation $\Delta f = 100$ kHz.
   (a) Determine the bandwidth of the resulting FM signal using the rule-of-thumb formula.
   (b) Suppose the analog signal was first converted into a PCM signal by quantizing it into 16 levels. Determine now the minimum bandwidth of the resulting FSK signal.

4.19. An FM modulator is used that can have a ± 10 kHz swing about a carrier frequency of 500 kHz. The bandwidth of $m(t)$ is 20 kHz.
   (a) Determine β_{max} and the bandwidth of the FM signal.
   (b) Repeat part (a) assuming that there are two frequency doubling stages at the transmitter after the FM modulator.

4.20. Our goal is to multiplex as many similar acoustical measurements through a 2-MHz cable as possible. Each message has a bandwidth of 18 kHz.
   (a) First assume frequency multiplexing using FM with a modulation index of 8 and 10% frequency guardbands. Determine the largest number of messages that can be multiplexed.
   (b) Assume now, TDM using PCM. Assume Nyquist sampling rates and $m = 1$. Further assume that 64 quantization levels are used. Determine the largest number of sources that can be multiplexed.

4.21. We wish to frequency multiplex, using FM, 100 messages each of bandwidth B_A into a channel whose bandwidth is B_C.
   (a) Solve for the largest value of the common modulation index β.
   (b) Evaluate β if the messages are speech signals as used in telephony and the channel is the TD2 telephone line and $A = 100$.

4.22. You wish to multiplex a number of voice channels ($B = 3.5$ kHz) through a cable whose bandwidth is 550 kHz.
   (a) You first wish to consider frequency multiplexing using standard DSB-AM. Your system must include guardbands of at least 20% of the message bandwidth. What is the largest number of messages that can be multiplexed?
   (b) You now wish to consider time multiplexing. Assume a sampling rate of 1.5 times the Nyquist rate and 32 levels of quantization. Assume that a bandwidth equal to the pulse rate is adequate. How many channels can be multiplexed?

4.23. Suppose it is desired to multiplex five messages, each having a bandwidth of 20 kHz by first separating the messages using single-sideband AM and then double-sideband AM modulating a 1-MHz carrier. What would be the minimum bandwidth? How would this compare with a system in which the carrier was 10 MHz, which was frequency modulated with a frequency deviation of 320 kHz?

4.24. Ten acoustic measurements are transmitted simultaneously through a micro-wave channel whose bandwidth is 2 MHz by first passing them through ideal low-pass filters (BW = B Hz), then modulating them, using DSB-AM, on subcarriers at frequencies of 0, 10, 20, . . . kHz, and finally frequency modulating the entire multiplexed signal.
  (a) What is the largest possible value for the bandwidth B of the initial filters?
  (b) What is the largest possible value for the FM modulation index β?
  (c) Sketch a receiver (a block diagram is adequate) whose output is the fourth measurement (i.e., the one modulated on a 30-kHz carrier).

4.25. Eight audio signals, whose bandwidths are each 5 kHz, are frequency division multiplexed on subcarriers at 50, 100, 150, . . . , and 400 kHz by using frequency modulation. The combined signal is then again frequency modulated for transmission through a microwave link whose band is centered at 30 MHz.
  (a) Find the largest possible β for each of the first stages of subcarrier modulation.
  (b) What is the overall bandwidth required for the microwave link if the frequency deviation $\Delta f = 1$ MHz?
  (c) Sketch (in block diagram form only) the receiver whose output is the audio message that used the 200-kHz subcarrier.

4.26. Multiplying a signal by a high-frequency carrier can be achieved by the balanced modulator shown below, where the input device is a center-tapped transformer that results in an inverted input signal as well as the input:

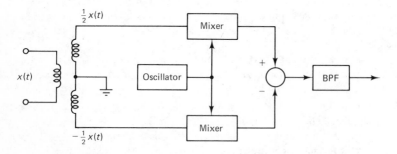

  (a) Assuming the mixer is the same as the one considered in Problem 4.8, prove that the output is $\alpha x(t) \cos 2\pi f_c t$ where $\cos 2\pi f_c t$ is the oscillator output.
  (b) Prove that the multiplier in a coherent receiver, such as Figure 3.40, can be replaced by this balanced modulator.

REFERENCES

1. L. R. Rabiner and R. W. Schafer, *Digital Processing of Speech Signals*, Prentice-Hall, Englewood Cliffs, NJ, 1978.
2. D. A. Huffman, "A Method for the Construction of Minimum Redundancy Codes," *Proc. IRE* 40:1098–1101 (1952).
3. J. G. Proakis, *Digital Communications*, McGraw-Hill, New York, 1983.
4. M. Schwartz, W. R. Bennet, and S. Stein, *Communication Systems and Techniques*, McGraw-Hill, New York, 1966.

5. M. Schwartz, *Information, Transmission, Modulation, and Noise*, McGraw-Hill, New York, 1980.
6. K. S. Shanmugam, *Digital and Analog Communication Systems*, Wiley, New York, 1979.
7. E. Jahnke and F. Emde, *Tables of Functions*, Dover, New York, 1945.
8. K. K. Clarke and D. J. Hess, *Communication Circuits: Analysis and Design*, Addison-Wesley, Reading, MA, 1971.

Chapter 5

Statistical Modeling of Random Signals

5.1 INTRODUCTION

In order to analyze the communication systems that we have considered, some modeling of random signals based on probability concepts is necessary. For example, the analysis of many of the discrete receivers involves the determination of the percentage of time that samples taken every T s of a random signal $x(t)$ exceed some threshold voltage V. These percentages $(100/P_T)$ can be different for processes with the same power spectrum, such as the shot noise and impulse noise processes shown in Figure 2.34. A measurement, called the *relative frequency* $_T r_N$, which is related to P_T, is defined by

$$_T r_N = \frac{1}{N} \sum_{n=1}^{N} u[x(nT) - V] \tag{5.1}$$

where $u(t)$ is the unit step function. The percentage P_T can be uniquely defined by

$$P_T = \lim_{N \to \infty} {_T r_N} \tag{5.2}$$

provided the limit exists, which means that $_T r_N$ converges as N increases to the same value every time the measurement is repeated. In this situation $_T r_N$ is an estimate of P_T that improves as N (the number of samples) increases.

If the relative frequency does not depend on the sampling time T, then Equation 5.2 can be replaced by

$$P = \lim_{T' \to \infty} \frac{1}{T'} \int_0^{T'} u[x(t) - V] \, dt \tag{5.3}$$

For this case, P can be thought of as the probability that any sample of $x(t)$ exceeds the voltage V. We seek probability models of random signals so that expressions, such as Equation 5.3, can be evaluated analytically rather than by measurements in the laboratory. Much of this chapter develops the concepts and theorems of probability theory that we will need. The final sections will utilize these concepts for the purpose of modeling random signals.

5.2 PROBABILITY MEASURE

Let us call the outcome of any experiment an *event* (denoted by capital letters, such as A). We further define the collection of all possible events as the *sample space* (denoted Ω) and an impossible event as the null event (denoted $\varnothing$). A *probability measure* P can be defined axiomatically in terms of the following properties:

   1. $0 \leq P(A) \leq 1$, for all A.
   2. $P(\Omega) = 1$.
   3. $P(\varnothing) = 0$.

The measure must conform to our intuitive notions of probability. If one performs an experiment n times and the event A occurs n_A times, the relative frequency of the event A is defined by

$$r_A = \frac{n_A}{n} \tag{5.4}$$

This relative frequency must converge to the probability of the event:

$$\frac{n_A}{n} \to P(A) \quad \text{as} \quad n \to \infty \tag{5.5}$$

for the probability measure to be useful. For example, if the probability of observing a head when flipping a coin is assumed to be 0.5, then the relative frequency of a head must converge to 0.5.

   Consider two arbitrary events A and B, which may or may not be mutually exclusive (i.e., they may or may not occur simultaneously). For example, the event of drawing a heart from a deck of cards (A) and the event of drawing an ace (B) may both occur at the same time and are not mutually exclusive. First, let us define the terminology

   $A \cup B$ means that either the event A or the event B occurred.
   $A \cap B$ or AB means that both events A and B occurred.

The symbols $\cup$ and $\cap$ are called union and intersection, respectively. There are four possible outcomes of an experiment relative to two events A and B.

A only, which occurs n_1 times.

B only, which occurs n_2 times.

A and *B*, which occurs n_3 times.

neither *A* nor *B*, which occurs n_4 times.

Here $n_1 + n_2 + n_3 + n_4 = n$ (the total number of experiments). We define $P(A)$ to be the probability of *A*, $P(B)$ as the probability of *B*, $P(A \cap B)$ the probability of *A* and *B*, and $P(A \cup B)$ the probability of *A* or *B*. If these measures are to conform to our notions of relative frequency, then

$$r_A = \frac{n_A}{n} = \frac{n_1 + n_3}{n} \to P(A) \qquad \text{as} \quad n \to \infty$$

$$r_B = \frac{n_B}{n} = \frac{n_2 + n_3}{n} \to P(B) \qquad \text{as} \quad n \to \infty$$

$$r_{A \cap B} = \frac{n_{A \cap B}}{n} = \frac{n_3}{n} \to P(A \cap B) \qquad \text{as} \quad n \to \infty$$

$$r_{A \cup B} = \frac{n_{A \cup B}}{n} = \frac{n_1 + n_2 + n_3}{n} \to P(A \cup B) \qquad \text{as} \quad n \to \infty$$

Since $n_1 + n_2 + n_3 = (n_1 + n_3) + (n_2 + n_3) - n_3$, it follows that

$$r_{A \cup B} = r_A + r_B - r_{A \cap B}$$

We therefore expect that

$$P(A \cup B) = P(A) + P(B) - P(A \cap B) \tag{5.6}$$

Equation 5.6 is often regarded as axiom 4 of the probability measure, and all other relationships can be proven formally from the four axioms. Equation 5.6 is conveniently visualized by a geometric figure (Figure 5.1), called a *Venn diagram*, where the probability of an event is associated with the area of a region. For mutually exclusive (or disjoint) events, the *A* region and *B* region do not overlap and $P(A \cap B) = 0$. Equation 5.6 can be generalized to *m* events. Using the notation

$$\bigcup_{i=1}^{m} A_i = A_1 \cup A_2 \cup \cdots \cup A_m$$

a convenient form known as the union bound is developed:

$$P\left(\bigcup_{i=1}^{m} A_i \right) \le \sum_{i=1}^{m} P(A_i) \tag{5.7}$$

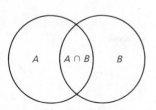

Figure 5.1 Venn diagram.

Equality is achieved only if the events are disjoint; that is,

$$P(A_i \cap A_j) = P(\varnothing) = 0 \quad \text{for any } i \neq j$$

Let us now consider the event that A has occurred given that the event B has occurred and denote its probability by $P(A/B)$. This is called a *conditional probability* because the probability is conditioned on the fact that either B occurred by itself or it occurred together with A; hence only these $n_2 + n_3$ experiments are to be considered. Thus the relative frequency of A given B is $n_3/(n_2 + n_3)$ and we expect that

$$\frac{n_3}{n_2 + n_3} \to P(A/B) \quad \text{as} \quad n \to \infty$$

This can be written $(n_3/n)/[(n_2 + n_3)/n]$, where n_3/n is the relative frequency of $A \cap B$ and $(n_2 + n_3)/n$ is the relative frequency of B, so we expect that

$$P(A/B)P(B) = P(A \cap B) \tag{5.8}$$

This relationship is called the *probability chain rule*. Since $P(A \cap B) = P(B \cap A)$, it follows that

$$P(A/B)P(B) = P(B/A)P(A)$$

or

$$P(A/B) = \frac{P(B/A)P(A)}{P(B)} \tag{5.9}$$

Equation 5.9, which is a form of the chain rule, is sometimes called *Bayes's rule*.

Two events are said to be *statistically independent* if $P(A \cap B) = P(A)P(B)$. For two such independent events, it is seen from Equation 5.8 that

$$P(A/B) = \frac{P(A \cap B)}{P(B)} = \frac{P(A)P(B)}{P(B)} = P(A)$$

Thus, if A is independent of B, the probability of A does not change by knowing that event B has occurred.

Finally, suppose one of m disjoint events A_i, $i = 1, 2, \ldots, m$, must occur with the event B. For example, one of the four disjoint events, corresponding to drawing a club, diamond, heart, or spade from a deck of cards, must occur with the event of drawing an ace. For this case

$$P(B) = \sum_{i=1}^{m} P(B \cap A_i) = \sum_{i=1}^{m} P(B/A_i)P(A_i) \tag{5.10}$$

where $P(A_i \cap A_j) = 0$ for any $i \neq j$. From our example, the probability of drawing an ace is equal to $P(\text{ace of clubs}) + P(\text{ace of diamonds}) + P(\text{ace of hearts}) + P(\text{ace of spaces}) = 4 \times \frac{1}{52} = \frac{1}{13}$. Sometimes Bayes's rule is given as

$$P(A_j/B) = \frac{P(B/A_j)P(A_j)}{\sum_{i=1}^{m} P(B/A_i)P(A_i)} \tag{5.11}$$

where $P(A_i \cap A_j) = 0$ for $i \neq j$.

5.3 RANDOM VARIABLES AND DISTRIBUTIONS

Let us assume that a sample space Ω of experiments is a collection of simple disjoint events w. Any event A defined on Ω consists of one or more w elements. Any real-valued function defined on the elements w of Ω is called a *random variable*, denoted $X(w)$. $X(w)$ maps Ω onto the real line:

$$\Omega \xrightarrow{X(w)} R$$

where R represents the real line. As an example, consider the experiment of drawing a card from a deck of 52 cards. The elements w are the 52 possible outcomes of the experiment. If the cards are ordered, we can assign to each card a number from 1 to 52 that corresponds to its position in the ordering. The function that relates the experiment to the first 52 integers is a random variable. Thus if the ordering of bridge is used, $X(2 \text{ of clubs}) = 1$, $X(2 \text{ of diamonds}) = 2$, etc.

Another example involves measuring the "value" of a sample of an electrical signal. By assigning a real value (in volts) to the electric force, we are defining a random variable that is defined on a continuous interval of the real line. All random variables have real values. They are called *discrete random variables* when defined on a set of discrete points on the line, and *continuous random variables* when defined on one or more continuous intervals of the real line.

We define the *probability function $p(x)$* by

$$p(x) = \Pr(X(w) = x) \tag{5.12}$$

This probability function characterizes a discrete random variable since the probability of any event A defined on Ω can be related to $p(x)$. For a continuous random variable, $p(x)$ could be zero for all $X(w)$ since there are an infinite number of elements w. On the other hand, a function that can be defined for any random variable which completely characterizes that random variable is called the *cumulative distribution function* (CDF), defined by

$$F(x) = \Pr(X(w) \leq x) = P(B) \tag{5.13}$$

where B consists of all elements w such that $X(w) \leq x$. For discrete random variables, the CDF is related to the probability function by

$$F(x) = \sum_{x_i \leq x} p(x_i) \tag{5.14}$$

In every case, the CDF satisfies the following properties:

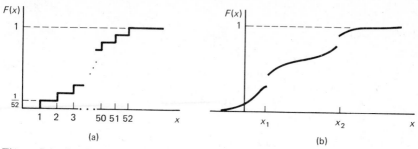

Figure 5.2 Legitimate CDFs.

1. $\lim_{x \to -\infty} F(x) = 0$.
2. $\lim_{x \to \infty} F(x) = 1$.
3. $F(x)$ is a monotonic, nondecreasing function of x that is continuous on the right for every x [i.e., $F(x) = \lim_{\epsilon \to 0} F(x + \epsilon)$]. Examples are shown in Figure 5.2.

Consider the class of events that map onto a right closed interval (i.e., the rightmost point is included but not the leftmost point) denoted $(a, b]$. The probability of such an event is

$$P((a, b]) = F(b) - F(a) \tag{5.15}$$

Since a point x can be characterized by

$$x = \lim_{\epsilon \to 0} (x - \epsilon, x]$$

it follows that

$$p(x) = \lim_{\epsilon \to 0} [F(x) - F(x - \epsilon)] \tag{5.16}$$

For the random variable characterized by the CDF of Figure 5.2a, $p(x) = \frac{1}{52}$ for the positive integers between 1 and 52, and there is no probability associated with the rest of the real line. This is a discrete random variable that could model the experiment of drawing a card from a deck. For the random variable characterized by the CDF of Figure 5.2b, except for the points x_1 and x_2, $p(x) = 0$. This is a continuous random variable.

5.4 DISCRETE PROBABILITY DISTRIBUTIONS

The probability function of discrete random variables must satisfy all the properties of the probability measure that were discussed in Section 5.2. We use the notation

$$p(x, y) = \Pr(X(w) = x \cap Y(w) = y) \tag{5.17}$$

where $p(x, y)$ is called the *bivariate* or *joint probability function*. It follows directly from Equations 5.8–5.11 that

$$p(x,y) = p(x/y)p(y) \tag{5.18}$$

$$p(x/y) = \frac{p(y/x)p(x)}{p(y)} \tag{5.19}$$

$$p(x) = \sum_i p(x,y_i) = \sum_i p(x/y_i)p(y_i) \tag{5.20}$$

$$p(y_j/x) = \frac{p(x/y_j)p(y_j)}{\sum_i p(x/y_i)p(y_i)} \tag{5.21}$$

and if x and y are statistically independent random variables,

$$p(x,y) = p(x)p(y) \tag{5.22}$$

We can extend many of these ideas to three or more random variables where $p(x_1, x_2, \ldots, x_n)$ is called a *multivariate probability function*. Equations 5.18, 5.20, and 5.22 can be logically extended to

$$p(x_1, x_2, \ldots, x_n) = p(x_1/x_2, x_3, \ldots, x_n)p(x_2, x_3, \ldots, x_n) \tag{5.23}$$

$$p(x_1) = \sum_i \sum_j \cdots \sum_k p(x_1, x_{2_i}, x_{3_j}, \ldots, x_{n_k}) \tag{5.24}$$

$$p(x_1, x_2, \ldots, x_n) = p(x_1)p(x_2) \cdots p(x_n) \tag{5.25}$$

if $x_1, x_2, \ldots, x_n$ are mutually statistically independent random variables.

EXAMPLE 5.1 ■

Let us consider a binary communication system which utilizes two pulses $s_1(t)$ and $s_0(t)$. If we map these outcomes onto the integers 1 and 0, we define a discrete random variable S that can have the value 1 or 0. Assume that a receiver is available, which may or may not be making decisions logically, that prints the symbols a and b. Thus the output is also characterized by a discrete random variable with two values. Upon experimentation, we discover that, when $s_0(t)$ is transmitted, 80% of the time the output is b and, when $s_1(t)$ is transmitted, 60% of the time the output is a. Furthermore, typical input sequences are found to be composed of $s_0(t)$ pulses 80% of the time and $s_1(t)$ pulses 20% of the time. We characterize this information in the manner shown in Figure 5.3. Let us assume that an a corresponds to $s_1(t)$ and a b to $s_0(t)$. This is called a *decision rule*. In this example the decision rule may be denoted

$$a \to 1 \qquad b \to 0$$

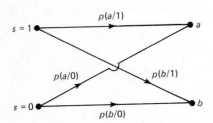

Figure 5.3 Characterization of Example 5.1. $p(0) = 0.8$, $p(1) = 0.2$, $p(a/1) = 0.6$, $P(b/1) = 0.4$, $P(b/0) = 0.8$, $p(a/0) = 0.2$.

Given this rule, two disjoint events correspond to an error: the event $s_0(t)$ is transmitted and a is received, and the event $s_1(t)$ is transmitted and b is received. Thus, from Equation 5.6, the probability of error $P(e)$ is

$$P(e) = p(0,a) + p(1,b) \qquad (5.26)$$

From Equation 5.18,

$$P(e) = p(a/0)p(0) + p(b/1)p(1)$$
$$= 0.2 \times 0.8 + 0.4 \times 0.2 = 0.24 \qquad (5.27)$$

Let us now calculate $p(1/b)$ and $p(0/b)$ using Bayes's rule or Equation 5.19.

$$p(1/b) = \frac{p(b/1)p(1)}{p(b)} = \frac{0.08}{p(b)}$$

$$p(0/b) = \frac{p(b/0)p(0)}{p(b)} = \frac{0.64}{p(b)}$$

The fact that $p(0/b) > p(1/b)$ means that it is more likely that $s_0(t)$ was transmitted than $s_1(t)$ given that the receiver prints a b. Since our decision rule assumes $s_0(t)$ when a b is received, it seems quite logical. Similarly,

$$p(1/a) = \frac{p(a/1)p(1)}{p(a)} = \frac{0.12}{p(a)}$$

$$p(0/a) = \frac{p(a/0)p(0)}{p(a)} = \frac{0.16}{p(a)}$$

Because $p(0/a) > p(1/a)$, it is logical to assume that $s_0(t)$ was transmitted when the receiver prints an a as well. It seems that a better decision rule than the one we have analyzed is to ignore the receiver entirely and always assume that $s_0(t)$ was transmitted. For this rule, the probability of error is the probability that $s_1(t)$ was transmitted which is 0.2. This rule is better since it results in a smaller probability of error. The decision rule based on determining the most likely signal is said to be the optimal decision rule.

To complete this example, let us calculate $p(a)$ and $p(b)$ from Equation 5.20:

$$p(a) = p(a/1)p(1) + p(a/0)p(0)$$
$$= 0.6 \times 0.2 + 0.2 \times 0.8 = 0.28$$
$$p(b) = p(b/1)p(1) + p(b/0)p(0)$$
$$= 0.4 \times 0.2 + 0.8 \times 0.8 = 0.72$$

As we expect, $p(a) + p(b) = 1$. We can now calculate the values of the four conditional probabilities we examined from Bayes's rule. For example,

$$p(1/b) = 0.08/p(b) = 0.08/0.72 = \tfrac{1}{9}$$

The procedure we just followed is explicitly indicated by Equation 5.21.

■ ■

5.5 CONTINUOUS PROBABILITY DISTRIBUTIONS

In this section we consider only absolutely continuous distributions defined by $p(x) = 0$ for all x. For such distributions we can define a *probability density function* (PDF), denoted $f(x)$, by

$$f(x) = \frac{d}{dx} F(x) \tag{5.28}$$

or

$$F(x) = \int_{-\infty}^{x} f_X(\xi) \, d(\xi) \tag{5.29}$$

where the random variable label X is used as a subscript for identification purposes when dummy variables are used. For $f(x)$ to be a density function of the probability measure, it must satisfy the following properties:

1. $f(x) \geq 0$ for all x. $\tag{5.30}$

2. $\displaystyle\int_{-\infty}^{\infty} f(x) \, dx = 1.$ $\tag{5.31}$

3. $\displaystyle\int_{a}^{b} f(x) \, dx = P((a,b]).$ $\tag{5.32}$

From the definition

$$\int_{a}^{b} f(x) \, dx = \int_{-\infty}^{b} f(x) \, dx - \int_{-\infty}^{a} f(x) \, dx$$
$$= F(b) - F(a) = P((a,b]) \tag{5.33}$$

(see Equation 5.15). Also,

$$\int_{-\infty}^{\infty} f(x) \, dx = \lim_{A \to \infty} \int_{-A}^{A} f(x) \, dx = \lim_{A \to \infty} [F(A) - F(-A)] = 1$$

from the properties of the CDF. Finally, since $F(x)$ is monotonically nondecreasing, it follows that $dF(x)/dx = f(x) \geq 0$.

EXAMPLE 5.2 ■
Let us reconsider the example of measuring the voltage of a sample of an electrical signal. Let us assume that this continuous random variable can only have values between $-A$ and $+A$ V. A distribution that

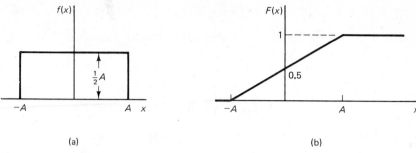

Figure 5.4 Uniform distribution.

can only have values between $-A$ and $+A$ V. A distribution that guarantees that the probability of any interval is proportional to the length of the interval is the uniform distribution characterized by Figure 5.4.

The height of $f(x)$ must be $A/2$ in order to ensure that

$$\int_{-\infty}^{\infty} f(x)\, dx = 1$$

■ ■

Bivariate and *multivariate PDFs* can be defined for continuous random variables in a manner analogous to that for discrete random variables.[1] If we start with

$$F(x_1, x_2, x_3) = \Pr(X_1 \le x_1, X_2 \le x_2, X_3 \le x_3) \tag{5.34}$$

then we can define

$$F(x_1, x_2, x_3) = \int_{-\infty}^{x_3} \int_{-\infty}^{x_2} \int_{-\infty}^{x_1} f_{X_1 X_2 X_3}(u_1, u_2, u_3)\, du_1\, du_2\, du_3 \tag{5.35}$$

Analogous to Equations 5.20 and 5.24,

$$f_{X_1}(x) = \int_{-\infty}^{\infty} \int_{-\infty}^{\infty} f_{X_1 X_2 X_3}(x, u_2, u_3)\, du_2\, du_3 \tag{5.36}$$

We must, however, approach conditional PDFs with caution since Equations 5.9 and 5.19 involve probabilities which are zero for absolutely continuous random variables. We can circumvent this problem by defining the sets A and B as follows:

A consists of all points w such that $Y(w) \le y$.

B consists of all points w such that $x - \Delta x < X(w) \le x$.

From Equation 5.8 we can write

---

[1] For convenience, we shall drop the argument of the random variable $X_1(w)$ from this point on. Capital letters will continue to refer to the random variables, small letters to their values.

$$P((-\infty,y]/(x - \Delta x, x])P((x - \Delta x, x]) = P((-\infty, y], (x - \Delta x, x]) \qquad (5.37)$$

If X and Y are continuous random variables, this can be written

$$P((-\infty, y]/(x - \Delta x, x]) = \int_{-\infty}^{y} \int_{x-\Delta x}^{x} f_{XY}(u_1, u_2) \, du_1 \, du_2 \Big/ \int_{x-\Delta x}^{x} f_X(u) \, du \qquad (5.38)$$

If we assume that Δx is very small, Equation 5.38 can be approximated by

$$P((-\infty, y]/(x - \Delta x, x]) \approx \int_{-\infty}^{y} f_{XY}(x, u) \, du \, \Delta x \Big/ f(x) \, \Delta x \qquad (5.39)$$

We now define

$$F(y/x) = \lim_{\Delta x \to 0} \Pr(Y \le y/(x - \Delta x, x]) \qquad (5.40)$$

Combining Equations 5.39 and 5.40, we get

$$F(y/x) = \int_{-\infty}^{y} f_{XY}(x, u) \, du \Big/ f(x) \qquad (5.41)$$

or, equivalently,

$$f(y/x) = f(x, y)/f(x) \qquad (5.42)$$

Since the chain rule holds for PDFs of continuous random variables, Bayes's rule also holds, so

$$f(y/x) = \frac{f(x/y)f(y)}{f(x)} \qquad (5.43)$$

and

$$f(y/x) = \frac{f(x/y)f(y)}{\int_{-\infty}^{\infty} f(x/y)f(y) \, dy} \qquad (5.44)$$

If Y is a continuous random variable and X a discrete one, then if A were defined as $(y - \Delta y, y]$ and B as x, Equation 5.8 would lead to (see Problem 5.11)

$$f(y/x) = \frac{p(x/y)f(y)}{p(x)} \qquad (5.45)$$

or, on rearranging,

$$p(x/y) = \frac{f(y/x)p(x)}{f(y)} \qquad (5.46)$$

These equations are called the mixed form of Bayes's rule.

Finally n continuous random variables are defined to be *statistically independent* if their multivariate PDFs factor, or

$$f(x_1, x_2, \ldots, x_n) = f(x_1)f(x_2) \cdots f(x_n) \qquad (5.47)$$

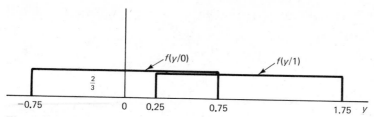

Figure 5.5 Conditional, uniform PDFs.

Similar to the discrete case, for independent random variables, Equation 5.42 becomes

$$f(y/x) = \frac{f(x, y)}{f(x)} = \frac{f(x)f(y)}{f(x)} = f(y)$$ (5.48)

Thus, knowing the value of X does not change the distribution of Y.

EXAMPLE 5.3 ■

Let us reconsider Example 5.1 but with a different receiver. Now the receiver is not a decision device, and its output is a continuous random variable Y whose value lies between -0.75 V and 1.75 V. On experimentation we discover that when $s_0(t)$ is transmitted, the PDF of Y is uniform over the interval $(-0.75, 0.75]$. When $s_1(t)$ is transmitted, the PDF of Y is uniform over the interval $(0.25, 1.75]$. We characterize this information by Figure 5.5. As before, $p(0) = 0.8$ and $p(1) = 0.2$. From Bayes's rule (the mixed form of Equation 5.46),

$$p(1/y) = \frac{f(y/1)p(1)}{f(y)} = \frac{0.2f(y/1)}{f(y)}$$

$$p(0/y) = \frac{f(y/0)p(0)}{f(y)} = \frac{0.8f(y/0)}{f(y)}$$

The numerators are plotted in Figure 5.6 (the denominators are the same). We see from Figure 5.6 that $s_0(t)$ was most likely to have been transmitted $[p(0/y) > p(1/y)]$ when the output $y < 0.75$. When $y >$

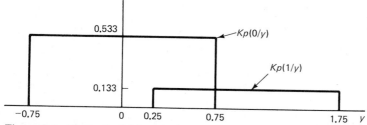

Figure 5.6 Plots of $p(1/y)$ and $p(0/y)$.

0.75, $s_1(t)$ is more likely to have been transmitted. Thus the optimal decision rule is

$$-0.75 < y < +0.75 \quad \rightarrow \quad 0$$
$$0.75 < y < 1.75 \quad \rightarrow \quad 1$$

The probability of error $P(e)$ for this rule is

$$
\begin{aligned}
P(e) &= P(y > 0.75, 0) + P(y < 0.75, 1) \\
&= P(y > 0.75/0)p(0) + P(y < 0.75/1)p(1) \\
&= 0.8 \int_{0.75}^{\infty} f(y/0)\, dy + 0.2 \int_{-\infty}^{0.75} f(y/1)\, dy \\
&= 0 + 0.2 \int_{0.25}^{0.75} \tfrac{2}{3} dy \approx 0.067
\end{aligned}
\tag{5.49}
$$

■ ■

EXAMPLE 5.4 ■

Consider as a candidate for a joint PDF

$$f(x,y) = Ke^{-a|x|-b|y|}$$

To be a legitimate PDF, it must be nonnegative and must integrate to 1:

$$1 = \int_{-\infty}^{\infty} \int_{-\infty}^{\infty} f(x, y)\, dx\, dy = K \int_{-\infty}^{\infty} e^{-a|x|}\, dx \int_{-\infty}^{\infty} e^{-b|y|}\, dy = K\frac{2}{a}\frac{2}{b}$$

If $K = ab/4$, the joint PDF is legitimate. From Equation 5.36

$$f(x) = \int_{-\infty}^{\infty} f_{XY}(x, u)\, du = \tfrac{1}{4}abe^{-a|x|} \int_{-\infty}^{\infty} e^{-b|y|}\, dy = \tfrac{1}{2}ae^{-a|x|}$$

Similarly $f(y) = \tfrac{1}{2}be^{-b|y|}$. Finally we note by inspection that

$$f(x,y) = f(x)f(y)$$

It follows from Equation 5.47 that the random variables characterized by the joint PDF of this example are statistically independent. ■ ■

5.6 STATISTICAL AVERAGES

The *expected value* $E\{X\}$, also called the *mean value* m_x, of a random variable X is defined as

$$E\{X\} = m_x = \sum_i x_i p(x_i) \tag{5.50}$$

for the discrete case, and

$$E\{X\} = m_x = \int_{-\infty}^{\infty} xf(x)\, dx \tag{5.51}$$

for the continuous case. This statistical average of a random variable is analogous to the time average of a random signal. We can extend the notion of expected value to arbitrary functions of one or more random variables

$$E\{g(X, Y)\} = \int_{-\infty}^{\infty} \int_{-\infty}^{\infty} g(x, y)f(x, y) \, dx \, dy \tag{5.52}$$

and to conditional expected values

$$E\{X/y\} = \int_{-\infty}^{\infty} xf(x/y) \, dx \tag{5.53}$$

Theorem 5.1. The expected value of the weighted sum of two or more random variables is the weighted sum of the expected values.

Proof: If we can prove this theorem for two random variables, it can be extended to n random variables by induction.

$$
\begin{aligned}
E\{aX + bY\} &= \int_{-\infty}^{\infty} \int_{-\infty}^{\infty} (ax + by)f(x, y) \, dx \, dy \\
&= \int_{-\infty}^{\infty} \int_{-\infty}^{\infty} axf(x, y) \, dy \, dx + \int_{-\infty}^{\infty} \int_{-\infty}^{\infty} byf(x, y) \, dx \, dy \\
&= \int_{-\infty}^{\infty} axf(x) \, dx + \int_{-\infty}^{\infty} byf(y) \, dy \\
&= am_x + bm_y
\end{aligned}
$$

Theorem 5.2. The expected value of the product of two statistically independent random variables is the product of their expected values.

Proof

$$
\begin{aligned}
E\{XY\} &= \int_{-\infty}^{\infty} \int_{-\infty}^{\infty} xyf(x, y) \, dx \, dy = \int_{-\infty}^{\infty} \int_{-\infty}^{\infty} xyf(x)f(y) \, dx \, dy \\
&= \int_{-\infty}^{\infty} xf(x) \, dx \int_{-\infty}^{\infty} yf(y) \, dy = m_x m_y
\end{aligned}
$$

The *rth moment* μ_r of a random variable X is defined as

$$\mu_r = E\{(X - m_x)^r\} \tag{5.54}$$

The second moment, or μ_2, is called the *variance* and usually denoted σ^2. σ is often called the *standard deviation*. Joint moments (μ_{ij}) can be defined in a similar manner:

$$\mu_{ij} = E\{(X - m_x)^i(Y - m_y)^j\} \tag{5.55}$$

The *correlation coefficient* ρ of two random variables is defined by

$$\rho = \frac{\mu_{11}}{\sqrt{\mu_{20}\mu_{02}}} = \frac{\mu_{11}}{\sigma_x \sigma_y} \tag{5.56}$$

Two random variables are said to be uncorrelated if $\rho = 0$.

EXAMPLE 5.5 ■

Let us find the relationship between σ^2 and $E\{X^2\}$:

$$\sigma^2 = E\{(X - m_x)^2\} = E\{X^2 - 2m_x X + m_x^2\}$$
$$= E\{X^2\} - 2m_x E\{X\} + m_x^2 \quad \text{(from Theorem 5.1)}$$
$$= E\{X^2\} - m_x^2 \tag{5.57} \quad ■ ■$$

EXAMPLE 5.6 ■

Let us find the mean and variance of a random variable X that is uniformly distributed between 0 and A [i.e., $f(x) = 1/A$ for $0 < x \le A$ and 0 elsewhere]:

$$m_x = \frac{1}{A} \int_0^A x \, dx = \frac{1}{A} \frac{A^2}{2} = \frac{A}{2}$$

$$E\{X^2\} = \frac{1}{A} \int_0^A x^2 \, dx = \frac{1}{A} \frac{A^3}{3} = \frac{A^2}{3}$$

From Equation 5.57

$$\sigma_x^2 = E\{X^2\} - m_x^2 = A^2/3 - A^2/4 = A^2/12 \quad ■ ■$$

EXAMPLE 5.7 ■

Let $Y_2 = a_1 X_1 + a_2 X_2$, where we are given the means m_1, m_2, the variances σ_1^2, σ_2^2, and the correlation coefficient ρ of the random variables X_1 and X_2. Let us compute the mean and variance of Y_2:

$$E\{Y_2\} = E\{a_1 X_1 + a_2 X_2\} = a_1 m_1 + a_2 m_2 \quad \text{(from Theorem 5.1)}$$
$$\sigma_y^2 = E\{(Y_2 - m_y)^2\} = E\{[a_1(X_1 - m_1) + a_2(X_2 - m_2)]^2\}$$
$$= a_1^2 E\{(X_1 - m_1)^2\} + a_2^2 E\{(X_2 - m_2)^2\}$$
$$+ 2a_1 a_2 E\{(X_1 - m_1)(X_2 - m_2)\}$$
$$= a_1^2 \sigma_1^2 + a_2^2 \sigma_2^2 + 2a_1 a_2 \mu_{11}$$
$$= a_1^2 \sigma_1^2 + a_2^2 \sigma_2^2 + 2a_1 a_2 \sigma_1 \sigma_2 \rho \tag{5.58} \quad ■ ■$$

Theorem 5.3. Statistically independent random variables are uncorrelated.

Proof

$$\mu_{11} = \int_{-\infty}^{\infty} \int_{-\infty}^{\infty} (x - m_x)(y - m_y) f(x, y) \, dx \, dy$$

$$= \int_{-\infty}^{\infty} \int_{-\infty}^{\infty} (x - m_x)(y - m_y) f(x) f(y) \, dx \, dy$$

$$= \int_{-\infty}^{\infty} (x - m_x) f(x) \, dx \int_{-\infty}^{\infty} (y - m_y) f(y) \, dy$$

$$= \left[\int_{-\infty}^{\infty} x f(x) \, dx - m_x \int_{-\infty}^{\infty} f(x) \, dx \right] \left[\int_{-\infty}^{\infty} y f(y) \, dy - m_y \int_{-\infty}^{\infty} f(y) \, dy \right]$$

$$= (m_x - m_x)(m_y - m_y) = 0$$

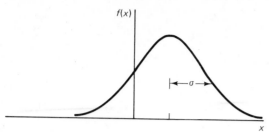

Figure 5.7   Gaussian PDF.

5.7 GAUSSIAN OR NORMAL DISTRIBUTIONS

A probability density function of central importance in communication theory is the *Gaussian* or *normal* PDF (see Figure 5.7). It is a bell-shaped function centered at the value m and having a width determined by the parameter σ:

$$f(x) = \frac{1}{\sqrt{2\pi\sigma^2}} \exp\left[-\frac{(x-m)^2}{2\sigma^2}\right] \tag{5.59}$$

Integrals such as $P(a < x \le b) = \int_a^b f(x)\, dx$ cannot be evaluated in closed form for the Gaussian PDF, and tables or approximations are needed. Even the following definite integrals, which we shall assume are correct without demonstration,[2] are difficult to evaluate:

$$\int_0^\infty e^{-u^2/2}\, du = \sqrt{\frac{\pi}{2}} \tag{5.60}$$

$$\int_0^\infty u e^{-u^2/2}\, du = 1 \tag{5.61}$$

$$\int_0^\infty u^{2n} e^{-u^2/2}\, du = 1 \cdot 3 \cdot 5 \cdots (2n-1)\sqrt{\frac{\pi}{2}} \tag{5.62}$$

This PDF is properly normalized, which follows from Equation 5.60:

$$\int_{-\infty}^\infty \frac{1}{\sqrt{2\pi\sigma^2}} e^{-(x-m)^2/2\sigma^2}\, dx = \int_{-\infty}^\infty \frac{1}{\sqrt{2\pi}} e^{-u^2/2}\, du$$

$$= \frac{2}{\sqrt{2\pi}} \int_0^\infty e^{-u^2/2}\, du = 1 \tag{5.63}$$

where in the first equality we used the transformation $x - m = u\sigma$. Using Equations 5.61 and 5.62, we can show that m is the expected value and σ^2 is the variance of a Guassian PDF (see Problem 5.18). Thus the Gaussian or normal PDF is specified by its mean value and variance and is often denoted by $N(m, \sigma^2)$.

   To facilitate computations of $P(a < x \le b)$, the integrals are often transformed to make the integrand a *standard normal PDF*, defined by $N(0, 1)$. This is achieved by the transformation $t = (x - m)/\sigma$: Thus

---

[2]See mathematical tables such as those from the *Handbook of Chemistry and Physics* [1].

$$\int_a^b \frac{1}{\sqrt{2\pi\sigma^2}} e^{-(x-m)^2/2\sigma^2} \, dx = \int_{(a-m)/\sigma}^{(b-m)/\sigma} \frac{1}{\sqrt{2\pi}} e^{-t^2/2} \, dt \tag{5.64}$$

These can be evaluated from tabulated integrals such as

$$Q(\alpha) = \int_\alpha^\infty \frac{1}{\sqrt{2\pi}} e^{-t^2/2} \, dt \tag{5.65}$$

which are given in many handbooks.[3]

EXAMPLE 5.8 ■

A plant manufactures 100-Ω resistors that are guaranteed to be within ±10% of the specified value (all resistors whose values are less than 90 or more than 110 Ω are discarded). Assume the values of the resistors satisfy a Gaussian PDF with mean 100 and variance 16. (This PDF cannot model the resistances exactly, since negative values are impossible. With this model, however, the probability of negative values is quite small.) Let us determine the percentage of resistors that are discarded:

$$\Pr(\text{discard}) = P_d = \int_{-\infty}^{90} f(x) \, dx + \int_{110}^\infty f(x) \, dx$$

where

$$f(x) = \frac{1}{\sqrt{2\pi \times 16}} \exp\left[-\frac{(x-100)^2}{32}\right]$$

By symmetry, for $t = (x - 100)/4$,

$$P_d = 2 \int_{110}^\infty \frac{1}{\sqrt{32\pi}} e^{-(x-100)^2/32} \, dx = 2 \int_{(110-100)/4}^\infty \frac{1}{\sqrt{2\pi}} e^{-t^2/2} \, dt = 2Q(2.5)$$

From available tables $Q(2.5) \approx 0.0062$; hence $P_d \approx 0.0124$ (or 1.24%).
■ ▪

Approximations for $Q(\alpha)$ are also used, particularly for large arguments, where the tables are usually incomplete. Setting $t = u + \alpha$ in Equation 5.65, changing variables, and using the property that $e^{a+b} = e^a e^b$ give

$$Q(\alpha) = \int_0^\infty \frac{1}{\sqrt{2\pi}} e^{-(u+\alpha)^2/2} \, du = \frac{e^{-\alpha^2/2}}{\sqrt{2\pi}} \int_0^\infty e^{-u^2/2} e^{-u\alpha} \, du$$

Since $1 > \exp(-\frac{1}{2}u^2) > 1 - \frac{1}{2}u^2$, it follows that

---

[3]Often integrals similar to that of Equation 5.65 are tabulated. The two most commonly found functions are the standard normal CDF [$\phi(\alpha)$] and the error function [erf(α)], which are defined by

$$\phi(\alpha) = \int_{-\infty}^\alpha \frac{1}{\sqrt{2\pi}} e^{-t^2/2} \, dt \quad \text{where} \quad Q(\alpha) = 1 - \phi(\alpha)$$

$$\text{erf}(\alpha) = \frac{2}{\sqrt{\pi}} \int_0^\alpha e^{-t^2} \, dt \quad \text{where} \quad Q(\alpha) = \frac{1}{2}\left[1 - \text{erf}\left(\frac{\alpha}{\sqrt{2}}\right)\right]$$

$$\frac{e^{-\alpha^2/2}}{\sqrt{2\pi}}\int_0^\infty (1 - \tfrac{1}{2}u^2)e^{-\alpha u}\,du < Q(\alpha) < \frac{e^{-\alpha^2/2}}{\sqrt{2\pi}}\int_0^\infty e^{-\alpha u}\,du$$

From tables of definite integrals, it can be found that

$$\int_0^\infty u^n e^{-\alpha u}\,du = n!/\alpha^{n+1}$$

Thus

$$\frac{e^{-\alpha^2/2}}{\sqrt{2\pi}\,\alpha}\left(1 - \frac{1}{\alpha^2}\right) < Q(\alpha) < \frac{e^{-\alpha^2/2}}{\sqrt{2\pi}\,\alpha} \tag{5.66}$$

The upper bound $[\exp(-\tfrac{1}{2}\alpha^2)/\sqrt{2\pi}\,\alpha]$ is an approximation that becomes quite good for large α since the lower bound becomes increasingly close to the upper bound as α increases. Another convenient upper bound that is not as close an approximation as that of Equation 5.66 but is simple and well behaved at α near zero is

$$Q(\alpha) < \tfrac{1}{2}e^{-\alpha^2/2} \tag{5.67}$$

$Q(\alpha)$ and the two upper bounds are shown in Figure 5.8.

EXAMPLE 5.9 ■

Let us reexamine Example 5.8. First we estimate P_d from the upper bound:

$$P_d = 2Q(2.5) < \frac{2}{\sqrt{2\pi}(2.5)}e^{-(2.5)^2/2} = 0.0140$$

For $\alpha > 2$ the upper bound can be used as an approximation. Let us now calculate the probability that the resistance could be negative (which is impossible) according to the Gaussian model:

$$\Pr(x < 0) = \int_{-\infty}^0 \frac{1}{\sqrt{32\pi}}e^{-(x-100)^2/32}\,dx = \int_{-\infty}^{-100/4}\frac{1}{\sqrt{2\pi}}e^{-t^2/2}\,dt$$

$$= \int_{25}^\infty \frac{1}{\sqrt{2\pi}}e^{-t^2/2}\,dt \quad \text{(by symmetry)}$$

$$= Q(25)$$

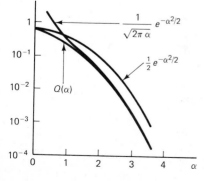

Figure 5.8 Approximations to $Q(\alpha)$.

$$Q(25) \approx \frac{1}{\sqrt{2\pi}25} e^{-(25)^2/2} = \frac{1}{\sqrt{2\pi}25} 10^{-\log_{10} e\,(625)/2}$$

$$= \frac{1}{\sqrt{2\pi}25} 10^{-0.21715(625)} \approx 8 \times 10^{-138.7} \qquad \blacksquare\;\blacksquare$$

The *joint* (or bivariate) *Gaussian* (or normal) *PDF* is defined by

$$f(x, y) = \frac{1}{2\pi\sigma_1\sigma_2\sqrt{1-\rho^2}} \exp\left\{-\frac{1}{2(1-\rho^2)}\right.$$

$$\left.\times \left[\left(\frac{x-m_1}{\sigma_1}\right)^2 - 2\rho\left(\frac{x-m_1}{\sigma_1}\right)\left(\frac{y-m_2}{\sigma_2}\right) + \left(\frac{y-m_2}{\sigma_2}\right)^2\right]\right\} \qquad (5.68)$$

where the constant ρ must satisfy $0 \le \rho < 1$. For the terminology to be consistent, $f(x)$ and $f(y)$, as determined from Equation 5.36, must both be Gaussian PDFs. Defining $x_1 = (x - m_1)/\sigma_1$ and $x_2 = (y - m_2)/\sigma_2$, we have

$$f(y) = \int_{-\infty}^{\infty} f(x, y)\, dx$$

$$= \frac{1}{2\pi\sigma_2\sqrt{1-\rho^2}} \int_{-\infty}^{\infty} \exp\left[-\frac{1}{2(1-\rho^2)}(x_1^2 - 2\rho x_1 x_2 + x_2^2)\right] dx_1$$

Since $x_1^2 - 2\rho x_1 x_2 + x_2^2 = (x_1 - \rho x_2)^2 + (1 - \rho^2)x_2^2$,

$$f(y) = \frac{e^{-x_2^2/2}}{2\pi\sigma_2\sqrt{1-\rho^2}} \int_{-\infty}^{\infty} \exp\left[-\frac{1}{2(1-\rho^2)}(x_1 - \rho x_2)^2\right] dx_1$$

Defining $t = (x_1 - \rho x_2)/\sqrt{1-\rho^2}$, we obtain

$$f(y) = \frac{e^{-x_2^2/2}}{\sqrt{2\pi}\sigma_2^2} \int_{-\infty}^{\infty} \frac{1}{\sqrt{2\pi}} e^{-t^2/2}\, dt = \frac{1}{\sqrt{2\pi}\sigma_2^2} e^{-(y-m_2)^2/2\sigma_2^2}$$

Similarly,

$$f(x) = \frac{1}{\sqrt{2\pi}\sigma_1^2} e^{-(x-m_1)^2/2\sigma_1^2}$$

It can also be shown (Problem 5.19) that the correlation coefficient is the constant ρ. The bell shape one obtains by rotating the Gaussian PDF of Figure 5.7 is a bivariate Gaussian PDF for $\sigma_1 = \sigma_2$ and $\rho = 0$ (see Figure 5.9a). The general shape is more oval and tilted relative to the x and y axes, as seen in Figure 5.9b. In Figure 5.9 the ellipses are equal probability contours.

Theorem 5.4. Two normal (or Gaussian) random variables that are uncorrelated are statistically independent.

Proof: Since (from Problem 5.19) ρ of Equation 5.68 is the correlation coefficient, setting $\rho = 0$ simplifies this expression to

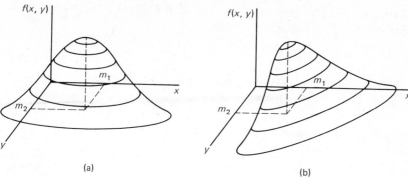

Figure 5.9 Bivariate Gaussian PDFs: (a) $\rho = 0$; (b) $\rho > 0$.

$$f(x, y) = \frac{1}{2\pi\sigma_1\sigma_2} \exp\left\{-\frac{1}{2}\left[\left(\frac{x - m_1}{\sigma_1}\right)^2 + \left(\frac{y - m_2}{\sigma_2}\right)^2\right]\right\}$$

$$= \frac{1}{\sqrt{2\pi\sigma_1^2}} e^{-(x-m_1)^2/2\sigma_1^2} \frac{1}{\sqrt{2\pi\sigma_2^2}} e^{-(y-m_2)^2/2\sigma_2^2}$$

$$= f(x)f(y)$$

Theorem 5.4 is an important property of Gaussian random variables. While it is often desirable to assume that two random variables are independent, this cannot be verified by a measurement. It may be possible to estimate the correlation coefficient however. For Gaussian random variables, if ρ is zero, statistical independence can be assumed.

The bivariate Gaussian PDF of Equation 5.68 can be generalized to a *multivariate Gaussian PDF*. For the case where all means are zero, the multivariate Gaussian PDF is defined by

$$f(x_1, x_2, \ldots, x_n) = \frac{1}{\sqrt{(2\pi)^n \det \mathbf{R}}} \exp\left(-\frac{1}{2}\sum_{i=1}^{n}\sum_{j=1}^{n} x_i x_j q_{ij}\right) \tag{5.69}$$

where $\mathbf{R}$ is an $n \times n$ matrix, q_{ij} is the ijth component of the inverse matrix $\mathbf{R}^{-1}$, and det $\mathbf{R}$ is the determinant of the matrix $\mathbf{R}$. Let us verify that this definition is consistent with Equations 5.59 and 5.68 with $m = 0$. For $n = 1$, $\mathbf{R}$ is a scalar. If we set this scalar equal to σ^2, Equation 5.68 reduces to $(1/\sqrt{2\pi\sigma^2}) \exp(-\frac{1}{2}x^2/\sigma^2)$, which agrees with Equation 5.59. For $n = 2$, $\mathbf{R}$ is a 2×2 matrix. If we define

$$\mathbf{R} = \begin{bmatrix} \sigma_1^2 & \sigma_1\sigma_2\rho \\ \sigma_1\sigma_2\rho & \sigma_2^2 \end{bmatrix} \tag{5.70}$$

then det $\mathbf{R} = \sigma_1^2\sigma_2^2(1 - \rho^2)$ and

$$\mathbf{R}^{-1} = \frac{1}{1 - \rho^2}\begin{bmatrix} 1/\sigma_1^2 & -\rho/\sigma_1\sigma_2 \\ -\rho/\sigma_1\sigma_2 & 1/\sigma_2^2 \end{bmatrix}$$

Substituting into Equation 5.68 results in Equation 5.69, where $m_1 = m_2 = 0$.

In both cases, the ijth element of the matrix $\mathbf{R}$ (r_{ij}) turns out to be the μ_{11} moment of the ith and jth random variable:

$$r_{ij} = E\{(X_i - m_i)(X_j - m_j)\} \tag{5.71}$$

Equation 5.71 holds for n Gaussian random variables; hence the multivariate Gaussian PDF is completely specified by this set of pairwise moments.

Theorem 5.5. If n Gaussian random variables are pairwise uncorrelated, they are all statistically independent.

    Proof: From Equation 5.71, for pairwise uncorrelated random variables, the matrix $\mathbf{R}$ becomes

$$\mathbf{R} = \begin{bmatrix} \sigma_1^2 & & & 0 \\ & \sigma_2^2 & & \\ & & \ddots & \\ 0 & & & \sigma_n^2 \end{bmatrix}$$

Thus det $\mathbf{R} = \sigma_1^2 \sigma_2^2 \cdots \sigma_n^2$, and

$$\mathbf{R}^{-1} = \begin{bmatrix} \sigma_1^{-2} & & & 0 \\ & \sigma_2^{-2} & & \\ & & \ddots & \\ 0 & & & \sigma_n^{-2} \end{bmatrix}$$

For this case, Equation 5.69 becomes

$$f(x_1, x_2, \ldots, x_n) = \frac{1}{\sqrt{(2\pi)^n \sigma_1^2 \sigma_2^2 \cdots \sigma_n^2}} \exp\left(-\frac{1}{2}\sum_{i=1}^{n}\frac{x_i^2}{\sigma_i^2}\right)$$

$$= \prod_{i=1}^{n}\frac{1}{\sqrt{2\pi\sigma_i^2}} e^{-x_i^2/2\sigma_i^2} = \prod_{i=1}^{n} f(x_i)$$

where $\Pi_{i=1}^{n}$ denotes the n-fold product.

5.8 THE SUM OF RANDOM VARIABLES

Let us define the random variable Z by $Z = X + Y$, where $f(x, y)$ is known. The probability that $Z \le z$ $[F(z)]$ can be identified as the volume under a portion of $f(x, y)$. Thus

$$F(z) = \Pr(X + Y \le z) = \iint_R f(x, y)\, dx dy \tag{5.72}$$

where the region R is the half-plane indicated in Figure 5.10. Using standard integration procedures, Equation 5.72 can be written

$$F(z) = \int_{-\infty}^{\infty} \left[\int_{-\infty}^{z-y} f(x, y)\, dx\right] dy \tag{5.73}$$

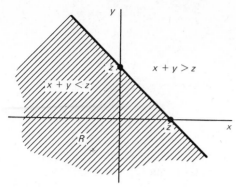

Figure 5.10 Region of integration for $F(z)$.

Suppose that X and Y are statistically independent random variables. It follows that

$$F(z) = \int_{-\infty}^{\infty} \left[\int_{-\infty}^{z-y} f(x)\, dx \right] f(y)\, dy$$

$$= \int_{-\infty}^{\infty} F_X(z - y) f_Y(y)\, dy$$

Since $f(z) = dF(z)/dz$, taking derivatives of both sides of the last equation results in

$$f(z) = \int_{-\infty}^{\infty} f_X(z - y) f_Y(y)\, dy \tag{5.74}$$

Thus, if X and Y are independent, the PDF of Z is the convolution of the PDF of X with the PDF of Y.

EXAMPLE 5.10 ■

Let X and Y be statistically independent random variables that are both uniformly distributed over the interval $(0, 1]$. Let us find the PDF of Z, defined by

$$Z = X + Y$$

Let us sketch the integrand of the convolution integral of Equation 5.74 for a fixed value of z that lies in the interval $(0, 1]$. We see, from Figure 5.11, that the product of $f_Y(y)$ and $f_X(z - y)$ is the crosshatched region whose area is z. Thus, for $0 < z < 1$, $f(z) = z$. If we repeat this procedure for $z < 0$, $1 < z < 2$, and $z > 2$, we determine $f(z)$, which is plotted in Figure 5.12. From symmetry considerations, the means of X and Y are both $\frac{1}{2}$ $\left(m_x = m_y = \frac{1}{2}\right)$. The variances are

$$\sigma_y^2 = \sigma_x^2 = \int_{-\infty}^{\infty} \left(x - \tfrac{1}{2}\right)^2 f(x)\, dx = \int_0^1 \left(x^2 - x + \tfrac{1}{4}\right) dx = \tfrac{1}{12}$$

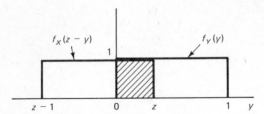

Figure 5.11 Evaluating the integrand of Equation 5.74.

From Theorem 5.1 $m_z = m_x + m_y = 1$, which also follows directly from Figure 5.12 using symmetry arguments. From Equation 5.58 ($\rho = 0$ for independent random variables)

$$\sigma_z^2 = \sigma_x^2 + \sigma_y^2 = \tfrac{1}{6}$$

To check this result,

$$\sigma_z^2 = \int_{-\infty}^{\infty} (z-1)^2 f(z)\, dz = \int_0^1 (z-1)^2 z\, dz + \int_1^2 (z-1)^2 (2-z)\, dz = \tfrac{1}{6}$$

■ ■

Observe that if we extend Example 5.10 to

$$Z_N = \sum_{i=1}^{N} X_i \tag{5.75}$$

where the X_i are statistically independent random variables with means m_i and variances σ_i^2, then by induction

$$m_{z_N} = \sum_{i=1}^{N} m_i \quad \text{and} \quad \sigma_{z_N}^2 = \sum_{i=1}^{N} \sigma_i^2 \tag{5.76}$$

To determine the PDF, we find $f(z_3)$ by convolving a uniform PDF with the PDF of Z_2 shown in Figure 5.12 and then continue this procedure. Surprisingly, as N increases, $f(z_N)$ approximates a Gaussian random variable:

$$f(z_N) \to N\left(\sum_{i=1}^{N} m_i, \sum_{i=1}^{N} \sigma_i^2\right) \quad \text{as} \quad N \to \infty \tag{5.77}$$

Theorem 5.6. If X_i are Gaussian random variables (not necessarily statistically independent), then $Z_N = \sum_{i=1}^{N} a_i X_i$ is Gaussian for any N.

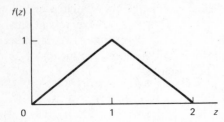

Figure 5.12 PDF of the sum of independent, uniform random variables.

Proof: It is enough to show that $Z_2 = a_1 X_1 + a_2 X_2$ is Gaussian. The general result follows by induction. With adjustments for the a_i weights, Equation 5.73 becomes

$$F(z_2) = \int_{-\infty}^{\infty} \left[\int_{-\infty}^{(z_2 - a_2 x_2)/a_1} f(x_1, x_2)\, dx_1 \right] dx_2 \tag{5.78}$$

From Theorem 5.1 and Equation 5.57,

$$m_{z_2} = a_1 m_1 + a_2 m_2$$

and $\hspace{8cm}$ (5.79)

$$\sigma_{z_2}^2 = a_1^2 \sigma_1^2 + a_2^2 \sigma_2^2 + 2 a_1 a_2 \sigma_1 \sigma_2 \rho$$

Substituting the bivariate Gaussian PDF of Equation 5.68, with the simplifying transformations

$$x = \frac{x_1 - m_1}{\sigma_1} \qquad y = \frac{x_2 - m_2}{\sigma_2} \qquad z = \frac{z_2 - m_{z_2}}{\sigma_{z_2}} \tag{5.80}$$

yields

$$F(z) = \frac{1}{2\pi \sqrt{1 - \rho^2}}$$
$$\times \int_{-\infty}^{\infty} \left\{ \int_{-\infty}^{(z\sigma_{z_2} - a_2 \sigma_2 y)/a_1 \sigma_1} \exp\left[-\frac{1}{2(1 - \rho^2)}(x^2 - 2\rho xy + y^2) \right] dx \right\} dy$$

$$\tag{5.81}$$

Because $x^2 - 2\rho xy + y^2 = (x - \rho y)^2 + (1 - \rho^2) y^2$, this becomes

$$F(z) = \frac{1}{2\pi \sqrt{1 - \rho^2}} \int_{-\infty}^{\infty} \exp\left(-\frac{y^2}{2}\right)$$
$$\times \left\{ \int_{-\infty}^{(z\sigma_{z_2} - a_2 \sigma_2 y)/a_1 \sigma_1} \exp\left[-\frac{1}{2(1 - \rho^2)}(x - \rho y)^2 \right] dx \right\} dy \tag{5.82}$$

Changing variables again with $u = (x - \rho y)/\sqrt{1 - \rho^2}$, Equation 5.82 becomes

$$F(z) = \frac{1}{\sqrt{2\pi}} \int_{-\infty}^{\infty} \exp\left(-\frac{y^2}{2}\right)$$
$$\times \left[\int_{-\infty}^{[z\sigma_{z_2} - (a_2 \sigma_2 + a_1 \sigma_1 \rho) y]/a_1 \sigma_1 \sqrt{1 - \rho^2}} \frac{1}{\sqrt{2\pi}} \exp\left(-\frac{u^2}{2}\right) du \right] dy$$

or

$$F(z) = \frac{1}{\sqrt{2\pi}} \int_{-\infty}^{\infty} \exp\left(-\frac{y^2}{2}\right) \phi\left[\frac{z\sigma_{z_2} - (a_2 \sigma_2 + a_1 \sigma_1 \rho) y}{a_1 \sigma_1 \sqrt{1 - \rho^2}} \right] dy \tag{5.83}$$

where $\phi(\alpha)$ is the CDF of the standard Gaussian distribution. Since

$$\frac{d}{dz} \phi(az + b) = \frac{a}{\sqrt{2\pi}} \exp\left[-\frac{(az + b)^2}{2} \right]$$

taking derivatives of both sides of Equation 5.83 yields

$$f(z) = \frac{1}{2\pi} \frac{\sigma_{z_2}}{a_1 \sigma_1 \sqrt{1 - \rho^2}}$$

$$\times \int_{-\infty}^{\infty} \exp\left(-\frac{y^2}{2}\right) \exp\left\{-\frac{1}{2}\left[\frac{z\sigma_{z_2} - (a_2\sigma_2 + a_1\sigma_1\rho)y}{a_1\sigma_1\sqrt{1 - \rho^2}}\right]^2\right\} dy \qquad (5.84)$$

The exponentials can be combined and the argument manipulated into the form

$$\exp\left(-\frac{1}{2}\left\{z^2 + \left[\frac{y\sigma_{z_2} - (a_2\sigma_2 + a_1\sigma_1\rho)z}{a_1\sigma_1\sqrt{1 - \rho^2}}\right]^2\right\}\right)$$

Thus Equation 5.84 becomes

$$f(z) = \frac{e^{-z^2/2}}{\sqrt{2\pi}} \frac{\sigma_{z_2}}{\sqrt{2\pi}\, a_1\sigma_1\sqrt{1 - \rho^2}} \int_{-\infty}^{\infty} \exp\left\{-\frac{1}{2}\left[\frac{y\sigma_{z_2} - (a_2\sigma_2 + a_1\sigma_1\rho)z}{a_1\sigma_1\sqrt{1 - \rho^2}}\right]^2\right\} dy$$

With one last change of variables,

$$t = \frac{y\sigma_{z_2} - (a_2\sigma_2 + a_1\sigma_1\rho)z}{a_1\sigma_1\sqrt{1 - \rho^2}}$$

this becomes

$$f(z) = \frac{e^{-z^2/2}}{\sqrt{2\pi}} \int_{-\infty}^{\infty} \frac{1}{\sqrt{2\pi}} e^{-t^2/2}\, dt = \frac{e^{-z^2/2}}{\sqrt{2\pi}} \qquad (5.85)$$

where $z = (z_2 - m_{z_2})/\sigma_{z_2}$.

Induction completes the proof.

We have seen in Example 5.10 that the PDF of the sum of N independent non-Gaussian random variables changes with N. For random variables with uniform PDFs, the PDF of the sum seems to approach a Gaussian PDF as N increases. On the other hand, the PDF of the sum of N Gaussian random variables (not necessarily independent) remains Gaussian. These results are consistent with the following theorem, known as the *central limit theorem*, which will be stated without proof.[4]

Theorem 5.7. If X_k are mutually independent random variables with finite variances that have the same probability distribution, then $Z_N = \sum_{k=1}^{N} X_k$ approaches a normal distribution as N increases without bound.

EXAMPLE 5.11 ■

Let us estimate the probability that $Y > 16$, where $Y = \sum_{i=1}^{25} X_i$ and where X_i are independent random variables uniformly distributed in the interval $(0, 1]$. We have seen in Example 5.6 that for the X_i random

[4]See Feller [2] for a proof of this theorem.

variables, $m_i = \frac{1}{2}$ and $\sigma_i^2 = \frac{1}{12}$. From Equation 5.76, $m_y = 25m_i = \frac{25}{2}$ and $\sigma_y^2 = 25\sigma_i^2 = \frac{25}{12}$. If we assume that Y can be approximated by a Gaussian random variable, $N\left(\frac{25}{2}, \frac{25}{12}\right)$, then

$$\Pr(Y > 16) \cong \int_{16}^{\infty} \frac{1}{\sqrt{2\pi\frac{25}{12}}} \exp\left[-\frac{1}{2}\frac{\left(y - \frac{25}{2}\right)^2}{\frac{25}{12}}\right] dy$$

$$= \int_{(16-12.5)/\sqrt{25/12}}^{\infty} \frac{1}{\sqrt{2\pi}} e^{-t^2/2}\, dt = Q(2.425)$$

$$\approx \frac{1}{\sqrt{2\pi}\,2.425} e^{-(2.425)^2/2} \approx 0.0087 \qquad \blacksquare\ \blacksquare$$

The last four theorems are an indication of the importance of the Gaussian distribution.

5.9 MODELING RANDOM SIGNALS

A process is said to be *stationary* if, for any function $g(\ )$ that is continuous or has a finite number of continuous segments,

$$\lim_{T' \to \infty} \frac{1}{T'} \int_0^{T'} g[x(t)]\, dt \qquad \text{exists} \tag{5.86}$$

A stationary process must have, for example, a measurable time average, $\langle x(t) \rangle$, an average power, $\langle x^2(t) \rangle$, indeed any arbitrary time average moment $\langle x^n(t) \rangle$. A PDF can be associated with a stationary process by equating statistical moments with time average moments. Thus we associate a PDF $f(x)$ with a stationary process $x(t)$ if

$$\int_{-\infty}^{\infty} x^n f(x)\, dx = \lim_{T' \to \infty} \frac{1}{T'} \int_0^{T'} x^n(t)\, dt, \tag{5.87}$$

or $E\{x^n\} = \langle x^n(t) \rangle$.

Any time average that can be expressed by Equation 5.86 can be modeled by a statistical average based on the appropriate PDF. For example, the percentage of time that a stationary process $x(t)$ exceeds a voltage V, or

$$\lim_{T' \to \infty} \frac{1}{T'} \int_0^{T'} u[x(t) - V]\, dt \tag{5.88}$$

where $u(t)$ is the unit step function, can be modeled by the probability that X exceeds V, or

$$\int_V^{\infty} f(x)\, dx = 1 - F_X(V) \tag{5.89}$$

where $F_X(\ )$ is the CDF. Equations 5.88 and 5.89 suggest a method of estimating the PDF model. Equation 5.88 can be estimated in the laboratory using the system of Figure 5.13. If C_n is the number of times the comparator output is high (as determined by the counter) out of n samples, then C_n/n is a

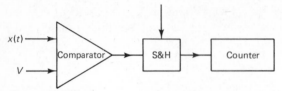

Figure 5.13   System for estimating the PDF model.

relative frequency that converges to $1 - F_X(V)$ as n increases. Thus by varying V one can estimate the CDF (hence the PDF).

An important question remains, however. Do averages, such as

$$\lim_{N\to\infty} \frac{1}{N} \sum_{n=1}^{N} g[x(nT)]$$ (5.90)

exist for stationary processes, and do they converge to the averages of Equation 5.86 for any T? If the answer is no, then the PDF model applies only in a global sense and may not be applicable for samples of the process. If the answer is yes, then every sample of the stationary process can be thought of as a random variable that is characterized by the PDF model. Let us consider two examples.

All periodic, deterministic signals are stationary since the averages of Equation 5.86 all exist for periodic signals. Thus a PDF model that applies, at least in the global sense, can be identified for periodic signals. If these signals are sampled once per period, the resulting values will all be the same. These values will depend on the time of the first sample (i.e., the location of the time origin). Thus whereas averages such as those of Equation 5.90 converge, the values depend on the location of the time origin and on T. Thus samples of a periodic deterministic signal should not be thought of as random variables.

Consider the impulse and shot noise processes that were discussed in Section 2.13. These are stationary processes that have the same power spectrum but different PDF models. We shall show that we can think of samples of these random processes as random variables characterized by the PDF model. Recall the distinction between these signals. If the pulses overlap a great deal, the process is called *shot noise*; otherwise, it is called *impulse noise*. For the shot noise process, a large number of independent pulses contribute to the value of the process at each instant of time. As we would expect intuitively from the central limit theorem, it can be proven [3] that a Gaussian PDF is an appropriate model for shot noise but not for impulse noise. Let us treat the average value separately as a dc signal, where from Equation 2.100

$$\langle x(t) \rangle = \gamma |H(0)|$$ (5.91)

Here γ is the average number of pulses per second and $H(f)$ is the Fourier transform of one of the pulses. The remaining shot noise is a zero-mean process whose power spectrum is $\gamma |H(f)|^2$ (see Equation 2.100). A sample of this process can be thought of as zero-mean Gaussian with variance

$$\sigma^2 = \int_{-\infty}^{\infty} \gamma |H(f)|^2 \, df \tag{5.92}$$

EXAMPLE 5.12 ■

Let us estimate the percentage of time that the shot noise of Example 2.13 exceeds one tenth its dc component. In Example 2.13 the pulse shape was given as

$$h(t) = \frac{q}{d} e^{-t/d} u(t)$$

The process is considered shot noise if $\gamma d \gg 1$. From Example 2.13, the dc component is γq, and the spectrum of the remaining shot noise is $\gamma q^2/[1 + (2\pi fd)^2]$. Let us model this shot noise with the distribution $N(0, \sigma^2)$, where

$$\sigma^2 = \int_{-\infty}^{\infty} \frac{\gamma q^2}{1 + (2\pi fd)^2} \, df = \frac{(\gamma q)^2}{2d\gamma}$$

The percentage of time that this shot noise exceeds $\gamma q/10$ is estimated as

$$\int_{\gamma q/10}^{\infty} \frac{1}{\sqrt{2\pi\sigma^2}} e^{-x^2/2\sigma^2} \, dx = \int_{\sqrt{2d\gamma}/10}^{\infty} \frac{1}{\sqrt{2\pi}} e^{-t^2/2} \, dt = Q\left(\frac{\sqrt{2d\gamma}}{10}\right)$$

For $\gamma d \gg 1$ this can be closely approximated by

$$Q\left(\frac{\sqrt{2d\gamma}}{10}\right) \approx \frac{10}{\sqrt{4\pi\gamma d}} e^{-\gamma d/100} \qquad\qquad ■ ■$$

It should be noted that stationary processes are not necessarily wide-sense stationary and vice versa. Any process that is neither stationary nor wide-sense stationary is called *nonstationary*.

If samples of a process that are sufficiently far apart can be considered as nearly independent, the process is said to be strong mixing. Here is a more formal definition [4, 5]: Let A, B be any events determined by conditions on the samples $y(kT)$, $k \leq m$, and $y(kT)$, $k \geq n$, respectively, with $n > m$. The process is said to satisfy the *strong mixing condition* if

$$|P(A \cap B) - P(A)P(B)| \leq a[(n - m)T] \tag{5.93}$$

for all such events A, B and some function $a(\tau)$ that decreases to zero as $\tau \to \infty$.

Deterministic processes, such as periodic signals, cannot be strong mixing, for the dependence between samples continues indefinitely. Intuitively, however, we expect most random processes to be strong mixing. Shot noise, for example, was modeled as the response of a filter to a series of independently located impulses. It follows that samples further apart than the "width" of the filter impulse response should be independent; hence shot noise is strong mixing. All communication signals that are modeled as the response of a filter to a random impulse train are strong mixing. Speech signals are also strong mixing.

A process that is both stationary and strong mixing is *ergodic*.[5] The following, *Birkhoff's ergodic theorem*, is a direct consequence of the definition [6–8]. The proof is complex, however, and is omitted.

Theorem 5.8.   For any ergodic process $x(t)$,

1. $(1/N) \sum_{n=1}^{N} g_1[x(nT)]$ converges to a unique value for any T, where $g_1(\ )$ is any piecewise continuous function of one variable; and
2. $(1/N) \sum_{n=1}^{N} g_2[x(nT), x(nT - \tau)]$ converges to a unique function of τ for any T and τ, where $g_2(\ )$ is any piecewise continuous function of two variables.

This result can be extended to functions of three or more variables. As a corollary to this theorem, we can make the following statements for any ergodic process $x(t)$:

1. Any sample of $x(t)$ can be modeled by a unique PDF.
2. An ergodic process is wide-sense stationary:

$$\frac{1}{N} \sum_{n=1}^{N} x(nT)x(nT - \tau) \to \lim_{T' \to \infty} \frac{1}{T'} \int_{0}^{T'} x(t)x(t - \tau)\, dt = R_x(\tau)$$

3. Any two samples of $x(t)$, spaced τ s apart, can be modeled by a bivariate PDF where

$$\langle x^n(t)x^m(t - \tau) \rangle = \int_{-\infty}^{\infty} \int_{-\infty}^{\infty} x_1^n x_2^m \, f(x_1, x_2; \tau)\, dx_1 \, dx_2 \tag{5.94}$$

4. Any n samples of an ergodic process can be modeled by an nth-order multivariate PDF that depends only on the times between the samples.

We saw in Chapter 2 that if we treat the average value separately [$x(t) = x'(t) + x_{dc}$, where $\langle x'(t) \rangle = 0$], then $R_x(\tau) = R_{x'}(\tau) + x_{dc}^2$. In each example $R_{x'}(\tau) \to 0$ as $\tau \to \infty$. We can now formalize this result for ergodic processes.

Theorem 5.9.   For a stationary and strong mixing process, $R_x(\tau) \to \langle x(t) \rangle^2$ as $\tau \to \infty$.

Proof

$$\langle x(t)x(t - \tau) \rangle = \int_{-\infty}^{\infty} \int_{-\infty}^{\infty} x_1 x_2 \, f(x_1, x_2; \tau)\, dx_1 \, dx_2$$

$$\to \int_{-\infty}^{\infty} x_1 f(x_1)\, dx_1 \int_{-\infty}^{\infty} x_2 f(x_2)\, dx_2 \qquad \text{as} \quad \tau \to \infty$$

$$= m_{x_1} m_{x_2} = \langle x(t) \rangle^2$$

---

[5] Ergodicity is actually defined [6–8] with a mixing condition that is slightly less restrictive than strong mixing. It is possible to invent a process that is ergodic by the standard definition but is not strong mixing [5, 9]. We are unlikely, however, to encounter such processes.

Observe the power of equating time averages and statistical averages. Let us emphasize this point by noting that for zero-mean ergodic processes the following expressions are all equivalent and well defined:

$$\langle x^2(t) \rangle = R_x(0) = \int_{-\infty}^{\infty} S_x(f)\, df = E\{X^2\} = \sigma_x^2 = \int_{-\infty}^{\infty} x^2 f(x)\, dx \qquad (5.95)$$

Two processes $x(t)$ and $y(t)$ are said to be *jointly ergodic* if they are both stationary and any pooled set of samples satisfies the strong mixing condition. Thus any n samples of $x(t)$ and m samples of $y(t)$ can be modeled by an $(n + m)$th-order multivariate PDF $f(x_1, x_2, \ldots, x_n, y_1, y_2, \ldots, y_m; \boldsymbol{\tau})$, where $\boldsymbol{\tau}$ is an $(n + m - 1)$-vector corresponding to the times between the samples. Two jointly ergodic processes are said to be *statistically independent* if

$$f(x_1, x_2, \ldots, x_n, y_1, \ldots, y_m; \boldsymbol{\tau}) = f(x_1, \ldots, x_n, \boldsymbol{\tau}_1) f(y_1, \ldots, y_m; \boldsymbol{\tau}_2) \qquad (5.96)$$

Thus the samples of $x(t)$ are statistically unrelated to any of the samples of $y(t)$ but not necessarily to each other.

Theorem 5.10. If two zero-mean jointly ergodic random processes are statistically independent, their cross-correlation function is zero for all τ.

Proof

$$R_{xy}(\tau) = \langle x(t)y(t - \tau) \rangle = \int_{-\infty}^{\infty} \int_{-\infty}^{\infty} xy\, f(x, y; \tau)\, dx\, dy$$

$$= \int_{-\infty}^{\infty} xf(x)\, dx \int_{-\infty}^{\infty} yf(y)\, dy = m_x m_y = 0$$

5.10 GAUSSIAN PROCESSES

An ergodic process is said to be a *Gaussian process* if any n samples can be modeled by a multivariate Gaussian PDF. (Equation 5.69 states the zero-mean case.) There are two reasons for the common use of this model. The model appears to be valid for many of the processes that we encounter, such as shot noise and thermal noise. Just as important, there are many unique properties of this model that make it mathematically tractable. Thus, even if the model is only approximately valid, we are able to obtain results which we hope are approximately correct. We shall state and prove some of these unique properties under the assumption that the average value or mean is zero. These results are general since we can (and usually do) treat the average value separately if it is not zero.

Property 5.1. A Gaussian process is completely characterized by its autocorrelation function or power spectrum.

Proof: From its definition (Equation 5.69), a multivariate Gaussian PDF is completely specified by the correlation matrix $\mathbf{R}$. The ijth element of this matrix is, from Equation 5.71, $r_{ij} = E\{(X_i - m_i)(X_j - m_j)\}$. If the process has zero mean, and the samples are spaced τ s apart:

$$r_{ij} = E\{x(t)x(t - \tau)\} = \int_{-\infty}^{\infty} \int_{-\infty}^{\infty} x_1 x_2 f(x_1, x_2; \tau) \, dx_1 dx_2$$

$$= \langle x(t)x(t - \tau) \rangle = R_x(\tau)$$

Thus all of the elements of the correlation matrix $\mathbf{R}$ are determined from the autocorrelation function $R_x(\tau)$ and the times between samples.

EXAMPLE 5.13 ■

Reconsider the Gaussian shot noise of Example 5.10, where $S_n(f) = \gamma q^2 / [1 + (2\pi f d)^2]$ and $\sigma^2 = \gamma q^2 / 2d$. Let us determine the PDF of two samples spaced d s apart, $f(x_1, x_2; d)$. From the Fourier transforms of Table 2.1

$$R_n(\tau) = \frac{\gamma q^2}{2d} e^{-|\tau|/d}$$

$$R_n(0) = \frac{\gamma q^2}{2d} = \sigma_1^2 = \sigma_2^2$$

$$\rho = \frac{E\{x_1 x_2\}}{\sigma^2} = \frac{R_n(d)}{R_n(0)} = e^{-1}$$

From Equation 5.68

$$f(x_1, x_2; d) = \frac{2d}{2\pi \gamma q^2 \sqrt{1 - e^{-2}}}$$

$$\times \exp\left[-\frac{2d}{2(1 - e^{-2})\gamma q^2} (x_1^2 - 2e^{-1}x_1 x_2 + x_2^2) \right]$$

Note that, as $\tau \to \infty$, $\rho \to 0$ and $f(x_1, x_2; \tau) \to f(x_1) f(x_2)$. ■ ■

Property 5.2. If samples of a process can be modeled by a zero-mean Gaussian PDF and $R(\tau) \to 0$ as $\tau \to \infty$, then the process is ergodic [5, 8].

Proof: As we saw in Example 5.13, $\rho = R(\tau)/R(0) \to 0$ as $\tau \to \infty$. From Theorem 5.4 or 5.5, if Gaussian random variables are uncorrelated, they are statistically independent. Thus this process is strong mixing and hence ergodic.

Property 5.3. Any linear operation on a Gaussian process results in another Gaussian process.

Proof: A general linear time-invariant operation is modeled by the convolution integral,

$$y(t) = \int_{-\infty}^{\infty} h(\xi)x(t - \xi)\, d\xi$$

We can approximate this with a discrete convolution sum:

$$y(kT) = \sum_{j=-\infty}^{\infty} a_j x[(k - j)T]$$

where the samples $x(kT)$ are Gaussian random variables. From Theorem 5.6 the output samples $y(kT)$ are Gaussian random variables whose means are zero:

$$E\{Y\} = \int_{-\infty}^{\infty} h(\xi)E\{x(t - \xi)\}\, d\xi = 0 \qquad \text{if} \quad E\{x(t)\} = 0 \tag{5.97}$$

From property 5.1, in order to characterize a set of output samples we need to know the autocorrelation function or power spectrum of $y(t)$. From Chapter 2 we know that

$$S_y(f) = |H(f)|^2 S_x(f) \tag{5.98}$$

EXAMPLE 5.14 ■

Suppose the input to a linear, causal filter, whose impulse response is $h(t) = e^{-t}u(t)$, is a zero-mean white (spectral level $\eta_0/2$ V^2/Hz) Gaussian process. From properties 5.3 and 5.1, the output is a zero-mean Gaussian process that is characterized by its autocorrelation function $R_y(\tau)$. The filter transfer function $H(f)$ is the Fourier transform of $h(t)$:

$$H(f) = \int_0^{\infty} e^{-t} e^{-j2\pi ft}\, dt = \frac{1}{1 + j2\pi f}$$

From Equation 5.98

$$S_y(f) = \frac{\eta_0}{2} \frac{1}{1 + (2\pi f)^2}$$

From Table 2.1

$$R_y(\tau) = \tfrac{1}{4}\eta_0 e^{-|\tau|}$$

We are now in position to determine the PDF for any n samples of the output provided we know the times between the samples. ■ ■

Property 5.4. Two jointly ergodic Gaussian processes $x(t)$ and $y(t)$ are completely specified by their autocorrelation functions $R_x(\tau)$ and $R_y(\tau)$ and the cross-correlation function $R_{xy}(\tau)$.

Proof: The PDF of any pooled set of n samples from $x(t)$ and m samples from $y(t)$ is completely specified by the correlation matrix $\mathbf{R}$. If we assume that the first n random variables are the samples from $x(t)$, $\mathbf{R}$ can be partitioned in the following way, where $r_{jk} = E\{x_j x_k\}$, $r_{rs} = E\{y_r y_s\}$, $r_{js} = E\{x_j y_s\}$, and $r_{rk} = E\{y_r x_k\}$:

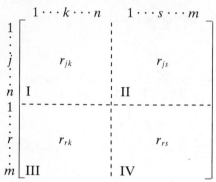

We see that the moments in block I are determined from $R_x(\tau)$, block IV is determined from $R_y(\tau)$, block II is determined from $R_{xy}(\tau)$, and block III from $R_{yx}(\tau)$. Since $R_{yx}(\tau) = R_{xy}(-\tau)$, only one of the cross-correlation functions need be specified.

Property 5.5. If the cross correlation of two zero-mean Gaussian processes is identically zero [i.e., $R_{xy}(\tau) = 0$ for all τ], then the processes are statistically independent.

  Proof: If $R_{xy}(\tau) = 0$ for all τ, all the elements in blocks II and III of the correlation matrix are zero. Thus

$$\mathbf{R} = \begin{array}{c} \\ 1 \\ \vdots \\ n \\ 1 \\ \vdots \\ m \end{array} \begin{array}{cc} 1 \cdots n & 1 \cdots m \\ \left[\begin{array}{c:c} \mathbf{R}_x & \phi \\ \hdashline \phi & \mathbf{R}_y \end{array}\right] \end{array}$$

For such matrices, it can be determined that (proof omitted) det $\mathbf{R} = $ det $\mathbf{R}_x$ det $\mathbf{R}_y$ and

$$\mathbf{R}^{-1} = \left[\begin{array}{cc} \mathbf{R}_x^{-1} & \phi \\ \phi & \mathbf{R}_y^{-1} \end{array}\right]$$

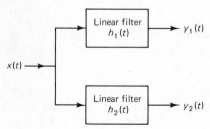

Figure 5.14 Generation of statistically related processes.

Substituting this into Equation 5.69 gives

$$f(x_1, \ldots, x_n, y_1, \ldots, y_m) = \frac{1}{\sqrt{(2\pi)^{n+m} \det \mathbf{R}_x \det \mathbf{R}_y}}$$

$$\times \exp\left[-\frac{1}{2}\left(\sum_{i=1}^{n}\sum_{j=1}^{m} x_i x_j q_{x_{ij}} + \sum_{i=1}^{n}\sum_{j=1}^{m} y_i y_j q_{y_{ij}}\right)\right]$$

$$= \frac{1}{\sqrt{(2\pi)^n \det \mathbf{R}_x}} \exp\left(-\frac{1}{2}\sum_{i=1}^{n}\sum_{j=1}^{m} x_i x_j q_{x_{ij}}\right)$$

$$\times \frac{1}{\sqrt{(2\pi)^m \det \mathbf{R}_y}} \exp\left(-\frac{1}{2}\sum\sum y_i y_j q_{x_{ij}}\right)$$

$$= f(x_1, \ldots, x_n) f(y_1, \ldots, y_m)$$

In order to gain further insight into the power of the Gaussian model, let us address ourselves to a question that will be of interest to us in later chapters. Consider the two processes $y_1(t)$ and $y_2(t)$ generated by a zero-mean ergodic process $x(t)$ from the circuit of Figure 5.14, where

$$y_i(t) = \int_{-\infty}^{\infty} h_i(\xi)x(t-\xi)\,d\xi$$

Under what circumstances, if any, are the processes $y_1(t)$ and $y_2(t)$ statistically independent? There is no mechanism to answer, or even address, this question if the input is not Gaussian. If the input is Gaussian, both $y_1(t)$ and $y_2(t)$ are Gaussian and their statistical properties are related to their autocorrelation functions and the cross-correlation function. From property 5.5, if $R_{y_1y_2}(\tau) = 0$ for all τ, then $y_1(t)$ and $y_2(t)$ are statistically independent. If we calculate an expression for $R_{y_1y_2}(\tau)$, we might be able to answer the question:

$$R_{y_1y_2}(\tau) = \langle y_1(t)y_2(t-\tau)\rangle$$

$$= \left\langle \int_{-\infty}^{\infty}\int_{-\infty}^{\infty} h_1(\xi)h_2(\eta)\,x(t-\xi)x(t-\tau-\eta)\,d\xi\,d\eta \right\rangle$$

$$= \int_{-\infty}^{\infty}\int_{-\infty}^{\infty} h_1(\xi)h_2(\eta)\,\langle x(t-\xi)x(t-\tau-\eta)\rangle\,d\xi\,d\eta$$

$$= \int_{-\infty}^{\infty}\int_{-\infty}^{\infty} h_1(\xi)h_2(\eta)R_x(\xi-\eta-\tau)\,d\xi\,d\eta \qquad (5.99)$$

If the input is white $[R_x(\xi - \eta - \tau) = R_x(0)\delta(\xi - \eta - \tau)]$, this equation reduces to

$$R_{y_1y_2}(\tau) = R_x(0)\int_{-\infty}^{\infty} h_1(\xi)h_2(\xi-\tau)\,d\xi \qquad (5.100)$$

Let us obtain another expression by taking Fourier transforms of both sides of Equation 5.99:

$$\mathscr{F}[R_{y_1 y_2}(\tau)] = \int_{-\infty}^{\infty} e^{-j2\pi f\tau} \int_{-\infty}^{\infty} \int_{-\infty}^{\infty} h_1(\xi) h_2(\eta) R_x(\xi - \eta - \tau)\, d\xi\, d\eta\, d\tau$$

$$= \int_{-\infty}^{\infty} \int_{-\infty}^{\infty} h_1(\xi) h_2(\eta) \left[\int_{-\infty}^{\infty} R_x(\xi - \eta - \tau) e^{-j2\pi f\tau}\, d\tau \right] d\xi\, d\eta$$

Changing variables $\rho = \tau + \eta - \xi$, this equation becomes

$$\mathscr{F}[R_{y_1 y_2}(\tau)] = \int_{-\infty}^{\infty} \int_{-\infty}^{\infty} h_1(\xi) h_2(\eta)\, e^{-j2\pi f(\xi - \eta)} \left[\int_{-\infty}^{\infty} R_x(\rho)\, e^{-j2\pi f\rho}\, d\rho \right] d\xi\, d\eta$$

$$= S_x(f) \int_{-\infty}^{\infty} h_1(\xi) e^{-j2\pi f\xi}\, d\xi \int_{-\infty}^{\infty} h_2(\eta) e^{j2\pi f\eta}\, d\eta$$

$$= S_x(f)\, H_1(f)\, H_2(-f) \tag{5.101}$$

EXAMPLE 5.15 ■

Suppose the filters are bandpass filters that are not overlapping.

$$|H_1(f)||H_2(f)| = 0 \qquad \text{for all } f$$

Then for any Gaussian input, $\mathscr{F}[R_{y_1 y_2}(\tau)] = 0$ or $R_{y_1 y_2}(\tau) = 0$. Hence $y_1(t)$ and $y_2(t)$ are statistically independent. ■ ■

5.11 CONCLUSIONS

In the next three chapters we shall use the concepts developed in this chapter to analyze the performance of both discrete and analog communication systems when the received signals are imbedded in noise. One obvious tool is the ability to model noise statistically. This will enable us to evaluate the performance of a receiver in terms of the probability of making a decision error. A less obvious tool is the ability to determine a decision rule that is optimum in the sense of minimizing the error probability. This ability was suggested by Examples 5.1 and 5.3. This concept will enable us ultimately to determine the "best" receivers, as well as to evaluate the performance of both optimal and suboptimal receivers.

PROBLEMS

5.1. (a) Calculate the probability that if a die is thrown twice, the first time it will have the value 2 and the second time 4.
   (b) Calculate the probability that when two dice are thrown together the sum is a 6.
   (c) Calculate the probability that the sum of two dice is 5 given that one has value 2.
   (d) Calculate the probability that the sum of two dice is at least 5 given that one has value 2.

5.2. (a) Determine the probability function of the random variable corresponding to the sum of the values of two dice.
   (b) Calculate and plot the CDF of this distribution.

5.3. Assume that you have a biased coin where the probability of a head is $p = 0.6$.
- (a) Calculate the probability that the first four flips come up heads and the next three are tails.
- (b) Prove that the probability that any r out of n flips are heads is $C_r^n p^r (1 - p)^{n-r}$ where $C_r^n = n!/(n - r)!r!$. (This is called the binomial distribution.)
- (c) What is the probability that four out of seven flips are heads?
- (d) What is the probability that at most four out of seven flips are heads?

5.4. Suppose that you are a sonar operator trying to decide whether or not there is a submarine nearby and you hear a peculiar sound. You know from experience that the probability of hearing this sound as a result of a submarine is 0.8. On the other hand, the probability of hearing this sound as a result of some natural water turbulence is 0.2. Also from experience, the a priori probability of finding a submarine in your present location is 0.15. What would be the optimum preliminary decision?

5.5. A certain communication system transmits only positive and negative pulses, which occur with equal probabilities. Because of noise in the channel, errors are made. If the probability is $\frac{1}{8}$ that a transmitted positive pulse will be received as a negative pulse and $\frac{1}{2}$ that a transmitted negative pulse will be received as a positive pulse, determine what fraction of the time the received positive pulses are correct.

5.6. Consider the problem of communicating one of six messages, each represented by an electrical signal s_i, $i = 1, 2, \ldots, 6$. Suppose we receive a noisy signal r and our knowledge of the noise is such that we can calculate the probabilities $P(r/s_i)$ as given below. Determine the best decision that we can make as to which message was sent.

	s_1	s_2	s_3	s_4	s_5	s_6
$p(s_i)$	0.1	0.2	0.3	0.2	0.1	0.1
$p(r/s_i)$	0.8	0.6	0.3	0.5	0.7	0.9

5.7. Assume that we are communicating three messages with a binary code (not PCM), and the received signal (after a decision is made on each pulse) is 10011. The transmitted signals that could have been set are $s_1 = 10111$, $s_2 = 11001$, $s_3 = 00011$. If $p(s_1) = 0.2$, $p(s_2) = 0.6$, $p(s_3) = 0.2$, and if the probability of mistaking a 1 for a 0 and vice versa is 0.3, determine which of the signals was most likely to have been sent. [*Hint:* The probability that any specific r out of n pulses are in error is $p^r (1 - p)^{n-r}$, where p is probability of error in any pulse.]

5.8.

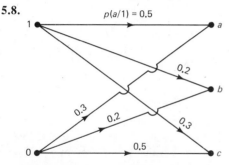

Here $p(1) = 0.5$, $p(0) = 0.5$.

(a) Determine the optimum decision rule; that is, when an a (or b or c) is received, should that be interpreted as a 1 or a 0?

(b) Calculate the probability of error $P(e)$, for this rule.

5.9. Consider a random variable X, whose distribution is characterized by

$$P((-\infty, 0]) = 0 \qquad P((0, a]) = a/a + 1$$

(a) Determine and sketch the CDF.

(b) Determine and sketch the PDF.

(c) What is the probability that $X > 2$?

5.10. Consider the random variable Y, whose CDF is given below:

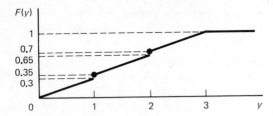

(a) Determine the probability that $Y \geq 2$, $Y < 2$, and $1 \leq Y < 2$.

(b) Find a PDF for Y using impulse functions at $y = 1$ and 2.

(c) Define a continuous random variable Y_1 and a discrete random variable Y_2 such that $Y = a_1 Y_1 + a_2 Y_2$. What are the distributions of Y_1 and Y_2?
[*Hint:* for any $g(\ )$, $E\{g(y)\} = \int g(y) f_Y(y)\, dy = a_1 \int g(y) f_{Y_1}(y)\, dy + a_2 \sum_{y_i} g(y_i) p_{Y_2}(y_i)$.]

5.11. Use Equation 5.8 and define A as the interval $(y - \Delta y, y]$, where Y is a continuous random variable, and B as the point x, where X is a discrete random variable. Prove the mixed form of Bayes's rule (Equation 5.45).

5.12. Consider again the random variable discussed in Problem 5.2.

(a) Calculate its mean value.

(b) Calculate its variance.

(c) Suppose 100 dice were thrown (50 pairs). Calculate the mean value and variance of the sum.

(d) Using the central limit theorem, estimate the probability that the sum of 100 dice will be greater than 260.

5.13. Consider the binomial distribution discussed in Problem 5.3 where r is the random variable and n is fixed.

(a) Calculate the mean value. (*Hint:* Prove first that $rC_r^n = nC_{r-1}^{n-1}$.)

(b) Calculate the variance.

(c) Evaluate (a) and (b) if $p = 0.6$ and $n = 100$.

5.14. Consider the random variable X of Problem 5.9. Prove that both the mean and the variance of X are infinite.

5.15. Consider the random variable Y of Problem 5.10.

(a) Calculate the mean values of Y_1, Y_2, and Y.

(b) Calculate the variances of Y_1, Y_2, and Y.

5.16. Determine the mean and variance of each of the random variables whose probability density functions are given. First evaluate b in terms of a so that the PDF is valid (i.e., $\int_{-\infty}^{\infty} f(x)\, dx = 1$).

      (a) $f(x) = b/2a$ if $|x| \leq a$, 0 otherwise.
      (b) $f(x) = be^{-a|x|}$.
      (c) $f(x) = be^{-ax}$ if $x \geq 0$, 0 otherwise.
      (d) $f(x) = b/[1 + (ax)^6]$.
      Hint:

$$\int_0^\infty x^n e^{-bx} \, dx = \frac{n!}{b^{n+1}}$$

$$\int_0^\infty \frac{x^{m-1}}{1 + x^n} \, dx = \frac{\pi}{n \sin(m\pi/n)}$$

5.17. Let X_1, X_2, and X_3 be statistically independent, zero-mean Gaussian random variables with variance σ^2. Define $Y_1 = X_1 + 0.6X_2$ and $Y_2 = 0.4X_2 + X_3$.
    (a) Determine the PDFs of Y_1 and Y_2 as well as the joint PDF of Y_1 and Y_2.
    (b) Calculate the variance of $Y_1 + Y_2$.

5.18. Show from the definition of the Gaussian PDF that the mean value is m and the variance is σ^2. That is, show that the terminology is consistent.

5.19. Show from the definition of the bivariate Gaussian PDF that for two jointly Gaussian random variables the correlation coefficient $\mu_{11}/\sigma_1\sigma_2$ is equal to the parameter ρ.

5.20. If X_1 and X_2 are independent zero-mean Gaussian random variables with $\sigma_1^2 = \sigma_2^2 = 1$, and if **X** is a vector in the Euclidean sense (i.e., $|\mathbf{X}| = \sqrt{X_1^2 + X_2^2}$), determine the mean and variance of the magnitude of **X**. (*Hint:* Convert the bivariate Gaussian PDF into polar form.)

5.21. Consider the random process $y(t)$, whose measured average value is zero and autocorrelation function is $R_y(\tau) = 10 \, \text{sinc}^2 \, 10^{+4}\tau$. Assuming the process is Gaussian, determine the percentage of time that $y(t)$ exceeds 7 V.

5.22. Repeat Problem 5.21, only now assume that the process is more impulsive and can be modeled by the PDF of Problem 5.16b. (*Hint:* Evaluate the parameter a in terms of σ_y^2 first.)

5.23. Consider a signal that is composed of a $+1$-V dc component plus a zero-mean stationary random component that can be modeled with a Gaussian PDF. The spectrum of the random component is given by

$$S_x(f) = \begin{cases} \eta_0/2 & \text{V/Hz} \quad |f| < B \\ 0 & \text{elsewhere} \end{cases}$$

Determine an expression for the probability that the signal lies in the interval between -2 and $+2$ V in terms of $Q(\ )$.

5.24. Repeat Problem 5.23, but now assume that the spectrum is given by

$$S_x(f) = \frac{n_0/\pi}{1 + (f/10^4)^2} \quad \text{V}^2/\text{Hz}$$

5.25. Referring to the output of Problem 2.22 (Chapter 2), write a PDF model for a sample. Also write the multivariate PDF for two samples spaced A s apart.

5.26. Consider now the output of Problem 2.23 (Chapter 2), when the input is a zero-mean Gaussian process.

(a) Write the PDF model for a sample of the output at any time t.

(b) Write the joint PDF for two samples spaced τ s apart.

(c) For what values of τ does the joint PDF factor into the product of the marginal PDFs?

5.27. Let $x_1(t)$, $x_2(t)$, and $x_3(t)$ be statistically independent zero-mean Gaussian processes with autocorrelation functions $R_1(\tau)$, $R_2(\tau)$, and $R_3(\tau)$, respectively. Define $y_1(t) = x_1(t) + 0.6x_2(t)$ and $y_2(t) = 0.4x_2(t) + x_3(t)$. Specify the statistical properties of the two processes $y_1(t)$ and $y_2(t)$ and determine $R_{y_1}(\tau)$, $R_{y_2}(\tau)$, and $R_{y_1 y_2}(\tau)$.

5.28. We have claimed that the system below is useful for measuring the autocorrelation function of a wide-sense stationary process $x(t)$:

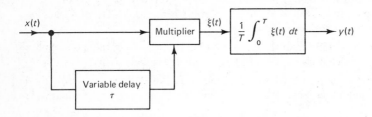

(a) Show that $E\{y\} = R_x(\tau)$.

(b) Find an expression for σ_y^2 if $x(t)$ is Gaussian with zero mean. (*Hint:* If x_1, x_2, x_3, and x_4 are Gaussian with zero mean, it can be shown that

$$E\{x_1 x_2 x_3 x_4\} = E\{x_1 x_2\}E\{x_3 x_4\} + E\{x_1 x_3\}E\{x_2 x_4\}$$
$$+ E\{x_1 x_4\}E\{x_2 x_3\})$$

(c) Simplify the expression in part (b) using the relation (without proof) that

$$\int_0^T \int_0^T f(t_2 - t_1)\, dt_1\, dt_2 = T \int_{-T}^T f(\xi)(1 - |\xi|/T)\, d\xi$$

for any continuous function $f(\ )$.

(d) Argue that $\sigma_y^2 \to 0$ as $T \to \infty$. This is a proof that the measurement converges to $R_x(\tau)$. Convergence of this type is said to be convergence in mean square.

REFERENCES

1. *Handbook of Chemistry and Physics*, Chemical Rubber, Cleveland, 1954.

2. W. Feller, *An Introduction to Probability Theory and Its Applications*, Wiley, New York, 1957.

3. E. N. Gilbert and H. O. Pollak, "Amplitude Distribution of Shot Noise," *Bell Syst. Tech. J.* 39:335–350 (1960).

4. M. Rosenblatt, "A Central Limit Theorem and a Strong Mixing Condition," *Proc. Natl. Acad. Sci. USA* 42:43–47 (1956).

5. M. Rosenblatt, "Independence and Dependence," *1961 Proceedings of the Fourth Berkeley Symposium* 2:431–443 (1961).

6. G. D. Birkhoff, "Proof of the Ergodic Theorem," *Proc. Natl. Acad. Sci. USA* 17:650–656 (1931).

7. J. Von Neumann, "Proof of the Quasi-ergodic Hypothesis," *Proc. Natl. Acad. Sci. USA* 18:70 (1932).

8. Y. A. Yaglom, *An Introduction to the Theory of Stationary Random Functions*, Prentice-Hall, Englewood Cliffs, NJ, 1962.

9. Y. A. Rozanov, "An Application of the Central Limit Theorem," *1961 Proceedings of the Fourth Berkeley Symposium* 2:445–454 (1961).

Chapter 6

Analysis of Binary Communications

6.1 INTRODUCTION

In this chapter we examine the discrete communication techniques of pulse code modulation as well as the modulated binary techniques that were introduced in Chapter 3. We take a closer look at the hardware, determine the probability of decision errors caused by additive noise, and evaluate the distortion due to quantization.

Throughout this chapter we assume the receiver structure shown in Figure 6.1, which is very similar to the receivers discussed in Chapter 3. The input to the receiver consists of one of two possible signal pulses [$s_0(t)$ and $s_1(t)$ represent the 1 and 0 of PCM] plus additive noise:

$$r(t) = s_i(t) + n(t) \qquad i = 1, 0 \tag{6.1}$$

The noise will always be an ergodic, zero-mean process with a known power spectrum $S_n(f)$.

6.2 THE OPTIMUM DECISION RULE

Because the input filter is linear, superposition permits us to regard the output of the sample and hold as

$$R = S_i + N \tag{6.2}$$

where S_i is a discrete random variable that represents the output when the signal alone is the input and N is a continuous random variable that represents the output when noise alone is the input. S_i has two possible values

176

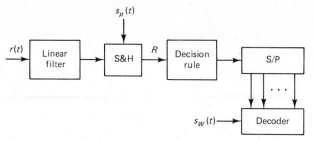

Figure 6.1 PCM receiver structure.

$(S_0$ and $S_1)$, which occur with probability P_0 and P_1. These probabilities are called *a priori probabilities* and may or may not be known. N is a zero-mean random variable whose variance σ_N^2 depends on the input spectrum $S_N(f)$ and the filter transfer function $H(f)$:

$$\sigma_N^2 = \int_{-\infty}^{\infty} S_N(f)|H(f)|^2 \, df \tag{6.3}$$

The decision rule is a device that, based on the value r of the random variable R and the a priori probabilities, if they are known, decides which signal was transmitted. For a given decision rule, the probability of error can be determined from

$$P(e) = \int_{-\infty}^{\infty} \Pr(\text{wrong decision}/r) f_R(r) \, dr \tag{6.4}$$

where $f_R(r)$ is the PDF of the random variable R. Suppose we can determine the conditional probabilities $p(S_0/r)$ and $p(S_1/r)$, which are called *a posteriori probabilities*. If for a given r the decision rule selects S_0, then $p(S_1/r)$ is the probability of a wrong decision given r. In order to minimize the probability of error, we require that for any r the decision rule select that signal whose a posteriori probability is the largest. In Examples 5.1 and 5.3 we called such a decision rule (namely, select the a posteriori "most likely" signal) the optimum rule. We now see that it minimizes the probability of error.

EXAMPLE 6.1 ■

Consider a noisy discrete binary communication channel that includes a decision device with three outputs as illustrated in Figure 6.2.

(a) If $P_0 = 0.7$ and $P_1 = 0.3$, determine the optimum decision rule (the assignment of a, b, c to S_0, S_1) and the resulting probability of error. The optimum rule maximizes the a posteriori probabilities. From Bayes's rule (Equation 5.9),

$$p(S_0/a) = \frac{p(a/S_0)P_0}{p(a)} = \frac{0.49}{p(a)}$$

$$p(S_1/a) = \frac{p(a/S_1)P_1}{p(a)} = \frac{0.09}{p(a)}$$

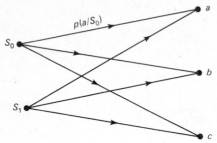

Figure 6.2   Example 6.1a: $P(a/S_0) = 0.7$, $P(a/S_1) = 0.3$, $P(b/S_0) = 0.2$, $P(b/S_1) = 0.2$, $P(c/S_0) = 0.1$, $P(c/S_1) = 0.5$.

When the receiver output is a, one should select S_0 as the signal. Following similar arguments, the optimum rule is seen to be $a \rightarrow S_0$, $b \rightarrow S_0$, $c \rightarrow S_1$. The probability of error is

$$P(e) = p(a, S_1) + p(b, S_1) + p(c, S_0)$$

$$= p(a/S_1)P_1 + p(b/S_1)P_1 + p(c/S_0)P_0$$

$$= 0.09 + 0.06 + 0.07 = 0.22$$

   (b)   Sketch the probability of error for each of the eight possible decision rules versus P_0. For the optimum rule (rule 1)

$$P(e) = p(a/S_1)(1 - P_0) + p(b/S_1)(1 - P_0) + p(c/S_0)P_0$$

$$= 0.5 - 0.4P_0$$

(see Figure 6.3). Consider the rule $a \rightarrow S_0$, $b \rightarrow S_1$, $c \rightarrow S_1$ (rule 2)

$$P(e) = p(a/S_1)(1 - P_0) + p(b/S_0)P_0 + p(c/S_0)P_0$$

$$= 0.3$$

The other six rules are formed in a similar way, and all are plotted in Figure 6.4. We observe that rule 1 is indeed optimum for $P_0 = 0.7$. We also observe that both rule 1 and rule 2 are optimum for $P_0 = 0.5$. Sometimes when the a priori probabilities are unknown a rule is selected that has the smallest maximum probability of error (called minimax). We see by inspection that rule 2 is minimax.   ■ ■

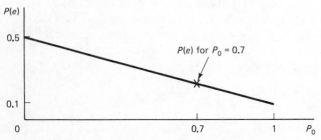

Figure 6.3   Example 6.1b: the optimal rule.

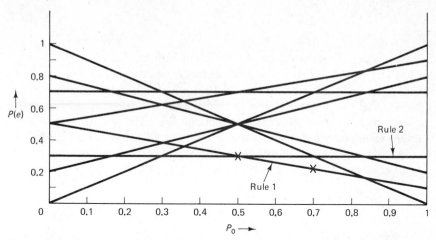

Figure 6.4   Example 6.1b: the eight possible rules.

From the mixed form of Bayes's rule (Equation 5.46), the a posteriori probabilities can be expressed

$$p(S_i/r) = \frac{f_R(r/S_i)P_i}{f_R(r)} \tag{6.5}$$

where $i = 0$ or 1. Since the denominator does not depend on the value of i, the optimum decision rule is to select S_0 if and only if

$$f_R(r/S_0)P_0 > f_R(r/S_1)P_1 \tag{6.6}$$

From Equation 6.2

$$\Pr(r_1 < R \le r_2/S_i) = \Pr(r_1 - S_i < N \le r_2 - S_i)$$

or

$$\int_{r_1}^{r_2} f_R(r/S_i)\, dr = \int_{r_1-S_i}^{r_2-S_i} f_N(n)\, dn$$

$$= \int_{r_1}^{r_2} f_N(r - S_i)\, dr \qquad (n = r - S_i)$$

It follows that

$$f_R(r/S_i) = f_N(r - S_i) \tag{6.7}$$

Thus the optimum decision rule is now to select S_0 if and only if

$$f_N(r - S_0)P_0 > f_N(r - S_1)P_1 \tag{6.8}$$

This rule can be generalized to include M signals ($M \ge 2$), where the optimum rule is then to select S_i if and only if

$$f_N(r - S_i)P_i > f_N(r - S_j)P_j \tag{6.9}$$

for $i, j = 0, 1, \ldots, M - 1$ and $j \neq i$.

Both sides of Equation 6.8 are plotted in Figure 6.5, where it is assumed that $S_1 > S_0$ and the PDF of the noise decreases monotonically from a single maximum at zero (called a *unimodal* PDF). It is clear from inspection of Figure 6.5 that the optimum decision rule is a simple threshold device: Select S_0 if and only if $r < a$; otherwise, select S_1. We have just proven that the optimum binary decision device is the comparator (which was used in Chapter 3 hardware discussions), provided the threshold voltage a corresponds to that value of r, where $P_0f_N(r - S_0) = P_1f_N(r - S_1)$, or

$$P_0f_N(a - S_0) = P_1f_N(a - S_1) \tag{6.10}$$

Let us now assume that the noise is zero-mean Gaussian. From properties 5.1 and 5.3 of Gaussian random processes, N is a zero-mean Gaussian random variable whose variance σ_N^2 is given by Equation 6.3. Thus Equation 6.10 becomes

$$\frac{P_0}{\sqrt{2\pi\sigma_N^2}} \exp\left[-\frac{1}{2\sigma_N^2}(a - S_0)^2\right] = \frac{P_1}{\sqrt{2\pi\sigma_N^2}} \exp\left[-\frac{1}{2\sigma_N^2}(a - S_1)^2\right]$$

or

$$\exp\left\{-\frac{1}{2\sigma_N^2}[(a - S_0)^2 - (a - S_1)^2]\right\} = \frac{P_1}{P_0} \tag{6.11}$$

Taking natural logarithms of both sides of Equation 6.11, we obtain

$$(a - S_1)^2 - (a - S_0)^2 = 2\sigma_N^2 \ln P_1/P_0$$

Solving for a yields

$$a = \frac{S_1 + S_0}{2} - \frac{\sigma_N^2}{S_1 - S_0} \ln \frac{P_1}{P_0} \tag{6.12}$$

If $P_0 = P_1 = \frac{1}{2}$ (i.e., if the 1s and 0s are equally likely a priori), which is a reasonable assumption for PCM, the optimum threshold lies halfway between the two values S_0 and S_1. If $P_0 \neq P_1$, the optimum threshold lies closer to the signal that is less likely a priori. Note that if the Gaussian PDF model is not

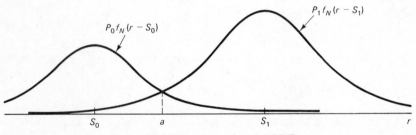

Figure 6.5 A posteriori probabilities for unimodal PDFs.

valid, Equation 6.12 does not define the location of the optimum threshold. If, however, $P_1 = P_0$ and the noise can be modeled by any zero-mean unimodal PDF, the optimum threshold lies halfway between the signal values.

When the a priori probabilities are unknown or the Gaussian assumption may not be valid, some strategy is necessary to set the threshold. The *maximum likelihood (ML) strategy* places the threshold halfway between the signal-only values. We shall use this strategy throughout, partly because the assumption of a priori equally likely signals seems reasonable for communication systems, partly because the formulas for the probability of error are simplified if this strategy is used. Thus, unless stated otherwise, devices called optimum decision devices are optimum in the sense of the maximum likelihood strategy.

6.3 ERROR PROBABILITY CALCULATIONS

For all binary communication systems we must distinguish between the probability that a pulse or bit is in error, $P_b(e)$, and the probability that an N-bit word is in error, $P_w(e)$. The relationship between them is

$$P_w(e) = 1 - \Pr(\text{none of the } N \text{ bits is in error})$$
$$= 1 - [1 - P_b(e)]^N \tag{6.13}$$

If $NP_b(e) \ll 1$, which is usually the situation, a simple approximation is

$$P_w(e) = 1 - [1 - NP_b(e) + \tfrac{1}{2}N(N-1)P_b^2(e) - \cdots]$$
$$\approx NP_b(e) \tag{6.14}$$

This approximation is equivalent to assuming that two or more bit errors are much less likely than a single bit error.

We can relate $P_b(e)$ to the joint probabilities $P_b(e, S_i)$ from

$$P_b(e) = \sum_{i=0}^{1} P_b(e, S_i)$$
$$= P_0 P_b(e/S_0) + P_1 P_b(e/S_1) \tag{6.15}$$

However,

$$P_b(e/S_0) = \Pr(r > a/S_0) = \Pr(n > a - S_0)$$
$$= \int_{a-S_0}^{\infty} f_N(x)\,dx \tag{6.16}$$

Similarly

$$P_b(e/S_1) = \int_{-\infty}^{a-S_1} f_N(x)\,dx \tag{6.17}$$

Thus Equation 6.15 becomes

$$P_b(e) = P_0 \int_{a-S_0}^{\infty} f_N(x)\,dx + P_1 \int_{-\infty}^{a-S_1} f_N(x)\,dx \tag{6.18}$$

If $f_N(x)$ is a zero-mean Gaussian PDF with variance σ_N^2, then after normalization

$$P_b(e) = P_0 \int_{(a-S_0)/\sigma_N}^{\infty} \frac{1}{\sqrt{2\pi}} e^{-x^2/2}\,dx + P_1 \int_{-\infty}^{(a-S_1)/\sigma_N} \frac{1}{\sqrt{2\pi}} e^{-x^2/2}\,dx \tag{6.19}$$

Because of the symmetry of the Gaussian PDF, this can be written

$$P_b(e) = P_0 \int_{(a-S_0)/\sigma_N}^{\infty} \frac{1}{\sqrt{2\pi}} e^{-x^2/2}\,dx + P_1 \int_{(S_1-a)/\sigma_N}^{\infty} \frac{1}{\sqrt{2\pi}} e^{-x^2/2}\,dx$$

or

$$P_b(e) = P_0 Q\left(\frac{a - S_0}{\sigma_N}\right) + P_1 Q\left(\frac{S_1 - a}{\sigma_N}\right) \tag{6.20}$$

where $Q(\ )$ was defined by Equation 5.65.

Substituting the optimum (ML) threshold $a = (S_1 + S_0)/2$ into Equation 6.20 yields

$$P_b(e) = (P_0 + P_1) Q\left(\frac{S_1 - S_0}{2\sigma_N}\right)$$

$$= Q\left(\frac{S_1 - S_0}{2\sigma_N}\right) \tag{6.21}$$

since the sum of the a priori probabilities must equal unity. For bipolar pulses, such as those used in the NRZ signal, the outputs due to signal alone have the same magnitude but different signs. Thus, if we define $S = S_1 = -S_0$, Equation 6.21 becomes

$$P_{b(bip)}(e) = Q\left(\frac{S}{\sigma_N}\right) = Q\left(\sqrt{\frac{S^2}{\sigma_N^2}}\right) < \frac{1}{2} \exp\left(-\frac{1}{2}\frac{S^2}{\sigma_N^2}\right) \tag{6.22}$$

where the bound of Equation 5.67 for $Q(\alpha)$ $[Q(\alpha) < \frac{1}{2} e^{-\alpha^2/2}]$ is used to show that $P_b(e)$ decreases exponentially with S^2/σ_N^2, called the *signal-to-noise ratio* (SNR). For monopolar pulses, $S_1 = S$, $S_0 = 0$, Equation 6.21 becomes

$$P_{b(mon)}(e) = Q\left(\sqrt{\frac{S^2}{4\sigma_N^2}}\right) < \frac{1}{2} \exp\left(-\frac{1}{8}\frac{S^2}{\sigma_N^2}\right) \tag{6.23}$$

In either case, $P_b(e)$ decreases rapidly with the SNR. When monopolar pulses are used, the SNR must be four times larger (6 dB) than when bipolar pulses are used in order to achieve the same $P_b(e)$.

EXAMPLE 6.2 ■

Impulse noise cannot be modeled by the Gaussian PDF. PDF models $[f(x)]$ for impulse noise should have a larger value at $x = 0$ than Gaussian PDFs and should fall off more slowly with an increase in its argument. One possible unimodal candidate is the double exponential or $f(x) = be^{-a|x|}$. Since $\int_{-\infty}^{\infty} f(x) \, dx$ must equal unity, this can be written $f(x) = \frac{1}{2}ae^{-a|x|}$ (see Problem 5.16). Setting the variance equal to the average noise power σ_N^2, we obtain

$$\sigma_N^2 = \int_{-\infty}^{\infty} x^2 f(x) \, dx = 2 \int_0^{\infty} x^2 (a/2) \, e^{-ax} \, dx = 2/a^2$$

Thus $a = \sqrt{2}/\sigma_N$, and the double-exponential PDF, which is useful for modeling impulse noise, is given by

$$f(x) = \frac{1}{\sqrt{2\sigma_N^2}} \exp\left(-\frac{\sqrt{2}}{\sigma_N}|x|\right) \tag{6.24}$$

From Equation 6.18,

$$P_b(e) = P_0 \int_{a-S_0}^{\infty} \frac{1}{\sqrt{2\sigma_N^2}} e^{-(\sqrt{2}/\sigma_N)|x|} \, dx + P_1 \int_{-\infty}^{a-S_1} \frac{1}{\sqrt{2\sigma_N^2}} e^{-(\sqrt{2}/\sigma_N)|x|} \, dx$$

Substituting the optimum (ML) threshold, one obtains

$$P_b(e) = (P_0 + P_1) \int_{(S_1-S_0)/2}^{\infty} \frac{1}{\sqrt{2\sigma_N^2}} e^{-(\sqrt{2}/\sigma_N)|x|} \, dx$$

or

$$P_b(e) = \frac{1}{2} \exp\left(-\frac{S_1 - S_0}{\sqrt{2\sigma_N^2}}\right) \tag{6.25}$$

For bipolar pulses this becomes

$$P_{b(bip)}(e) = \frac{1}{2} \exp\left(-\sqrt{\frac{2S^2}{\sigma_N^2}}\right)$$

which does not decrease with the SNR as fast as when the Gaussian model is valid (see Equation 6.22). Thus overcoming the occasional large spikes of impulse noise requires much more signal power than when the noise is Gaussian. ■ ■

6.4 QUANTIZATION ERROR

Let us assume that the signal is an ergodic bandlimited random process that is sampled at or faster than the Nyquist rate and quantized into M levels. We assume first that the signal lies within some finite interval $[-A, A]$ and is quantized as in Figure 6.6. If the signal sample S lies in the ith interval I_i, whose center is a_i and whose width is Δ_i, the quantized signal S_q has value a_i. The quantity $S - S_q$ is the distortion caused by quantization. The mean square error due to quanitzation, called the *quantization noise power* N_q, is defined by

$$N_q = E\{(S - S_q)^2\} \tag{6.26}$$

Because $S_q = a_i$ when S lies in I_i, N_q can be evaluated from

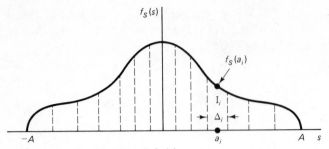

Figure 6.6   Quantization definition.

$$N_q = \sum_{i=1}^{M} \int_{a_i-\Delta_i/2}^{a_i+\Delta_i/2} (s - a_i)^2 f_S(s)\, ds \tag{6.27}$$

If we approximate $f_S(s)$ with straight line segments,

$$f_S(s) \approx f(a_i) + (s - a_i)f'(a_i) \qquad \text{for} \quad S \text{ in } I_i$$

then

$$N_q \approx \sum_{i=1}^{M} \left[f(a_i) \int_{a_i-\Delta_i/2}^{a_i+\Delta_i/2} (s - a_i)^2\, ds + f'(a_i) \int_{a_i-\Delta_i/2}^{a_i+\Delta_i/2} (s - a_i)^3\, ds \right] \tag{6.28}$$

where the approximation is good if M is large. Changing variables by $x = s - a_i$, we have

$$N_q \approx \sum_{i=1}^{M} \left[f(a_i) \int_{-\Delta_i/2}^{\Delta_i/2} x^2\, dx + f'(a_i) \int_{-\Delta_i/2}^{\Delta_i/2} x^3\, dx \right]$$

$$= \sum_{i=1}^{M} f(a_i)\, 2 \int_0^{\Delta_i/2} x^2\, dx = \frac{1}{12} \sum_{i=1}^{M} f(a_i)\Delta_i^3 \tag{6.29}$$

If p_i is the probability that S lies in I_i,

$$p_i = \int_{a_i-\Delta_i/2}^{a_i+\Delta_i/2} f_S(s)\, ds \approx f(a_i)\Delta_i \tag{6.30}$$

Equation 6.29 can also be expressed

$$N_q = \frac{1}{12} \sum_{i=1}^{M} p_i \Delta_i^2 \tag{6.31}$$

For equal-width quantization ($\Delta_i = 2A/M$), which is achieved by all standard A/D devices,

$$N_q = \frac{A^2}{3M^2} \sum_{i=1}^{M} p_i = \frac{A^2}{3M^2} \tag{6.32}$$

since $\Sigma_i\, p_i$ must equal 1. For other types of quantization we define γ_i from

$$\Delta_i = \gamma_i 2A/M \tag{6.33}$$

Hence

$$N_q = \frac{A^2}{3M^2} \sum_{i=1}^{M} \gamma_i^2 p_i \tag{6.34}$$

The effect of the quantization noise power depends on the signal power σ_s^2, defined by

$$\sigma_s^2 = \int_{-A}^{A} s^2 f_S(s) \, ds \tag{6.35}$$

For this reason it is more useful to evaluate the SNR due to quantization:

$$\text{SNR}_q = \frac{\sigma_s^2}{N_q} = cM^2 \tag{6.36}$$

where $c = 3\sigma_s^2/A^2$ for uniform quantization and in general

$$c = \frac{3\sigma_s^2}{A^2} \Bigg/ \sum_{i=1}^{M} \gamma_i^2 p_i \tag{6.37}$$

EXAMPLE 6.3 ■

Assume the signal has a uniform PDF, or

$$f_S(s) = \begin{cases} 1/2A & |s| \leq A \\ 0 & |s| > A \end{cases}$$

For this case uniform quantization ($\Delta_i = 2A/M$) seems logical. Note that $p_i = f(a_i)\Delta_i = 1/M$ for all i, and uniform quantization also corresponds to equal interval probability:

$$\sigma_s^2 = 2 \int_0^A s^2 \frac{1}{2A} \, ds = \frac{A^2}{3}$$

Thus, for a uniform PDF, $c = 1$ and

$$\text{SNR}_q = M^2 \tag{6.38} \quad \blacksquare\ \blacksquare$$

EXAMPLE 6.4 ■

Consider the triangular PDF shown in Figure 6.7, where $A = 2$ and $M = 8$.

$$\sigma_s^2 = 2 \int_0^2 s^2 \frac{1}{2}(1 - \frac{1}{2}s) \, ds = \frac{2}{3}$$

hence for uniform quantization $c = \frac{3}{4}\sigma_s^2 = \frac{1}{2}$, and $\text{SNR}_q = \frac{1}{2}M^2$. Let us now consider equal interval probability quantization ($p_i = \frac{1}{8}$ for all i) where the details are shown in Figure 6.7. From Equation 6.31, $N_q = \frac{1}{12}(\frac{1}{8}\sum_{i=1}^{8} \Delta_i^2)$, and, from Figure 6.7,

$$N_q = \frac{1}{12 \times 8} 2[(0.268)^2 + (0.313)^2 + (0.419)^2 + 1^2] = \frac{1}{12 \times 8} 2.69$$

Thus

$$\text{SNR}_q = \frac{2}{3}(12 \times 8)/2.69 = 0.3715 \times 8^2$$

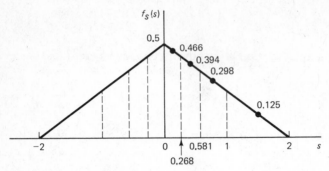

Figure 6.7 Example 6.4.

For this type of quantization $c = 0.3715$, which is slightly smaller than the value for uniform quantization ($c = 0.5$).   ■ ■

Let us now consider untruncated PDFs where $f_S(s) \neq 0$ for $|s| > A$ and where $2A$ is the maximum dynamic range that the quantizer can handle. For this case, Equation 6.29 should be replaced by

$$N_q = \frac{1}{12} \sum_{i=1}^{M} f(a_i)\Delta_i^3 + 2 \int_{A}^{\infty} (s - a_{max})^2 f_S(s)\, ds \qquad (6.39)$$

where $a_{max} = A(1 - 1/M)$, and the second term of Equation 6.39 is the effect of truncation. For uniform quantization ($\Delta_i = 2A/M$) the first term can be expressed

$$\frac{A^2}{3M^2} \sum_{i=1}^{M} f(a_i)\Delta_i \qquad (6.40)$$

where for large M

$$\sum_{i=1}^{M} f(a_i)\Delta_i + 2 \int_{A}^{\infty} f_S(s)\, ds \approx 1 \qquad (6.41)$$

Combining the last three equations results in

$$N_q = \frac{A^2}{3M^2} + \xi(A) \qquad (6.42)$$

where

$$\xi(A) = 2 \int_{A}^{\infty} \left[(s - a_{max})^2 - \frac{A^2}{3M^2} \right] f_S(s)\, ds$$

$$= 2 \int_{A}^{\infty} s^2 f_S(s)\, ds - 4A\left(1 - \frac{1}{M}\right) \int_{A}^{\infty} s f_S(s)\, ds$$

$$+ 2A^2 \left(1 - \frac{2}{M} + \frac{2}{3M^2}\right) \int_{A}^{\infty} f_S(s)\, ds \qquad (6.43)$$

The truncation effect (Equation 6.43) depends on the PDF of the signal.

Assume that the signal is zero-mean Gaussian with variance σ_S^2. It can be shown that

$$\int_A^\infty \frac{s^2}{\sqrt{2\pi\sigma_S^2}} e^{-s^2/2\sigma_S^2}\, ds = \frac{\sigma_S A}{\sqrt{2\pi}} e^{-A^2/2\sigma_S^2} + \sigma_S^2 Q\left(\frac{A}{\sigma_S}\right)$$

$$\int_A^\infty \frac{s}{\sqrt{2\pi\sigma_S^2}} e^{-s^2/2\sigma_S^2}\, ds = \frac{\sigma_S}{\sqrt{2\pi}} e^{-A^2/2\sigma_S^2}$$

$$\int_A^\infty \frac{1}{\sqrt{2\pi\sigma_S^2}} e^{-s^2/2\sigma_S^2}\, ds = Q\left(\frac{A}{\sigma_S}\right) \approx \frac{1}{\sqrt{2\pi}\, A/\sigma_S} e^{-A^2/2\sigma_S^2}$$

Substituting these equations into Equation 6.43 yields

$$\xi(A) = \sqrt{\frac{2}{\pi}}\frac{\sigma_S}{A}\left(\sigma_S^2 + \frac{2A^2}{3M^2}\right) e^{-A^2/2\sigma_S^2} \tag{6.44}$$

We now let $A = k\sigma_S$; that is, we assume the signal is truncated at k standard deviations. Equation 6.44 now becomes

$$\xi(k) = \sqrt{\frac{2}{\pi}}\frac{\sigma_S^2}{k}\left(1 + \frac{2k^2}{3M^2}\right) e^{-k^2/2} \tag{6.45}$$

$$N_q = \frac{k^2\sigma_S^2}{3M^2} + \xi(k) \tag{6.46}$$

Finally, the SNR becomes $\text{SNR}_q = cM^2$, where

$$c = \frac{3}{k^2}\bigg/\left[1 + \frac{3M^2}{k^3}\sqrt{\frac{2}{\pi}}\left(1 + \frac{2k^2}{3M^2}\right) e^{-k^2/2}\right] \tag{6.47}$$

Note that $3/k^2 = 3\sigma_S^2/A^2$ was the result for uniform quantization given that $f_S(s) = 0$ for $|s| > A$. Equation 6.47 is evaluated as a function of k in Table 6.1, where it is assumed that $M = 512$. We see that, when the signal is Gaussian, truncation at five standard deviations is optimum ($c \approx 0.118$). For the double-exponential PDF of Example 6.2 (see Problem 6.10), however, which is also evaluated in Table 6.1 (labeled *impulsive*), the optimum truncation is less sharp and occurs at eight standard deviations ($c \approx 0.041$).

The SNR due to quantization varies with the assumed dynamic range and the statistics of the signal. This problem is compounded greatly in that many analog signals are nonstationary; that is, σ_S^2 may vary considerably from one time to the next. One way to deal with both problems is to quantize at ten or more standard deviations of the "weak" signal passages. The cost of doing this, in terms of SNR or, equivalently, the large M needed to achieve a given SNR, is significant. In this situation nonuniform quantization can be very helpful.

6.5 COMPANDING ANALYSIS

In this section we assume that $A \geq 10\sigma_S$ and truncation effects can be ignored. Thus Equation 6.37 is valid or $\text{SNR}_q = cM^2$, where

TABLE 6.1   EVALUATION OF c ($SNR_q = cM^2$) for $M = 512$

k	$3/k^2$	c (Gaussian)	c (impulsive)
3	0.3333	0.0013	0.0004
4	0.1875	0.0437	0.0012
5	0.1200	0.1178	0.0043
6	0.0833	0.0833	0.0149
7	0.0612	0.0612	0.0338
8	0.0469	0.0469	0.0407
9	0.0370	0.0370	0.0360
10	0.0300	0.0300	0.0298

$$c \simeq \frac{3\sigma_s^2}{A^2} \bigg/ \sum_{i=1}^{M} \gamma_i^2 f_s(a_i)\Delta_i$$

$$\simeq \frac{3\sigma_s^2}{A^2} \bigg/ \int_{-\infty}^{\infty} \gamma^2(s) f_s(s)\, ds \tag{6.48}$$

In Chapter 4 we considered the use of companding, which employs a nonlinearity followed by equal-width quantization. We saw (Equations 4.3 and 4.5) that if the nonlinearity is modeled by $g[s(t)]$, then

$$\gamma_i = \frac{g(A)}{A} \bigg/ \frac{dg(s)}{ds}\bigg|_{a_i}$$

We specifically considered the μ-*law compander*,

$$g(s) = \begin{cases} \dfrac{\ln(1 + \mu s/A)}{\ln(1 + \mu)} & s \geq 0 \\[2mm] -\dfrac{\ln(1 - \mu s/A)}{\ln(1 + \mu)} & s < 0 \end{cases}$$

For this case (Equation 4.6)

$$\gamma(s) = \frac{(1 + \mu|s|/A)\ln(1 + \mu)}{\mu} \tag{6.49}$$

Thus for μ-law companding Equation 6.48 becomes

$$c = \frac{3\sigma_s^2}{A^2} \frac{\mu^2}{\ln^2(1 + \mu)} \bigg/ \int_{-\infty}^{\infty} \left(1 + 2\frac{\mu}{A}|s| + \frac{\mu^2}{A^2}s^2\right) f_s(s)\, ds \tag{6.50}$$

Since $\int_{-\infty}^{\infty} f_s(s)\, ds = 1$, $\int_{-\infty}^{\infty} s^2 f_s(s)\, ds = \sigma_s^2$, and $\int_{-\infty}^{\infty} |s| f_s(s)\, ds = E\{|s|\}$, Equation 6.50 becomes

$$c = \frac{3\sigma_s^2}{A^2} \bigg/ \ln^2(1 + \mu)\left(\frac{1}{\mu^2} + \frac{2}{\mu}\frac{E\{|s|\}}{A} + \frac{\sigma_s^2}{A^2}\right) \tag{6.51}$$

While the term $E\{|s|\}$ depends on the PDF model, it does not vary much with the model; thus σ_s^2/N_q is nearly independent of the PDF model for

TABLE 6.2 QUANTIZATION SNR VERSUS σ_S^2

		Uniform quantization				μ-law companding ($\mu = 255$)	
			SNR$_q$ (dB)				SNR$_q$ (dB)
$3\sigma_S^2/A^2$	c (dB)	9 bits	11 bits	13 bits	c (dB)		9 bits
3×10^{-2}	-15.2	39.0	51.0	63.0	-10.6		43.6
10^{-2}	-20	34.2	46.2	58.2	-10.9		43.3
10^{-3}	-30	24.2	36.2	48.2	-12.2		42.0
10^{-4}	-40	14.2	26.2	38.2	-14.9		39.3
10^{-5}	-50	4.2	16.2	28.2	-19.4		34.8
10^{-6}	-60	-5.8	6.2	18.2	-25.6		28.6

μ-law companding (provided $A \geq 10\sigma_S$). This is an important feature of this type of companding. If the signal is Gaussian,

$$E\{|s|\} = 2 \int_0^\infty \frac{s}{\sqrt{2\pi\sigma_S^2}} e^{-s^2/2\sigma_S^2} \, ds = \sqrt{\frac{2}{\pi}}\sigma_S \tag{6.52}$$

For impulsive noise $E\{|s|\}$ is similar (see Problem 6.11). For Gaussian inputs Equation 6.51 becomes

$$c = \frac{3\sigma_S^2}{A^2} \bigg/ \ln^2(1+\mu)\left(\frac{1}{\mu^2} + \frac{2}{\mu}\sqrt{\frac{2}{\pi}}\frac{\sigma_S}{A} + \frac{\sigma_S^2}{A^2}\right) \tag{6.53}$$

Both c and the SNR (SNR$_q$ = cM^2) are given in Table 6.2 for various input signal levels for $\mu = 255$, and $M = 512$ (nine bits per word). Observe that the quantization SNR for μ-law companding is nearly constant with variations in signal power. For $\sigma_S^2 = 10^{-2}A^2$, μ-law companding is 4.6 dB better than uniform quantization, whereas for very weak signals ($\sigma_S^2 \leq 10^{-5}A^2$) the improvement is more than 30 dB.

If our design criterion were a quantization SNR of 40 dB, when the signal power is $\geq 10^{-4}A^2$ this would require 13 bits (8192 levels) if uniform quantization is used but only 9 bits (512 levels) with μ-law companding. From this viewpoint, companding has achieved bit and bandwidth compression. Thus the required bit rate or bandwidth is only $\frac{9}{13}$ that required with uniform quantization.

6.6 PROBABILITY OF ERROR
AND QUANTIZATION TRADE-OFF

Increasing the number of quantization intervals M dramatically decreases the quantization noise. Increasing M, however, increases the signal bandwidth, which increases the probability of error $P_b(e)$. Thus, while the quantization noise N_q and $P_b(e)$ can be analyzed separately, they interact, and a trade-off may be necessary. The "best" value of M depends on the communication technique. We consider PCM in this section.

The bandwidth needed for a PCM signal is

$$B = mf_s \log_2 M$$

where f_s is the symbol or sampling rate and the factor m is on the order 1. Solving for M, we obtain

$$M = 2^{B/mf_s} = \exp\left(\frac{\ln 2}{m}\frac{B}{f_s}\right)$$

and

$$\text{SNR}_q = cM^2 = c \exp\left(\frac{2\ln 2}{mf_s}B\right) \tag{6.54}$$

where, for uniform quantization, $c = 3\sigma_s^2/A^2$. This exponential growth with bandwidth is sketched in Figure 6.8. We need a method of characterizing $P_b(e)$ that shows the trade-off explicitly.

Let us call the mean square distortion due to decision errors the decision noise N_d. If the dynamic range of the digitizer is $2A$ V, as assumed in the previous sections, then an error in the first or most significant bit of an N-bit word causes a distortion of A V (a mean square distortion of A^2 V^2). Similarly, a decision error in the second bit causes a mean square distortion of $(A/2)^2$ V^2, and the ith bit a mean square distortion of $(A/2^{i-1})^2$ V^2. Because all N bits have the same probability of error $P_b(e)$, it follows that

$$\begin{aligned} N_d &= \sum_{i=1}^{N} \left(\frac{A}{2^{i-1}}\right)^2 P_b(e) \\ &= P_b(e)A^2\left[1 + \left(\frac{1}{2}\right)^2 + \left(\frac{1}{4}\right)^2 + \cdots + \left(\frac{2}{M}\right)^2\right] \\ &\approx \tfrac{4}{3}A^2 P_b(e) \end{aligned} \tag{6.55}$$

where the approximation is quite good if M is large. The SNR we seek is

$$\sigma_s^2/N_d = c/4P_b(e) \tag{6.56}$$

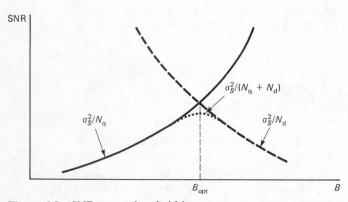

Figure 6.8 SNR versus bandwidth.

where $c = 3\sigma_S^2/A^2$ as before. We saw (Equation 6.22) that for bipolar pulses

$$P_b(e) = Q(\sqrt{S^2/\sigma_N^2}) < \tfrac{1}{2}e^{-S^2/2\sigma_N^2} \tag{6.57}$$

If the noise is white $[S_n(f) = \tfrac{1}{2}\eta_0 \ V^2/Hz]$ and the filter of Figure 6.1 is an ideal bandpass filter whose bandwidth is that of the signal (B), it follows from Equation 6.3 that

$$\sigma_N^2 = \eta_0 B \tag{6.58}$$

Combining the last three equations, we obtain

$$\frac{\sigma_S^2}{N_d} > \frac{c}{2}\exp\left(\frac{S^2}{2\eta_0}\frac{1}{B}\right) \tag{6.59}$$

This SNR decreases exponentially with the bandwidth B and is shown plotted as the dashed curve in Figure 6.8.

The combined or net SNR, $\sigma_S^2/(N_q + N_d)$ is shown as the dotted curve in Figure 6.8. For most values of B either the quantization noise N_q or the decision noise N_d dominates. The best or largest SNR is achieved when they are equal. This optimum bandwidth B_{opt} is determined by equating Equations 6.54 and 6.59. A similar situation exists for other communication techniques.

The preceding discussion neglects, however, the arguments of Section 3.7. Mathematically the mean square distortion due to quantization follows Equation 6.29 or 6.39, but for sufficiently large M the distortion may not be considered relevant. If the quantization noise is less than the measurement error of measured data, or less than an observable amount in a speech or music signal, it is not a real distortion. Two situations are shown in Figure 6.9, where N_m is the measurement noise, which does not depend on the bandwidth. When the situation 2 occurs, the best operating bandwidth is determined by quantizing "fine enough." It is only when the probability of decision error is relatively large, as in curve 1, that the question of optimizing the bandwidth by a trade-off between N_q and $P_b(e)$ arises.

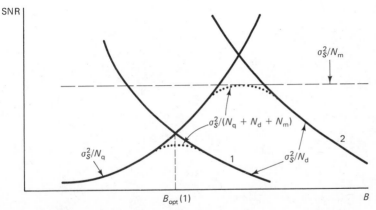

Figure 6.9 SNR (including measurement error) versus bandwidth.

6.7 REPEATERS

For most long-range communication schemes, hardware, called *repeaters*, is introduced along the communication channel to make up for the inevitable attenuation of signal power. For cw communications, these repeaters are amplifiers and filters. For discrete communications, however, the repeaters detect the signal pulses with the decision mechanism of a receiver and subsequently generate and transmit new noise-free pulses. Such a repeater, was shown for PSK in Section 3.11. This difference between repeaters represents a major advantage of discrete communications over cw techniques.

For convenience, assume that the repeaters are equidistant and that at the input to every repeater an independent zero-mean Gaussian noise of variance σ_N^2 is added to the signal. Suppose, first, that the repeaters are amplifiers whose voltage gain G exactly compensates for the attenuation between stages, as in Figure 6.10. At the input to the mth stage (or receiver) the noise power is $m\sigma_N^2$ since the components are all independent. If σ_S^2 is the signal power, the SNR becomes

$$\text{SNR} = \frac{1}{m}\frac{\sigma_S^2}{\sigma_N^2} \tag{6.60}$$

If the signal is a bipolar PCM signal and the last stage is a PCM receiver, then, from Equation 6.22,

$$P_b(e) = Q\left(\sqrt{\frac{1}{m}\frac{\sigma_S^2}{\sigma_N^2}}\right) < \tfrac{1}{2}\exp\left(-\frac{1}{2m}\frac{\sigma_S^2}{\sigma_N^2}\right) \tag{6.61}$$

If the signal is a PCM signal, however, the repeaters should be binary decision mechanisms followed by transmitters. Let P_m be the probability of a single pulse being in error after the mth stage, where

$$P_1 = Q(\sqrt{\sigma_S^2/\sigma_N^2}) \tag{6.62}$$

After two stages

$$P_2 = P_1(1 - P_1) + (1 - P_1)P_1 = 2P_1 - 2P_1^2$$

After m stages

$$P_m = P_1(1 - P_{m-1}) + (1 - P_1)P_{m-1}$$
$$= mP_1 + m(m - 1)P_1^2 - \cdots$$

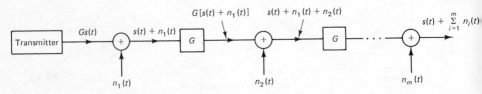

Figure 6.10 Compensating amplifiers.

Assuming $mP_1 \ll 1$ (otherwise the overall performance is inadequate), we have

$$P_m \approx mP_1 = mQ(\sqrt{\sigma_S^2/\sigma_N^2}) < \tfrac{1}{2}me^{-\sigma_S^2/2\sigma_N^2} \tag{6.63}$$

The performance of the amplifiers degrades exponentially with m (Equation 6.61), whereas the performance of the PCM repeaters degrades only linearly with m (Equation 6.63). This is a substantial difference.

EXAMPLE 6.5 ■

Compare two channels that have 100 repeaters such that the received SNR (σ_S^2/σ_N^2) is 25 for each stage. In one channel the repeaters are amplifiers only, while in the other channel they are PCM receivers and transmitters. Using decision-type repeaters, we obtain

$$P_1 = Q(5) \simeq \frac{1}{(\sqrt{2\pi})(5)}e^{-25/2} \simeq 3 \times 10^{-7}$$

After 100 stages

$$P_b(e) = P_{100} \simeq 100P_1 \simeq 3 \times 10^{-5}$$

Using amplifiers only,

$$P_b(e) = Q(\sqrt{25/100}) = Q(0.5) = 0.3085$$

This latter channel is virtually useless.   ■ ■

The use of decision devices followed by transmission of clean pulses is very efficient and represents a significant advantage of discrete communications relative to cw techniques.

6.8 MATCHED FILTERS

Let us now consider the linear filter of the PCM receiver shown in Figure 6.1. From Equations 6.22 and 6.23, we see that $P_b(e)$ decreases exponentially with the SNR (S^2/σ_N^2). It follows that the optimum filter should maximize this SNR at some appropriate sample time t_0. There is no advantage to preserving the shape of the signal.

The output of a causal linear filter at time t_0 as a result of the signal component $s(t)$ is given by

$$S = \int_0^\infty h(t)s(t_0 - t)\, dt \tag{6.64}$$

where $h(t)$ is the filter impulse response. The output signal power at t_0 is S^2. The output as a result of noise $n(t)$ at t_0 is

$$N(t_0) = \int_0^\infty h(t)n(t_0 - t)\, dt \tag{6.65}$$

The output noise power σ_N^2 is given by

$$\sigma_N^2 = \langle N^2(t_0) \rangle = \left\langle \int_0^\infty \int_0^\infty h(\xi)h(\eta)n(t_0 - \xi)n(t_0 - \eta) \, d\xi \, d\eta \right\rangle$$

$$= \int_0^\infty \int_0^\infty h(\xi)h(\eta)\langle n(t_0 - \xi)n(t_0 - \eta) \rangle \, d\xi \, d\eta$$

$$= \int_0^\infty \int_0^\infty h(\xi)h(\eta)R_n(\xi - \eta) \, d\xi \, d\eta \tag{6.66}$$

where $R_n(\tau)$ is the autocorrelation of the noise process $n(t)$. The output SNR is for time t_0:

$$\frac{S^2}{\sigma_N^2} = \left[\int_0^\infty h(t)s(t_0 - t) \, dt \right]^2 \Big/ \int_0^\infty \int_0^\infty h(\xi)h(\eta)R_n(\xi - \eta) \, d\xi \, d\eta \tag{6.67}$$

Let us define λ as the maximum SNR that is achievable among all possible filters. If there exists some filter $h_0(t)$ that achieves this maximum, it must satisfy

$$\lambda \int_0^\infty \int_0^\infty h_0(\xi)h_0(\eta)R_n(\xi - \eta) \, d\xi \, d\eta = \left[\int_0^\infty h_0(t)s(t_0 - t) \, dt \right]^2 \tag{6.68}$$

If we define $S_{max} = \int_0^\infty h_0(t)s(t_0 - t) \, dt$, or the largest possible signal output at the time t_0, then Equation 6.68 can be written

$$\lambda \int_0^\infty \int_0^\infty h_0(\xi)h_0(\eta)R_n(\xi - \eta) \, d\xi \, d\eta - S_{max} \int_0^\infty h_0(\xi)s(t_0 - \xi) \, d\xi = 0$$

or

$$\int_0^\infty h_0(\xi)\left[\lambda \int_0^\infty h_0(\eta)R_n(\xi - \eta) \, d\eta - S_{max}s(t_0 - \xi) \right] d\xi = 0 \tag{6.69}$$

A nontrivial solution[1] for $h_0(\eta)$—that is, a solution other than $h_0(\eta) = 0$—is obtained if $h_0(\eta)$ satisfies

$$\int_0^\infty h_0(\eta)R_n(\xi - \eta) \, d\eta = \frac{S_{max}}{\lambda}s(t_0 - \xi) \tag{6.70}$$

This equation is called a Wiener–Hopf equation. Unfortunately, it is not easy to solve in general.

Usually the spectrum of the noise is flat in the vicinity of the signal spectrum and a white noise model is valid. In this case, the Wiener–Hopf equation reduces to

$$h_0(\xi) = \begin{cases} Ks(t_0 - \xi) & \xi \geq 0 \\ 0 & \xi < 0 \end{cases} \tag{6.71}$$

where $K = 2S_{max}/\eta_0\lambda$. Actually, we shall soon see that the scale factor K can have any value without affecting the results. The solution of Equation 6.71 is

---

[1] A method known as the calculus of variations (see, for example, Hildebrand [1]) can be used to prove that this solution is the only possible nontrivial solution.

called a *matched filter* (MF) because of the way that the impulse response is related to the signal shape. For white noise Equation 6.67 becomes

$$S^2/\sigma_N^2 = \left[\int_0^\infty h(t)s(t_0 - t)\, dt \right]^2 \bigg/ \frac{\eta_0}{2} \int_0^\infty h^2(\xi)\, d\xi \tag{6.72}$$

Substituting the matched filter impulse response (Equation 6.71) yields

$$\lambda = K^2 \left[\int_0^\infty s^2(t_0 - t)\, dt \right]^2 \bigg/ \frac{\eta_0}{2} K^2 \int_0^\infty s^2(t_0 - \xi)\, d\xi$$

$$= \left[\int_{-\infty}^{t_0} s^2(\eta)\, d\eta \right]^2 \bigg/ \frac{\eta_0}{2} \int_{-\infty}^{t_0} s^2(\eta)\, d\eta$$

$$= \frac{2}{\eta_0} \int_0^{t_0} s^2(\eta)\, d\eta, \tag{6.73}$$

where the lower limit of the integral is changed to zero because the signal pulses are assumed to be zero for negative time. Observe that the scale factor K cancels and does not affect the SNR. If the pulse is of width T, then the sampling time which maximizes the SNR is at the trailing edge of the pulse or $t_0 = T$. At this sampling time,

$$S_{max} = \int_0^T s^2(t)\, dt = E_b \tag{6.74}$$

the total energy in the pulse, and

$$\frac{S^2}{\sigma_N^2}\bigg|_{max} = \frac{2}{\eta_0} E_b \tag{6.75}$$

It follows that the minimum probability of error for bipolar pulses is given by

$$P_{b(bip)}(e) = Q(\sqrt{2E_b/\eta_0}) < \tfrac{1}{2} e^{-E_b/\eta_0} \tag{6.76}$$

EXAMPLE 6.6 ■

Consider two types of bipolar pulses discussed in Chapter 3—RZ and NRZ (see Figure 3.1). Suppose that for the NRZ pulses $2E_b/\eta_0 = 25$. Then with matched filters in the receiver

$$P_b(e) = Q(\sqrt{25}) \simeq \frac{1}{(\sqrt{2\pi})(5)} e^{-25/2} \simeq 3 \times 10^{-7}$$

The energy per pulse E_b for the RZ pulses is smaller because of the zero-value interval between pulses. If this interval is 20% of the total time, then $2E_b/\eta_0 = 0.8(25) = 20$ for the RZ pulses, and

$$P_b(e) = Q(\sqrt{20}) \simeq \frac{1}{(\sqrt{2\pi})(20)} e^{-20/2} \simeq 4 \times 10^{-6}$$

The probability of error depends on the duration of the pulse as well as its amplitude.   ■ ■

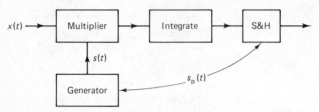

Figure 6.11   Matched filter implementation.

The response $y(t)$ of a linear filter to any input $x(t)$ that is zero for $t < 0$ is

$$y(t) = \int_0^t x(\xi)h(t - \xi)\,d\xi$$

For the matched filter $h(t) = Ks(T - t)$ for $t \geq 0$ and this becomes

$$y(t) = K \int_0^t x(\xi)s(T - t + \xi)\,d\xi$$

$$= K \int_0^t x(\xi)s[\xi - (t - T)]\,d\xi \qquad\qquad (6.77)$$

At time $t = T$

$$y(T) = K \int_0^T x(\xi)s(\xi)\,d\xi \qquad\qquad (6.78)$$

Equation 6.78 suggests a possible MF implementation, shown in Figure 6.11. Strictly, the sample and hold is not part of the MF but is needed for the output to be $y(T)$ rather than $y(t)$. When the input is signal only,

$$y(T) = S = K \int_0^T s^2(\xi)\,d\xi = KE_b \qquad\qquad (6.79)$$

When the input is zero-mean noise only, $y(T)$ is a zero-mean random variable whose variance σ_N^2 is

$$\sigma_N^2 = E\left\{\left[K \int_0^T n(\xi)s(\xi)\,d\xi\right]^2\right\}$$

$$= K^2 E\left\{\int_0^T \int_0^T n(\xi)n(\eta)s(\xi)s(\eta)\,d\xi\,d\eta\right\}$$

$$= K^2 \int_0^T \int_0^T s(\xi)s(\eta)E\{n(\xi)n(\eta)\}\,d\xi\,d\eta$$

since all $s(\xi)$ and $s(\eta)$ are deterministic. This now becomes[2]

---

[2]Observe that operating with expected values of random variables is much the same as taking the time average of the random signals that are being modeled.

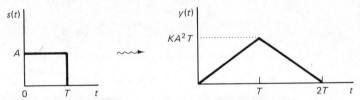

Figure 6.12 Matched filter response to a flat-topped pulse.

$$\sigma^2_N = K^2 \int_0^T \int_0^T s(\xi)s(\eta)R_n(\xi - \eta) \, d\xi \, d\eta$$

$$= K^2(\eta_0/2) \int_0^T \int_0^T s(\xi)s(\eta)\delta(\xi - \eta) \, d\xi \, d\eta \qquad \text{(for white noise)}$$

$$= K^2(\eta_0/2) \int_0^T s^2(\xi) \, d\xi = K^2(\eta_0/2)E_b \qquad (6.80)$$

As we saw, the SNR at $t = T$ is

$$\frac{S^2}{\sigma^2_N} = \frac{(KE_b)^2}{\frac{1}{2}K^2\eta_0 E_b} = \frac{2}{\eta_0}E_b$$

Let us look more carefully at the output without the sample and hold as a function of time. When the input is a signal pulse, Equation 6.77 becomes

$$y(t) = K \int_0^t s(\xi)s[\xi - (t - T)] \, d\xi \qquad (6.81)$$

The response to a flat-topped pulse is shown in Figure 6.12. Although the output at time T is KE_b, the response spreads into the next pulse, causing intersymbol interference if a sequence of decisions is needed, as in PCM. To avoid this problem, the integrator of Figure 6.11 must be replaced by an integrate and dump (I&D) circuit that discharges the integrator immediately after the output is sampled. The integrate and dump circuit requires pulse synchronization in order to know when to discharge and start integrating again. The MF response (with this modification) to an NRZ signal is shown in Figure 6.13.

When the noise is not white, it is generally assumed that the spectrum can be modeled as the spectrum of the response of a causal filter to white noise (see Section 2.9). Thus

$$S_n(f) = K|H(f)|^2 \qquad (6.82)$$

where $H(f)$ is a causal filter. If $1/H(f)$ is realizable,[3] it can be regarded as a prewhitening filter. If the input to this filter has the spectrum $S_n(f)$, the spectrum of its response is

---

[3] If $S_n(f)$ can be modeled by a ratio of polynomials in f^2, then a filter can always be found that satisfies Equation 6.82 whose inverse $1/H(f)$ is causal. This procedure, called spectral factorization, is discussed in many texts [2–5].

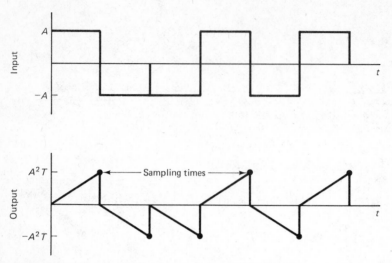

Figure 6.13 Matched filter response to an NRZ signal.

$$S_y(f) = S_n(f)\frac{1}{|H(f)|^2} = K \tag{6.83}$$

Hence the output is white. The signal $s(t)$ is changed by this prewhitening filter to a different signal $\hat{s}(t)$, and the problem has been reduced to that of detecting a known signal $\hat{s}(t)$ in white noise, a problem we have already solved. An implementation of the general matched filter is shown in Figure 6.14.

For this receiver to work for pulse sequences, such as PCM, the prewhitening filter must not cause intersymbol interference. If it does, the matched filter is a more complex, noncausal filter that can only be approximately realized. Throughout the rest of this text, however, we shall assume white noise, which does not require prewhitening filters.

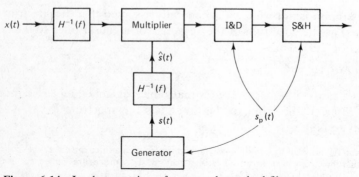

Figure 6.14 Implementation of a general matched filter.

6.9 COMPARISON OF MFs
WITH SUBOPTIMAL FILTERS

We first consider near-ideal low-pass filters that are sufficiently wide band to ensure that the shape of the pulses is not distorted, so that a sample in the middle of a pulse of amplitude $\pm A$ V is itself $\pm A$ V. If the pulse rate is f_p (or $1/T$) pulses/s, the bandwidth of such a filter is $B = mf_p = m/T$, where normally $m \geq 1$. If the input noise is white, so that $S_n(f) = \frac{1}{2}\eta_0$ V^2/Hz, the noise variance after the filter (Equation 6.3) is

$$\sigma_N^2 = \eta_0 B = \eta_0 m/T \tag{6.84}$$

The SNR for a sample of the output is

$$\frac{A^2}{\sigma_N^2} = \frac{A^2 T}{\eta_0 m} = \frac{1}{2m}\frac{2}{\eta_0}E_b \tag{6.85}$$

since $A^2 T$ is the energy in a flat-topped pulse of amplitude $\pm A$ V. Thus the relationship between the SNRs of the ideal low-pass filter (ILPF) and the MF is

$$\text{SNR}_{\text{ILPF}} = \frac{1}{2m}\text{SNR}_{\text{MF}} \tag{6.86}$$

For $m = 2$, for example, the SNR resulting from the matched filter is four times as large, or 6 dB better.

In Chapter 3 we considered the narrow-band sinc-shaped pulses, which, if sampled precisely at the pulse peaks, cause no intersymbol interference. In the frequency domain these signals are completely contained in a bandwidth of $\frac{1}{2}f_p$ and are used for bandwidth compression. All the assumptions in deriving Equation 6.85 are satisfied for this signal when $m = \frac{1}{2}$. Hence this minimum bandwidth signal achieves the maximum SNR, $(2/\eta_0)A^2 T$, in additive white noise. It should be remembered, however, that this signal is very sensitive to precise pulse shaping and pulse synchronization.

Let us now return to the flat-topped pulses and replace the MFs with first-order low-pass filter. For these filters the impulse response is $h(t) = (1/\tau)e^{-t/\tau}u(t)$ and the transfer function is $H(f) = 1/(1 + j2\pi f\tau)$, where τ is the time constant. The response of such a filter to a single flat-topped pulse of amplitude A and width T is

$$y(t) = \begin{cases} A(1 - e^{-t/\tau}) & t \leq T \\ A(1 - e^{-t/\tau})e^{-(t-T)/\tau} & t > T \end{cases} \tag{6.87}$$

As for MFs, the appropriate sampling time is at the trailing edge of the pulse, $t_0 = T$ (see Figure 6.15b).

Furthermore, as was the case with MFs, intersymbol interference occurs (see Figure 6.15b) unless the filter output is suddenly discharged immediately after it is sampled (see Figure 6.15c). Observe that, with discharging, the output looks similar to the MF output of Figure 6.13.

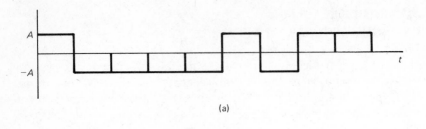

(a)

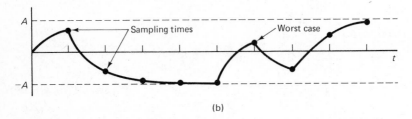

(b)

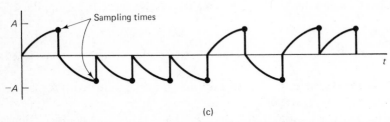

(c)

Figure 6.15   Response of first-order LPFs: (a) a NRZ signal, (b) the first-order LPF, and (c) the first-order LPF with sudden discharging.

The signal sampled values are found from Equation 6.87 with $t = T$:

$$S = y(T) = A(1 - e^{-T/\tau}) \tag{6.88}$$

where discharging (no intersymbol interference) is assumed. The output noise power is given by

$$\sigma_N^2 = \int_{-\infty}^{\infty} \frac{\eta_0/2}{1 + (2\pi f \tau)^2} \, df = \frac{\eta_0}{4\tau} \tag{6.89}$$

It follows that the SNR is

$$\frac{S^2}{\sigma_N^2} = \frac{A^2(1 - e^{-T/\tau})^2}{\eta_0/4\tau} = \frac{2}{\eta_0} A^2 T \frac{2(1 - e^{-x})^2}{x} \tag{6.90}$$

where $x = T/\tau$. This SNR is maximized (see Problem 6.19) when $x \approx 1.25$ or $\tau = 0.8T$. For this time constant,

$$\frac{S^2}{\sigma_N^2} = 0.814 \frac{2}{\eta_0} E_b \tag{6.91}$$

which corresponds to a cost of only 0.89 dB ($-10 \log_{10} 0.814$) relative to the MF.

When discharging is not used, the worst case occurs when the output is building to A V from an initial value of $-A$ V (see Figure 6.15b). For this case $S = A(1 - 2e^{-T/\tau})$, and, using this worst case analysis, we have

$$\frac{S^2}{\sigma_N^2} = \frac{2}{\eta_0}A^2 T\frac{2(1 - 2e^{-x})^2}{x} \leq 0.56\frac{2}{\eta_0}E_b \qquad (6.92)$$

where the maximum is achieved when $x \approx 2.5$ or $\tau \approx 0.4T$. For this case the cost is 2.5 dB, which means that 1.79 ($1/0.56$) times more signal power is needed to achieve the same $P_b(e)$ as would be achieved with an MF.

6.10 MFs FOR ASK AND PSK

Most of the results of this chapter apply directly to the modulated binary techniques of ASK and PSK. Equations 6.22 and 6.23 are general in that they do not depend on the shape of the signal pulses. Recognizing that ASK is monopolar and PSK is bipolar, we may conclude that

$$P_{b(PSK)}(e) = Q(\sqrt{S^2/\sigma_N^2}) < \tfrac{1}{2}e^{-S^2/2\sigma_N^2} \qquad (6.93)$$

and

$$P_{b(ASK)}(e) = Q(\sqrt{S^2/4\sigma_N^2}) < \tfrac{1}{2}e^{-S^2/8\sigma_N^2} \qquad (6.94)$$

Assuming similar hardware, ASK requires four times (6 dB) more signal power to achieve the same $P_b(e)$ as PSK. To implement matched filters it is necessary to generate a replica of the signal pulses $s(t)$. For the sinusoidal pulses of ASK and PSK this requires accurate phase synchronization. An optimum receiver, assuming white noise, for both ASK and PSK is shown in Figure 6.16, where only the comparator threshold voltage is different. For sinusoidal pulses of amplitude A V and of T-s duration,

$$E_b(\text{sinusoidal}) = \tfrac{1}{2}A^2 T \qquad (6.95)$$

For the matched filters, as implemented in Figure 6.16, Equations 6.93 and 6.94 become

$$P_{b(PSK)}(e) = Q\left(\sqrt{\frac{2}{\eta_0}E_b}\right) < \tfrac{1}{2}e^{-E_b/\eta_0} \qquad (6.96)$$

and

$$P_{b(ASK)}(e) = Q\left(\sqrt{\frac{1}{2\eta_0}E_b}\right) < \tfrac{1}{2}e^{-E_b/4\eta_0} \qquad (6.97)$$

The matched filter of Figure 6.16 looks very much like the PSK demodulating modem of Figure 3.40 except the LPF is replaced by an I&D

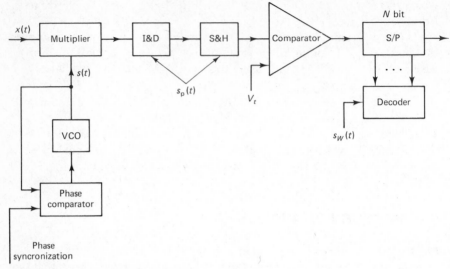

Figure 6.16 Optimal ASK and PSK receivers.

circuit. We saw in Section 6.9 that replacing the I&D with a simple low-pass filter can cause less than 1 dB degradation provided the filter output is discharged after every decision. Furthermore, note that the multiplier can be replaced with a mixer for ASK or PSK and a gate for baseband PCM. As discussed in Chapter 3, for ASK the matched filters can be replaced with envelope detectors that do not require phase synchronization. These devices are discussed in more detail in Chapter 8, where it is seen that, depending on how the envelope detectors are implemented, the cost relative to MFs is somewhere between 0 and 3 dB. This option is not available, however, for the more efficient PSK; DPSK, discussed briefly in Chapter 3, permits a method of self-phase-synchronization (see Figure 3.41) at a cost of 1–2 dB [6] relative to PSK.

The interaction between phase and pulse synchronization can be studied with the help of Equation 6.81, which gives the output of an MF (without sample and hold and for signal only) as a function of time. The response to a flat-topped pulse (Figure 6.12) gives some indication of the sensitivity of pulse synchronization for baseband PCM. If the phase is known accurately a priori and phase synchronization is not used, Equation 6.81 becomes for sinusoidal pulses

$$y_1(t) = A^2 \int_0^t \sin 2\pi f\xi \, \sin\{2\pi f[\xi - (t - T)] + \theta\} \, d\xi \tag{6.98}$$

where θ is a possible small phase error. If, on the other hand, phase synchronization is used (as in Figure 6.16), the output becomes

$$y_2(t) = A^2 \int_0^t \sin 2\pi f\xi \, \sin(2\pi f\xi + \theta) \, d\xi \tag{6.99}$$

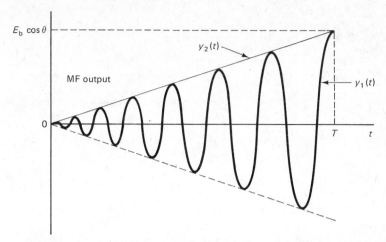

Figure 6.17 Output of MF with and without phase synchronization.

If we let $f = k/T$ (i.e., if there are an integral number of cycles per pulse), then for $t < T$ Equation 6.98 becomes, after straightforward calculations,

$$y_1(t) = E_b \frac{t}{T} \cos\left[2\pi k\left(1 - \frac{t}{T}\right) + \theta\right] + E_b \frac{\sin 2\pi f T(1 - t/T)}{2\pi k} \cos \theta$$

$$\approx E_b \frac{t}{T} \cos\left[2\pi k\left(1 - \frac{t}{T}\right) + \theta\right] \tag{6.100}$$

where the second term is small if k is large ($k > 5$) and $E_b = \frac{1}{2}A^2 T$. Similarly,

$$y_2(t) = E_b \frac{t}{T} \cos \theta + E_b \frac{\sin[4\pi k(1 - t/T) + \theta] - \sin \theta}{4\pi k}$$

$$\approx E_b \frac{t}{T} \cos \theta \tag{6.101}$$

for large k. Both of these outputs (with and without phase synchronization) are plotted in Figure 6.17. We see that even if the phase is known a priori without phase synchronization, pulse synchronization becomes a problem. For example, for $k > 8$ the sampling time must be accurate to better than 1%. Such accuracy is not needed when phase synchronization is used.

6.11 MFs FOR FSK

It will be shown in Chapter 7 that the FSK receiver of Figure 6.18 is the optimum receiver. Comparing this receiver with the receiver of Chapter 3 (Figure 3.44) makes the analogy with ASK evident. If the signal frequencies are sufficiently different, the MFs can be replaced with BPFs and envelope detectors with a reduction in SNR between 0 and 3 dB. If the MFs are used

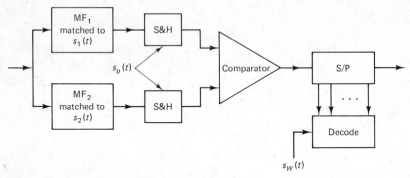

Figure 6.18   Optimum FSK receiver.

with separate signals for phase synchronization, however, the difference between the signal frequencies can be very small.

Let us replace the decision mechanism of the FSK receiver with the equivalent hardware of Figure 6.19. If both signals $s_1(t)$ and $s_2(t)$ have the same energy E_b, the output at the correct sampling time T is either $+KE_b$ [if $s_1(t)$ is present] or $-KE_b$ [if $s_2(t)$ is present]. Thus the input to the comparator is a sequence of bipolar pulses as in PSK. The noise input N is $n_1(T) - n_2(T)$, where $n_1(T)$ and $n_2(T)$ are both zero-mean Gaussian with variance $\frac{1}{2}K^2\eta_0 E_b$ (Equation 6.80). Hence N is zero-mean Gaussian with variance (Equation 5.58)

$$\sigma_N^2 = 2\left(\tfrac{1}{2}K^2\eta_0 E_b\right) + 2\left(\tfrac{1}{2}K^2\eta_0 E_b\right)\rho \tag{6.102}$$

where ρ is the correlation coefficient between $n_1(T)$ and $n_2(T)$. From Example 5.15, however, if the filters do not overlap, then $\rho = 0$. Hence $\sigma_N^2 = K^2\eta_0 E_b$ $(S^2/\sigma_N^2 = E_b/\eta_0)$ and

$$P_{b(FSK)}(e) = Q(\sqrt{E_b/\eta_0}) < \tfrac{1}{2}e^{-E_b/2\eta_0} \tag{6.103}$$

If we compare this result with Equations 6.96 and 6.97, we see that FSK requires 3 dB more energy than PSK but 3 dB less energy than ASK to achieve the same probability of error.

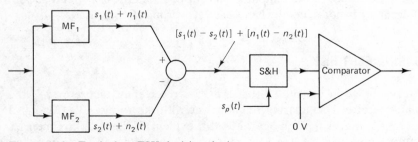

Figure 6.19   Equivalent FSK decision device.

EXAMPLE 6.7 ■

Let us compare the use of ASK with FSK given the following constraints: peak power (at receiver) $= 12 \text{ V}^2$, $f_p = 10^5$ Hz, $S_n(f) = 10^{-6}$ V^2/Hz. In both cases $E_b = \frac{1}{2}A^2T = \frac{1}{2} \times 12 \times 10^{-5} = 6 \times 10^{-5} \text{ V}^2 \cdot \text{s}$. Therefore, $(2/\eta_0)E_b = 6 \times 10^{-5} \times 10^6 = 60$.

1. Using matched filters

$$P_{b(\text{ASK})}(e) = Q\left(\sqrt{\frac{1}{4}\frac{2E_b}{\eta_0}}\right) = Q(\sqrt{15}) \approx \frac{1}{\sqrt{(2\pi)(15)}}e^{-15/2}$$

$$\approx 6 \times 10^{-5}$$

$$P_{b(\text{FSK})}(e) = Q\left(\sqrt{\frac{1}{2}\frac{2E_b}{\eta_0}}\right) = Q(\sqrt{30}) \approx \frac{1}{\sqrt{(2\pi)(30)}}e^{-30/2}$$

$$\approx 2 \times 10^{-8}$$

$$B_{\text{ASK}} = 2mf_p > 4 \times 10^5 \text{ Hz} \qquad (m > 2 \text{ to preserve pulse shape})$$

B_{FSK} is slightly larger.

Other advantages are that ASK requires that only one MF be implemented and that FSK is less susceptible to unknown variations in the signal strength (fading), because the comparator threshold (see Figure 6.18 or 6.19) does not depend on the signal strength.

2. By use of suboptimum BPFs and envelope detectors, the SNR (both ASK and FSK) could be degraded by up to 3 dB, depending on implementation of envelope detectors:

$$B_{\text{ASK}} = 2mf_p = 4 \times 10^5 \text{ Hz} \qquad (\text{assuming } m = 2)$$
$$B_{\text{FSK}} \geq 4mf_p \geq 8 \times 10^5 \text{ Hz} \qquad (\text{for BPFs not to overlap}) \qquad ■ ■$$

6.12 TIME SCALING

For all the binary techniques, when matched filters are used,

$$P_b(e) = Q(\sqrt{2KE_b/\eta_0}) \tag{6.104}$$

where $K = 1$ for baseband PCM and PSK, $\frac{1}{2}$ for FSK, and $\frac{1}{4}$ for ASK. The energy of the pulses can be increased either by increasing the power with a more powerful transmitter or by increasing the pulse width T. It follows that if one has the capability of storage at both transmitter and receiver, the pulses could be transmitted at a very slow rate, thereby making $P_b(e)$ as small as desirable. This procedure would introduce a significant time delay. Given no system real-time constraint, this results in an extremely efficient encoding/

decoding system. In fact, since it is relatively easy to store binary data, it is hard to imagine a better technique than PCM or some modulated version in the absence of a real-time constraint.

EXAMPLE 6.8 ■

In Chapter 1 we briefly discussed the communication problem of transmitting pictures from a satellite passing Mars. We are now in position to explain the techniques that were used. Using a grid of samples (200 × 200) and quantizing each sample into 64 levels allowed the analog picture to be converted to a discrete signal. The degree of sampling and quantization in this example is determined subjectively. It was decided to use a binary technique (six pulses per sample) and to use sufficient time scaling to result in an acceptable probability of error. Considering the time it took for the satellite to reach Mars and the time for the transmission of electromagnetic waves to Earth, there is clearly no reason whatsoever to impose a real-time constraint. Under these conditions there is no point to consider any technique other than some form of modulated PCM. Because the overall cost of the project is so great, there is virtually no cost constraint on the receiver. Furthermore, with the time scaling used, the pulse rate was quite small ($f_p < 10$ pulse/s), and maintaining accurate pulse and phase synchronization was not difficult. Under these conditions the logical choice is PSK with a receiver similar to Figure 6.16.   ■ ■

6.13 CONCLUSIONS

In this chapter the receiver structure of Figure 6.1 (or Figure 6.18 for FSK) was assumed and the filter and decision rule components optimized. We shall verify in Chapter 7 that the structure is optimum, and the optimum receiver for any signal set will be determined. In Chapter 7 we shall also find a geometric visualization and determine a general formula for the probability of error for any signal set. This will allow us to verify in an elegant way the main results of this chapter (Equations 6.22, 6.23, 6.76, 6.96, 6.97, and 6.103). We shall also be in position to compare the binary communication techniques with multilevel techniques such as QPSK and FPM.

As pointed out in Chapter 3, even when modulated and multilevel techniques are used, quantization (and time division multiplexing or time scaling when needed) is always carried out in connection with baseband PCM. Because of this, there will be no reason to reconsider quantization analysis and companding in connection with multilevel techniques.

PROBLEMS

6.1. A stationary signal is to be transmitted using a baseband PCM system that employs NRZ pulses. The additive noise can be modeled by a zero-mean

Gaussian PDF. Assume that the SNR at the receiver comparator (i.e., the ratio of S^2/σ_N^2, where S is the value of a pulse sample) is 6.

(a) Estimate the probability that a pulse is in error.

(b) Repeat if the SNR is 16.

6.2. Repeat Problem 6.1, assuming that the noise is a zero-mean impulse noise whose PDF model is that of Example 6.2.

6.3. Sixteen analog messages are to be time division multiplexed and then transmitted in real time over a cable using PCM with bipolar pulses. The pulses are not flat topped but rather sinc-shaped by filters whose rolloff factor $r = 0.2$. Each signal has a bandwidth of 4.5 kHz, is quantized uniformly into 256 levels, and is sampled at the Nyquist rate. At the front end of the receiver (which is an ideal low-pass filter), the peak value of the signal pulses is $+12$ V and the additive noise is white with spectral level $\eta_0/2 = 10^{-6}$ V^2/Hz.

(a) Determine the pulse rate going over the cable.

(b) What is the required bandwidth of the cable (and the ideal LPF of the receiver)?

(c) Calculate $P_b(e)$.

(d) Calculate $P_w(e)$.

(e) Sketch in block diagram form the receiver hardware; indicate where synchronization is needed.

6.4. Suppose the signals of Problem 6.3 are all uniformly distributed within ± 7 V. Determine the SNR due to quantization.

6.5. We wish to quantize a continuous signal that can be modeled by the PDF

$$f_S(s) = \begin{cases} \dfrac{1}{4(1 - e^{-2})} e^{-|s|/2} & |s| \leq 4 \\ 0 & |s| > 4 \end{cases}$$

(a) What is the signal power σ_S^2?

(b) Assuming equal width quantization and $M = 8$, determine SNR$_q$.

(c) What is c in the expression SNR$_q = cM^2$?

(d) Assuming c does not vary with M, how large must M be in order to achieve an SNR due to quantization of 40 dB?

6.6. We wish to transmit an analog signal using a baseband PCM code employing seven flat-topped bipolar pulses per sample. The signal has a bandwidth of 15 kHz and can be modeled by a zero-mean uniform PDF. The noise is white with a two-sided spectral level of 3×10^{-6} V^2/Hz.

(a) Estimate the SNR due to quantization assuming uniform quantization is used.

(b) Estimate the bandwidth of the encoded signal (assume $m = 1$).

(c) Assuming that the receiver filter is an ideal LPF with a bandwidth just wide enough to pass the signal, what is the received noise power?

(d) Estimate the probability that a pulse is in error, assuming ± 5 V pulses.

(e) What is the probability that a sample of the analog signal is in error?

6.7. You are to design a baseband PCM system using an NRZ signal that has the smallest possible probability of error. You have the following constraints:

1. The filter in the receiver is an ideal LPF whose bandwidth is twice the pulse rate.

2. The average received signal power is 40 V^2.
3. The received noise is white with spectral level of 2.5 × 10^{-6} V^2/Hz.
4. The original message is stationary, zero mean, and Gaussian with a bandwidth of 20 kHz.
5. SNR due to quantization must be at least 30 dB using uniform quantization.

    (a) Determine the number of quantization levels. (*Hint:* Use Table 6.1 to determine the largest value for c.)

    (b) Determine the bandwidth of your NRZ signal and the input noise power.

    (c) Calculate the probability that a sample of the original message will be in error.

6.8. Repeat Problem 6.7 if the SNR due to quantization must be at least 30 dB for a dynamic range of ±17.3 standard deviations. (Assume uniform quantization.)

6.9. Referring to Problem 6.8, determine which effect (the quantization noise N_q or the decision noise N_d) is dominant.

6.10. Evaluate the SNR due to quantization (σ_s^2/N_q) for the double-exponential PDF of Example 6.2 from Equations 6.42 and 6.43. Verify that

$$\text{SNR}_q = \frac{3}{k^2} M^2 \left/ \left\{ 1 + \frac{3M^2}{k^2} \left[1 + \frac{\sqrt{2}k}{M} + \frac{1}{3}\left(\frac{\sqrt{2}k}{M}\right)^2 \right] e^{-\sqrt{2}k} \right\} \right.$$

6.11. Compute $E\{|s|\}$ for the double-exponential PDF and compare the result with that of Equation 6.52 for Gaussian signals. Evaluate Equation 6.51 for this impulse PDF model.

6.12. Your goal is to design a system using μ-law companding ($\mu = 255$) that achieves an SNR due to quantization of at least 30 dB with a dynamic range of 173.2 standard deviations. How many bits per sample are required?

6.13. You are asked to design a quantizer that achieves at least 60 dB SNR when $3\sigma_s^2/A^2$ is −20 dB.

    (a) Determine the number of bits needed for uniform quantization and for μ-law companding ($\mu = 255$).

    (b) For both answers to part (a), determine the SNR when $3\sigma_s^2/A^2$ is −40 dB and −60 dB.

6.14. **(a)** Repeat Problem 6.8 under the assumption that μ-law companding ($\mu = 255$) is used.

    (b) Determine the received signal power that would be needed to achieve the same $P_w(e)$ as in Problem 6.8.

6.15. Reconsider Problem 6.6.

    (a) Calculate the probability of a sample being in error if matched filters are used.

    (b) Repeat if the pulse rate is slowed by a factor of 10.

6.16. Reconsider Problem 6.7 with an optimum filter replacing the ideal LPF.

    (a) Calculate the probability $P_w(e)$ of a sample being in error.

    (b) Repeat part (a) assuming that the signal and noise levels are at the input to repeaters and there are 12 decision devices (11 repeaters and one receiver) in the transmission path.

6.17. We wish to implement an optimum receiver for a PCM system employing the bipolar raised-cosine-shaped pulses of Figure 3.1. Assume a peak amplitude A, a pulse width T, and white noise (spectral level $\eta_0/2$).
(a) Determine an expression for $P_b(e)$ in terms of the given parameters.
(b) Sketch an implementation of the optimum receiver that is very similar to the PSK receiver of Figure 6.16.

6.18. In an effort to confound the enemy, a military PCM scheme (six bits per letter) is employed using a bipolar scheme, i.e., $s_i(t) = \pm s(t)$, where $s(t)$ is shown below:

(a) How would you implement an MF in your receiver? (Sketch a block diagram only.)
(b) If $E_b = \int_0^T s^2(t)\, dt = 2 \text{ V}^2 \cdot \text{s}$ and $\eta_0/2 = 10^{-5} \text{ V}^2/\text{Hz}$, what is the probability of error per letter?

6.19. A flat-topped pulse of width T, along with white noise, is the input to a first-order LPF with time constant τ. The output SNR is given by Equation 6.90. Determine the time constant which maximizes the SNR and verify Equation 6.91. (*Hint:* The equation is transcendental and must be solved by trial-and-error methods.)

6.20. Assume that the same flat-topped pulse of width T is the input to a finite-memory integrator $[y(t) = \int_{t-\tau}^t x(\xi)\, d\xi]$.
(a) What value of τ maximizes the SNR?
(b) Assuming the integrator is discharged at $t = T$, sketch the output (signal only) as a function of time for $\tau = T$ and $\tau = 0.8T$.

6.21. Twelve signals are to be transmitted by sampling, each at a 20-kHz rate, quantized into 256 levels, and encoded into a PCM code using NRZ signals. These encoded signals are time division multiplexed and then modulated by a modem into one of three possible signals: ASK, PSK, and FSK.
(a) Determine the bandwidth required for each scheme. Assume that $m = 2$ and that for FSK the difference between the tones is $2mf_p$.
(b) Assuming optimum receivers, determine $P_b(e)$ for each scheme in terms of E_b and $\eta_0/2$.
(c) Determine E_b if the signal amplitude is 10^{-2} V.
(d) Indicate a few ways E_b can be increased while still communicating in real time.

6.22. Consider the problem of transmitting data from a computer on board a satellite to some ground station by PSK. The pulse rate is 500 kHz and the peak value of the received sinusoidal pulses is 0.1 V. The signals are imbedded in white zero-mean Gaussian noise whose spectral level is 10^{-9} V^2/Hz. The receiver employs MFs.
(a) Determine the energy of one of the received signal pulses.
(b) Calculate $P_b(e)$.

(c) Recalculate $P_b(e)$ assuming the signal is slowed down, by time scaling, to a pulse rate of 100 kHz.

6.23. Consider the task of communicating multiplexed speech signals via satellite using PSK. The satellite, which acts like a repeater, is situated halfway between the transmitter and receiver. Assume that the received sinusoidal pulses (both at the satellite and receiver) have a peak value of 0.2 V and the pulse rate is 200 kHz. The noise is white zero-mean Gaussian noise with spectral level of 10^{-9} V^2/Hz at the satellite and 10^{-8} V^2/Hz at the receiver. Both satellite and receiver employ matched filters.
(a) Determine the energy of one of the received signal pulses (same at satellite and receiver).
(b) Calculate $P_b(e)$ at the satellite and at the receiver.
(c) Calculate the probability of error per pulse for the combined system.

6.24. Assume that a matched filter is implemented for a T-s pulse modeled by $A \sin(100\pi/T)t$. Determine in the following case the output of the MF if the pulse synchronization is off by $T/200$ s:
(a) No phase-locked loop is used but the phase error is only 10°, and
(b) A phase-locked loop is used ($\theta \simeq 0°$).

6.25. In Example 6.8 we discussed the problem of communicating a picture via the Mariner satellite. Suppose a picture is taken every minute as the satellite passes near Mars.
(a) Determine $P_b(e)$ if the pictures are communicated in real time (i.e., for a 1-min duration signal) and if the signal peak value at the receiver is 1 mV and the noise spectral level is 0.25×10^{-8} V^2/Hz.
(b) The signals are time scaled so that it takes 8.5 h to transmit one picture. Compute $P_b(e)$, assuming the same values for signal and noise power.

REFERENCES

1. F. B. Hildebrand, *Methods of Applied Mathematics*, Prentice-Hall, Englewood Cliffs, NJ, 1952.
2. A. Papoulis, *Probability, Random Variables, and Stochastic Processes*, McGraw-Hill, New York, 1965.
3. W. B. Davenport and W. L. Root, *An Introduction to the Theory of Random Signals and Noise*, McGraw-Hill, New York, 1958.
4. J. B. Thomas, *An Introduction to Statistical Communication Theory*, Wiley, New York, 1969.
5. E. Wong, *Introduction to Random Processes*, Springer-Verlag, New York, 1983.
6. J. G. Proakis, *Digital Communications*, McGraw-Hill, New York, 1983.

Chapter 7

Discrete Communication Theory

7.1 INTRODUCTION

In this chapter we consider the task of detecting one of M signals imbedded in additive, white, Gaussian noise. We are concerned with receiver hardware as well as error analysis. The analysis approach,[1] which is largely geometric, will enable us to visualize and evaluate the discrete encoding schemes discussed in Chapter 3 as well as the optimum (in the sense of minimum probability of error) encoding scheme.

7.2 FOURIER SERIES REPRESENTATION OF RANDOM SIGNALS

In Chapter 2 we saw that deterministic signals that repeat every T s can be represented by a Fourier series:

$$x(t) = \frac{a_0}{2} + \sum_{n=1}^{\infty} \left(a_n \cos \frac{n2\pi}{T} t + b_n \sin \frac{n2\pi}{T} t \right) \tag{7.1}$$

where the series is said to converge in the mean. By the orthogonality of sinusoids (Equations 2.35–2.37), the coefficients were seen to be found from

$$a_n = \frac{2}{T} \int_0^T x(t) \cos \frac{2\pi}{T} nt \, dt \qquad b_n = \frac{2}{T} \int_0^T x(t) \sin \frac{2\pi}{T} nt \, dt \tag{7.2}$$

We can represent with a Fourier series any deterministic signal that is bounded and continuous over a T-s interval $(0, T)$ by extending the signal

---

[1] This approach is based on that of Wozencraft and Jacobs [1].

periodically. Equation 7.1 is understood to apply only over the interval $0 < t < T$. If the signal is strictly bandlimited, the Fourier series can be truncated to

$$x(t) = \frac{a_0}{2} + \sum_{n=1}^{f_{max}T} \left(a_n \cos n\frac{2\pi}{T}t + b_n \sin n\frac{2\pi}{T}t \right) \tag{7.3}$$

for signals that are said to be low pass, and

$$x(t) = \sum_{n=f_{min}T}^{f_{max}T} \left(a_n \cos n\frac{2\pi}{T}t + b_n \sin n\frac{2\pi}{T}t \right) \tag{7.4}$$

for signals that are said to be bandpass. Thus a bandlimited signal can be represented by $2f_{max}T + 1$ coefficients if it is low pass and by $2(f_{max} - f_{min})T + 2$ coefficients if it is bandpass.[2] Let us define the bandwidth B by

$$B = \begin{cases} f_{max} + 1/2T & \text{for low-pass signals} \\ f_{max} - f_{min} + 1/T & \text{for bandpass signals} \end{cases} \tag{7.5}$$

A justification is given in Figure 7.1. With this definition we can say that a strictly bandlimited signal can be represented by $2BT$ coefficients.

For analytic convenience we can normalize the sinusoidal functions by defining

$$\phi_m(t) = \begin{cases} \dfrac{1}{\sqrt{T}} & m = 0 \\ \sqrt{\dfrac{2}{T}} \cos \dfrac{2\pi}{T} \dfrac{(m+1)}{2} t & m \text{ odd} \\ \sqrt{\dfrac{2}{T}} \sin \dfrac{2\pi}{T} \dfrac{m}{2} t & m \text{ even} \end{cases} \tag{7.6}$$

With this definition, the orthogonality principle of sinusoids becomes

$$\int_0^T \phi_n(t)\phi_m(t)\, dt = \delta_{nm} = \begin{cases} 1 & m = n \\ 0 & \text{otherwise} \end{cases} \tag{7.7}$$

and the $\phi_m(t)$ functions are said to be an *orthonormal* set of basis functions. Equations 7.1 and 7.2 can now be written compactly:

$$x(t) = \sum_{m=0}^{\infty} c_m \phi_m(t) \tag{7.8}$$

where

$$c_m = \int_0^T x(t)\phi_m(t)\, dt \tag{7.9}$$

Can Equations 7.6–7.9 be used to represent a random noiselike signal over some T-s interval? For such signals, Equation 7.9 defines not constant

---

[2]The coefficients represent the signal in the sense that the signal can be uniquely determined from them through Equation 7.1.

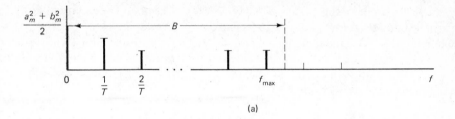

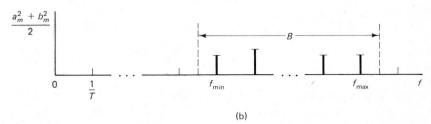

Figure 7.1 Bandwidth definitions for strictly bandlimited signals: (a) LP and (b) BP signal average-power distributions.

coefficients c_m but random variables N_m whose value n_m is determined from

$$n_m = \int_0^T n(t)\phi_m(t)\, dt \tag{7.10}$$

where $n(t)$ is a zero-mean Gaussian process. The N_m random variables can be interpreted, using Equation 7.10, as a sample of the output of a linear operation on the process $n(t)$. If $n(t)$ is an ergodic Gaussian process, it follows from property 5.3 of Gaussian processes that N_m is a Gaussian random variable. The relevant moments of N_m are

$$E\{N_m\} = \int_0^T E\{n(t)\}\phi_m(t)\, dt = 0 \tag{7.11}$$

and

$$\sigma_m^2 = E\{N_m^2\} = \int_0^T \int_0^T E\{n(t_1)n(t_2)\}\phi_m(t_1)\phi_m(t_2)\, dt_1\, dt_2$$

$$= \int_0^T \int_0^T R_n(t_2 - t_1)\phi_m(t_1)\phi_m(t_2)\, dt_1\, dt_2 \tag{7.12}$$

Similarly, the joint moments of the N_m random variables are determined from

$$E\{N_m N_k\} = \int_0^T \int_0^T E\{n(t_1)n(t_2)\}\phi_m(t_1)\phi_k(t_2)\, dt_1\, dt_2$$

$$= \int_0^T \int_0^T R_n(t_2 - t_1)\phi_m(t_1)\phi_k(t_2)\, dt_1\, dt_2 \tag{7.13}$$

Any sample of the model, $\sum_{m=1}^{\infty} N_m \phi_m(t_0)$, is a sum of Gaussian random variables and therefore is also a Gaussian random variable. Thus for ergodic Gaussian processes the model also has a Gaussian PDF. From property 5.1 of Gaussian processes, the model is statistically identical to the process $n(t)$ if the autocorrelation function of the model is equal to the autocorrelation function of the process. Thus, if

$$R_n(t_2 - t_1) = E\left\{ \sum_{m=0}^{\infty} \sum_{k=0}^{\infty} N_m N_k \phi_m(t_1)\phi_k(t_2) \right\}$$

$$= \sum_{m=0}^{\infty} \sum_{k=0}^{\infty} E\{N_m N_k\}\phi_m(t_1)\phi_k(t_2) \tag{7.14}$$

where the joint moments are determined from Equation 7.13, an ergodic Gaussian process is represented by the model—and hence by the N_m random variables.

Let us now assume that the process $n(t)$ is white, so that $R_n(t_2 - t_1) = (\eta_0/2)\delta(t_2 - t_1)$. It follows from Equations 7.12 and 7.13 that

$$\sigma_m^2 = \frac{\eta_0}{2} \int_0^T \int_0^T \delta(t_2 - t_1)\phi_m(t_1)\phi_m(t_2)\, dt_1\, dt_2$$

$$= \frac{\eta_0}{2} \int_0^T \phi_m^2(t_1)\, dt_1 = \frac{\eta_0}{2} \tag{7.15}$$

for all m, and

$$E\{N_m N_k\} = \frac{\eta_0}{2} \int_0^T \int_0^T \delta(t_2 - t_1)\phi_m(t_1)\phi_k(t_2)\, dt_1\, dt_2$$

$$= \frac{\eta_0}{2} \int_0^T \phi_m(t_1)\phi_k(t_1)\, dt_1 = \frac{\eta_0}{2}\delta_{mk} \tag{7.16}$$

Thus for white processes the coefficients are uncorrelated; if the process is Gaussian, they are statistically independent. The autocorrelation function of the model becomes

$$R_n(t_2 - t_1) = \sum_{m=0}^{\infty} \sum_{k=0}^{\infty} \frac{\eta_0}{2}\delta_{mk}\phi_m(t_1)\phi_k(t_2) = \sum_{m=0}^{\infty} \frac{\eta_0}{2}\phi_m(t_1)\phi_m(t_2) \tag{7.17}$$

Substituting Equation 7.6 into Equation 7.17, we obtain

$$R_n(t_2 - t_1) = \sum_{m=0}^{\infty} \frac{\eta_0}{2}\frac{2}{T}\left(\cos\frac{2\pi m}{T}t_1 \cos\frac{2\pi m}{T}t_2 + \sin\frac{2\pi m}{T}t_1 \sin\frac{2\pi m}{T}t_2\right)$$

$$= 2\sum_{m=0}^{\infty} \frac{\eta_0}{2}\cos\left[\frac{2\pi m}{T}(t_2 - t_1)\right]\frac{1}{T}$$

If we define $m/T = f$ and $1/T = \Delta f$, this sum is essentially

$$2\int_0^{\infty} \frac{\eta_0}{2}\cos 2\pi f(t_2 - t_1)\, df$$

which is by definition the Fourier transform of a white spectrum or the correct autocorrelation function of $n(t)$. We conclude that for white noise the model represents the process where the coefficients are all statistically independent zero-mean Gaussian random variables with variance $\eta_0/2$.

The received signal $r(t)$ consists of one of M transmitted signals plus additive noise, or

$$r(t) = s_i(t) + n(t) \tag{7.18}$$

where $i = 1, \ldots, M$. If the signal is strictly bandlimited,

$$r(t) = \sum_{m=\hat{m}}^{2BT+\hat{m}-1} s_{mi}\phi_m(t) + \sum_{m=0}^{\infty} n_m\phi_m(t) \tag{7.19}$$

where s_{mi} are constant coefficients, $\hat{m}$ is either 0 (for LP signals) or $2f_{\min}T$ (for bandpass signals), and n_m are the values of N_m that are independent Gaussian random variables. If we define

$$n(t) = \sum_{m=\hat{m}}^{2BT+\hat{m}-1} n_m\phi_m(t) + \hat{n}(t)$$

$$= \hat{n}(t) + \overset{\approx}{\hat{n}}(t) \tag{7.20}$$

then

$$r(t) = \sum_{m=\hat{m}}^{2BT+\hat{m}-1} (s_{mi} + n_m)\phi_m(t) + \overset{\approx}{\hat{n}}(t) \tag{7.21}$$

The $\hat{n}(t)$ terms of $n(t)$ are called the *relevant noise*. It is intuitively clear that $\overset{\approx}{\hat{n}}(t)$ is irrelevant, since it lies outside of the frequency band of the signal. More rigorously, since the coefficients of the irrelevant part of $r(t)$ can tell us nothing about either the relevant noise (since all coefficients are statistically independent) or the signal, they must be irrelevant to the decision. We conclude that if $\hat{\mathbf{N}}$ is the vector of $2BT$ random variables, the coefficients of the relevant noise model, it represents the relevant noise $\hat{n}(t)$ because $\hat{n}(t)$ can be uniquely determined from this vector through Equation 7.20.

This $2BT$-dimensional vector of random variables is characterized not by a vector value (as $\mathbf{s}_i$) but by a multivariate PDF. Since all of the random variables are statistically independent zero-mean Gaussian random variables with equal variances $\eta_0/2$, this PDF is given by

$$f_{\hat{\mathbf{N}}}(\mathbf{n}) = \prod_{m=\hat{m}}^{2BT+\hat{m}-1} \frac{1}{\sqrt{\pi\eta_0}} e^{-n_m^2/\eta_0} \tag{7.22}$$

This representation can be extended to the nonwhite case according to the theory of Karhunen and Loeve [2, 3]. The relevant noise can always be represented by a $2BT$-dimensional vector of zero-mean Gaussian independent random variables. Unlike the white noise case, however, the variances σ_m^2 vary with m. More significantly, the basis functions are not the sinusoids of a Fourier series but are an orthonormal set that depends on the spectrum $S_n(f)$ and on the duration of the time interval T. This theory is beyond the scope of this text, however, and is mentioned only for completeness.

7.3 GEOMETRIC ANALOGY

We wish to draw an analogy between the process

$$\sum_{m=\dot{m}}^{2BT+\dot{m}-1} c_m\phi_m(t) \qquad \text{subject to} \qquad \int_0^T \phi_m(t)\phi_n(t)\, dt = \delta_{mn}$$

with an N-dimensional vector

$$\mathbf{v} = \sum_{m=\dot{m}}^{N+\dot{m}-1} a_m\mathbf{r}_m \tag{7.23}$$

subject to

$$\mathbf{r}_m \cdot \mathbf{r}_n = \delta_{mn} \tag{7.24}$$

where the $\mathbf{r}_m$ are perpendicular unit-length vectors. If $N \le 3$ we can sketch the vector $\mathbf{v}$, as in Figure 7.2, where the a_m coefficients are the projections along the m axis. We know from the Pythagorean theorem that the distance of the vector from the origin, its magnitude $|\mathbf{v}_i|$, is given by

$$d_i^2 = |\mathbf{v}_i|^2 = \sum_{m=1}^{3} a_{mi}^2 \tag{7.25}$$

Furthermore, the distance between vectors is given by

$$d_{12}^2 = |\mathbf{v}_1 - \mathbf{v}_2|^2 = \sum_{m=1}^{3} (a_{m1} - a_{m2})^2 \tag{7.26}$$

Although we cannot draw N-dimensional vectors for $N > 3$, we can define them mathematically; in this case the Pythagorean theorem generalizes to

$$d_i^2 = |\mathbf{v}_i|^2 = \sum_{m=1}^{N} a_{mi}^2 \tag{7.27}$$

and

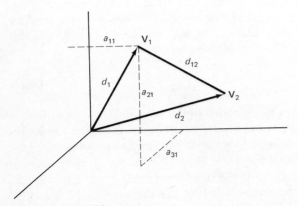

Figure 7.2 Euclidean vectors.

$$d_{12}^2 = |\mathbf{v}_1 - \mathbf{v}_2|^2 = \sum_{m=1}^{N} (a_{m1} - a_{m2})^2 \tag{7.28}$$

Fortunately, we almost always can generalize geometric results visualized in two or three dimensions into N dimensions.

By analogy between orthonormal functions $\phi_m(t)$ and unit-length perpendicular vectors $\mathbf{r}_m$, the vector representation $\mathbf{s}_j$ of a signal $s_j(t)$ can be visualized geometrically. An important property of signals is their energy E_j. Using the vector model,

$$\begin{aligned}
E_j &= \int_0^T s_j^2(t)\, dt = \int_0^T \sum_m \sum_k s_{mj} s_{kj} \phi_m(t) \phi_k(t)\, dt \\
&= \sum_m \sum_k s_{mj} s_{kj} \int_0^T \phi_m(t) \phi_k(t)\, dt = \sum_m \sum_k s_{mj} s_{kj} \delta_{mk} \\
&= \sum_{m=\hat{m}}^{2BT+\hat{m}-1} s_{mj}^2 \tag{7.29}
\end{aligned}$$

It follows that the magnitude of the signal vector $\mathbf{s}_j$ corresponds to the square root of its energy $\sqrt{E_j}$. Similarly, the energy difference ΔE_{ij} between two signals $s_i(t)$ and $s_j(t)$ becomes

$$\begin{aligned}
\Delta E_{ij} &= \int_0^T [s_i(t) - s_j(t)]^2\, dt \\
&= \int_0^T \left[\sum_m (s_{mi} - s_{mj}) \phi_m(t) \right]^2 dt \\
&= \sum_m \sum_k (s_{mi} - s_{mj})(s_{ki} - s_{kj}) \int_0^T \phi_m(t) \phi_k(t)\, dt \\
&= \sum_m (s_{mi} - s_{mj})^2 \tag{7.30}
\end{aligned}$$

It is clear that in the geometric analogy distance corresponds to the square root of energy. If we define the source constraints, that is, if we note that E_m is the maximum energy a signal can have, B is the maximum signal bandwidth, and T is the duration of the signal, then each of the signals $s_i(t)$ is represented as a point $\mathbf{s}_i$ in a $2BT$-dimensional hypersphere of radius $\sqrt{E_m}$. Figure 7.3 shows three dimensions of such a hypersphere. The axes are labeled according to the basis functions they represent. Keep in mind that the basis functions constitute a set of $2BT$ sinusoids from which the signals $s_i(t)$ can be reconstructed using Equation 7.8, where the coefficients, which are the components of $\mathbf{s}_i$, are the projections on the various axes.

The relevant noise vector $\mathbf{N}$ is not a point but rather is characterized by a $2BT$-dimensional PDF. The noise is visualized as a hyperspheric cloud centered at the origin. If white noise is assumed, resulting in the PDF of Equation 7.22, this cloud is perfectly symmetric in all directions (i.e., the variance of the components is the same, $\sigma_m^2 = \eta_0/2$, in all directions). The received signal $r(t)$ is represented according to Equation 7.21 by the vector

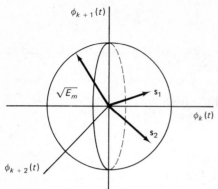

Figure 7.3 Geometric representation of signals.

$$\mathbf{R} = \mathbf{s}_i + \hat{\mathbf{N}} \tag{7.31}$$

and may be visualized as a cloud centered about the signal $\mathbf{s}_i$. A two-dimensional picture is shown in Figure 7.4.

EXAMPLE 7.1 ■

Let us visualize geometrically the flat-topped pulses often used in PCM. If a flat pulse of T-s duration is extended periodically, it becomes a constant and requires only the function $\phi_0(t)$ (see Equation 7.6)—or one dimension—for representation. Bipolar pulses of the maximum possible energy E_m are indicated by the squares in Figure 7.5. Because the absence of a pulse corresponds to the origin, monopolar pulses are indicated by the $\times$ marks. The distance between pulses is $2\sqrt{E_m}$ for bipolar pulses and $\sqrt{E_m}$ for monopolar pulses. Hence the energy difference is $4E_m$ for bipolar pulses and E_m for monopolar pulses.

■ ■

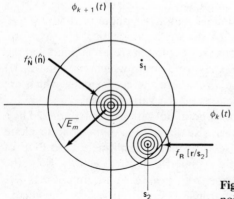

Figure 7.4 Geometric representation of noise and noisy signals.

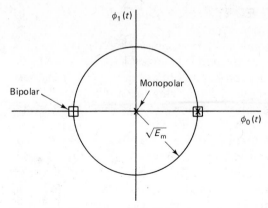

Figure 7.5 Representation of flat-topped pulses.

EXAMPLE 7.2 ■

Let us now visualize the bipolar raised-cosine-shaped pulses of Figure 3.1, which can be modeled for $0 < t < T$ by

$$s_i(t) = \pm\left(\frac{A}{2} + \frac{A}{2}\cos\frac{2\pi}{T}t\right)$$

or

$$s_i(t) = \pm[a\phi_0(t) + b\phi_1(t)]$$

These pulses can be visualized in two dimensions as in Figure 7.6, where $a = \sqrt{2E_b/3}$, $b = \sqrt{E_b/3}$, and $E_b = \frac{3}{8}E_m$, where $E_m = A^2T$ (see Problem 7.1). ■ ■

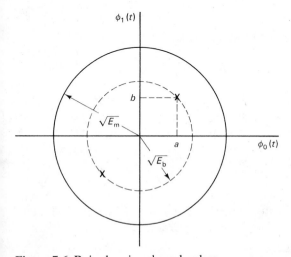

Figure 7.6 Raised cosine-shaped pulses.

7.4 OPTIMAL DECISION THEORY

We discussed in Section 6.2 the notion of a decision rule based on the random variable R $(R = S_i + N)$, which corresponds to a sample of the input to the decision rule device and the a priori probabilities P_i of the signals. We argued that the optimum rule (in the sense of minimizing the probability of error) selects that signal whose a posteriori probability $p(S_i/r)$, where r is the value of R, is a maximum. We now have the entire T-s input modeled by the vector $\mathbf{R}$ according to Equation 7.31. It follows, logically, that the optimum decision rule based on the received signal $r(t)$, where $0 < t < T$, and the a priori probabilities P_i, where $i = 1, \ldots, M$, selects that signal whose a posteriori probability $p(\mathbf{s}_i/\mathbf{r})$, where $\mathbf{r}$ is the value of $\mathbf{R}$, is a maximum. The following argument is directly analogous to that of Section 6.2 (see Equations 6.5-6.9).

From Bayes's rule, the optimum rule selects that $s_i(t)$ which maximizes

$$f_\mathbf{R}(\mathbf{r}/\mathbf{s}_i)P_i/f_\mathbf{R}(\mathbf{r}) \tag{7.32}$$

Since $f_\mathbf{R}(\mathbf{r})$ depends on the received signal only and not on i, this is equivalent to determining the maximum of $P_i f_\mathbf{R}(\mathbf{r}/\mathbf{s}_i)$. Since

$$\Pr(\mathbf{r}_1 < \mathbf{R} \le \mathbf{r}_2/\mathbf{s}_i) = \Pr(\mathbf{r}_1 - \mathbf{s}_i < \hat{\mathbf{N}} \le \mathbf{r}_2 - \mathbf{s}_i)$$

or

$$\int_{r_1}^{r_2} \cdots \int f_\mathbf{R}(\mathbf{r}/\mathbf{s}_i)\, d\mathbf{r} = \int_{r_1-s_i}^{r_2-s_i} \cdots \int f_\hat{\mathbf{N}}(\mathbf{n})\, d\mathbf{n} = \int_{r_1}^{r_2} \cdots \int f_\hat{\mathbf{N}}(\mathbf{r} - \mathbf{s}_i)\, d\mathbf{r}$$

then

$$f_\mathbf{R}(\mathbf{r}/\mathbf{s}_i) = f_\hat{\mathbf{N}}(\mathbf{r} - \mathbf{s}_i) \tag{7.33}$$

Thus the optimum decision rule selects that signal $s_i(t)$ that maximizes

$$P_i f_\hat{\mathbf{N}}(\mathbf{r} - \mathbf{s}_i) \tag{7.34}$$

or on substitution of the PDF of Equation 7.22,

$$P_i \prod_{m=\dot{m}}^{2BT+\dot{m}-1} \exp\left[-\frac{(r_m - s_{mi})^2}{\eta_0} \right] \tag{7.35}$$

where the multiplicative constant has been dropped. The natural logarithm is a monotonic function, which means that $x_j > x_i$ if and only if $\ln x_j > \ln x_i$ (see Figure 7.7). Thus the optimum decision rule selects that signal $s_i(t)$ that maximizes

$$\ln P_i - \frac{1}{\eta_0} \sum_{m=\dot{m}}^{2BT+\dot{m}-1} (r_m - s_{mi})^2 \tag{7.36}$$

Alternatively, we can say that the optimum decision rule selects that signal $s_i(t)$ that minimizes

$$\sum_{m=\dot{m}}^{2BT+\dot{m}-1} (r_m - s_{mi})^2 - \eta_0 \ln P_i \tag{7.37}$$

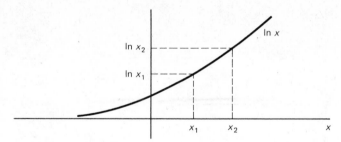

Figure 7.7 Natural logarithm.

If the signals all have equal a priori probabilities ($P_i = 1/M$ for all i), then $\ln P_i$ does not depend on i and can be dropped. When this assumption is not valid, the a priori probabilities are usually unknown. We shall use the maximum likelihood (ML) strategy throughout this chapter. With this strategy the optimum decision rule selects that signal $s_i(t)$ that minimizes

$$\sum_{m=\dot{m}}^{2BT+\dot{m}-1} (r_m - s_{mi})^2 \tag{7.38}$$

We observe from Equation 7.28 that the summation term is the geometric distance between the vector representing the received signal, $\mathbf{r}$, and the vector representing one of the possible transmitted signals, $\mathbf{s}_i$. This distance corresponds to the square root of the energy difference between $r(t)$ and $s_i(t)$ (see Equation 7.30). Hence the optimum rule may be interpreted (Figure 7.8) as finding that signal vector that is closest to the actual received vector $\mathbf{r}$. An alternative visualization involves dividing up the hypersphere with hyperplanes into zones that contain one transmitted-signal vector. The dotted lines of Figure 7.8 represent a two-dimensional version of these

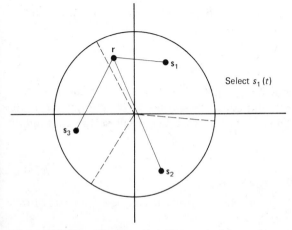

Figure 7.8 Visualization of optimum decision rule.

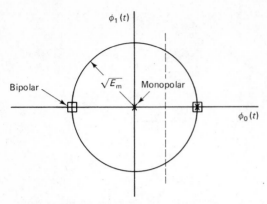

Figure 7.9 Flat-topped pulses and the optimum threshold hyperplanes.

hyperplanes. The selected signal is the one that lies in the same zone as the received signal. With the ML strategy these hyperplanes correspond to those points that are equidistant from adjacent signals.

EXAMPLE 7.3 ■

Let us reconsider the flat-topped pulses of Example 7.1, depicted in Figure 7.9. For the bipolar pulses, the decision hyperplane is the axes labeled $\phi_1(t)$. The other dashed line is the decision hyperplane for the monopolar pulses. Since the bipolar pulses are further apart than the monopolar pulses (and hence further away from the hyperplane thresholds), we see intuitively from geometrical considerations why $P_b(e)$ is smaller for the bipolar pulses than for the monopolar pulses.   ■ ■

7.5 THE OPTIMUM ML RECEIVER

The optimum decision rule (Equation 7.38) is equivalent to finding the signal that maximizes

$$-\sum_{m=\dot{m}}^{2BT+\dot{m}-1}(r_m-s_{mi})^2 = -\sum_{m=\dot{m}}^{2BT+\dot{m}-1}r_m^2 + 2\sum_{m=\dot{m}}^{2BT+\dot{m}-1}r_ms_{mi} - \sum_{m=\dot{m}}^{2BT+\dot{m}-1}s_{mi}^2 \qquad (7.39)$$

The first summation is the same for all signals (i.e., does not depend on i) and can be dropped. The last summation corresponds to E_i, the energy in the ith signal (see Equation 7.29). Thus the optimum decision rule selects that signal $s_i(t)$ that maximizes

$$\sum_{m=\dot{m}}^{2BT+\dot{m}-1}r_ms_{mi} - \tfrac{1}{2}E_i \qquad (7.40)$$

where E_i and $\mathbf{s}_i$ are known a priori and $\mathbf{r}$ represents the relevant part of the input signal.

The random variables r_m are the Fourier coefficients of $r(t)$, and from Equation 7.9

$$r_m = \int_0^T r(t)\phi_m(t)\, dt \tag{7.41}$$

Substituting Equation 7.41 into the summation of Equation 7.40, we obtain

$$\sum_{m=\hat{m}}^{2BT+\hat{m}-1} r_m s_{mi} = \int_0^T r(t) \sum_{m=\hat{m}}^{2BT+\hat{m}-1} s_{mi}\phi_m(t)\, dt$$

Now, $\sum_{m=\hat{m}}^{2BT+\hat{m}-1} s_{mi}\phi_m(t) = s_i(t)$ (see Equation 7.19). Hence the optimum ML decision rule selects the signal $s_i(t)$ that maximizes

$$\int_0^T r(t)s_i(t)\, dt - \tfrac{1}{2}E_i \tag{7.42}$$

The integral of Equation 7.42 is the output of a filter matched to the *i*th signal (see Equation 6.78). It follows that the optimum receiver for deciding among M signals is given by Figure 7.10, where the matched filters are implemented as in Figure 6.11. If all of the signals have the same energy, it is not necessary to include the biases $\tfrac{1}{2}E_i$, and the outputs of the MFs are compared directly.

EXAMPLE 7.4 ■

Let us consider the optimum receiver for binary pulses, both bipolar as in PSK and monopolar as in ASK. In the bipolar case, one MF is the inverse of the other; that is, the outputs are identical in magnitude but opposite in sign. The energies are equal, and therefore the bias terms are unnecessary. The largest term corresponds to the positive output. It follows that only one MF is required, and the decision is based on whether the output is positive or negative. Thus the PSK receiver of Figure 6.16 with one MF, whose output is compared with a zero

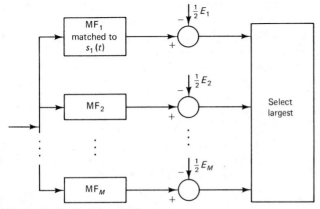

Figure 7.10 Optimum ML receiver.

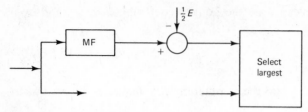

Figure 7.11 Example 7.4.

threshold, is optimum. For the monopolar case, the filter that is matched to the absence of a signal is an open circuit. If in Figure 7.10 we take into account the signal energy differences, Figure 7.11 results. Once again, this is the same as the receiver of Figure 6.16, where the output of one MF is compared with a threshold voltage of $\frac{1}{2}E$. ■ ■

EXAMPLE 7.5 ■

Let us consider the optimum receiver for deciding between bursts of different frequencies such as FSK and FPM. In this case, if there are M different signals, there must be M different MFs. The signal energies are all equal; hence the bias terms are unnecessary. Thus the FSK receiver of Figure 6.18, previously analyzed, is seen to be optimum. The FPM receiver of Figure 3.46 is near optimum, depending on how the envelope detectors are implemented. If the BPFs and envelope detectors of this receiver are replaced with MFs, it becomes the optimum detector. ■ ■

EXAMPLE 7.6 ■

Consider now the optimum receiver for PPM. Once again the bias terms are unnecessary. The MFs are implemented at different times but have the same structure. It follows that the optimum receiver can be implemented as in Figure 7.12. The multiplier part of the MF can be eliminated since the pulses are flat topped and we are multiplying by a constant. It is necessary to control the I&D and sample and hold circuits with the clock signal $s_c(t)$. Because the MF outputs are taken at different times, they must be stored through the use of an A/D circuit. The reader should compare this receiver with that suggested in Figure 3.15 for PPM. ■ ■

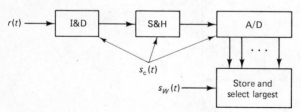

Figure 7.12 Optimum receiver for PPM.

There are other signal sets, such as QPSK (see Problem 7.7), for which the bank of MFs can be implemented with fewer than M filters.

7.6 GEOMETRIC CONSIDERATIONS FOR ERROR CALCULATIONS

When we attempt to determine the probability of error $P(e)$ of a signal set from its geometry, it is often convenient to be able to visualize moving signals around. Since the only important consideration for $P(e)$ calculations will be the distance between signals, we are free to move the signals as long as all the distances remain the same. We cannot move some of the signal points and keep others fixed without changing some of the distances and hence $P(e)$. For the white noise case the probability of error is invariant to any translation or rotation of the entire signal geometry. This concept is convenient since it often simplifies the computations needed to line up signal points on the axes (if possible) or some other convenient geometric locations.

We can use such geometric concepts to formulate interesting problems. Because the energy of a signal corresponds to the square of the distance to the origin, it follows that by translating a signal geometry we get a new signal set whose signals have different energies but whose performance in additive noise is the same. Can we translate an arbitrary signal geometry so that the average or mean energy of the signal set is a minimum? It is not at all surprising that the answer is the affirmative, and the translation is to move the centroid or center of gravity of the signal set to the origin of the axes [1].

In the rest of this chapter we shall limit ourselves for the most part to symmetric signal geometries. In general, the probability of error $P(e)$ of a signal set (or an encoding technique) is evaluated from (see Equation 6.15)

$$P(e) = \sum_{i=1}^{M} P(e/\mathbf{s}_i) P_i \qquad (7.43)$$

Thus we must consider each of the M possible signals and determine the probability of error if that signal was sent. These conditional probabilities of error are combined according to Equation 7.43, which requires knowledge of the a priori probabilities. A signal set is said to be *symmetric* if $P(e/\mathbf{s}_i)$ is the same for all M signals when an optimum ML receiver is implemented. Usually it can be determined by inspection whether a signal set is symmetric. Thus in Figure 7.13 the set indicated by xs is clearly symmetric while that indicated by 0s is not. Since for a symmetric signal set $P(e/\mathbf{s}_i)$ is a constant and $\sum_{i=1}^{M} P_i = 1$, Equation 7.43 becomes

$$P(e) = P(e/\mathbf{s}_i) \sum_{i=1}^{M} P_i = P(e/\mathbf{s}_i) \qquad (7.44)$$

Thus we arbitrarily select one signal and compute the probability of error given that that signal was sent. The a priori probabilities P_i do not affect the overall probability of error. It should be recognized that the computational

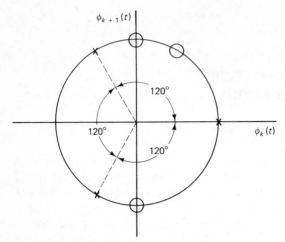

Figure 7.13 Symmetric versus nonsymmetric geometries.

simplicity of Equation 7.44 is a compelling reason for the restriction to symmetric geometries.

Intuitively, we can argue that symmetric signal sets are better than nonsymmetric sets. In Section 7.11 we shall see that the probability of error is strongly related to the smallest distance between signals. On this basis we conclude that the symmetric set of Figure 7.13 performs better than the nonsymmetric set. Indeed, it is hard to imagine three signal points further apart than those of the symmetric set even if we include a third dimension. We shall use just such arguments based on symmetry to define the optimum signal set. We conclude that since the best signal sets are symmetric, concentration on these symmetric geometries is not very restrictive.

7.7 BINARY SIGNAL SETS

As the first and most simple example, let us consider binary signals. Any binary signal set is trivially symmetric and can be visualized as the $\mathbf{x}_0$ and $\mathbf{x}_1$ points of Figure 7.14. The shape of the signals depends on the particular set of basis functions needed to characterize the signal. We shall soon verify that the performance of this signal set in additive white noise depends only on the distance d between them and not on their shapes. These two points can be freely translated and rotated without changing this distance and are symmetric by definition since $P(e/\mathbf{x}_i) = P(e/\mathbf{x}_0)$. If we translate and rotate, we can get the signal set $(\mathbf{s}_1, \mathbf{s}_0)$, which lies on the $\phi_1(t)$ axis only. These signals are equidistant $(d/2)$ from the origin. This new signal set, which performs equally as well as the original one, is easier to evaluate.

The ML receiver splits the hypersphere with a hyperplane that goes through the origin and lies along all of the axes except $\phi_1(t)$. In other words,

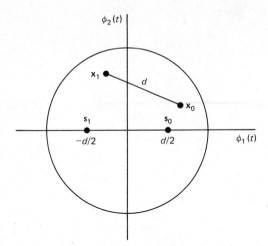

Figure 7.14 Binary signal sets.

only component n_1 of $\hat{\mathbf{n}}$ is relevant as far as this new $(\mathbf{s}_i, \mathbf{s}_0)$ set is concerned. From Equation 7.44 we see that

$$P(e) = P(e/\mathbf{s}_1) = \Pr(n_1 > d/2) \tag{7.45}$$

Thus, if $\mathbf{s}_1$ was sent and n_1 is more positive than $d/2$, the received signal will lie to the right of the hyperplane and be closer to $\mathbf{s}_0$. The optimum receiver will then select $\mathbf{s}_0$ as the transmitted signal and be in error. Since n_1 is a zero-mean Gaussian random variable with variance $\eta_0/2$, it follows that

$$
\begin{aligned}
P(e) &= \int_{d/2}^{\infty} \frac{1}{\sqrt{\pi \eta_0}} e^{-n_1^2/\eta_0}\, dn_1 \\
&= \int_{d/\sqrt{2\eta_0}}^{\infty} \frac{1}{\sqrt{2\pi}} e^{-x^2/2}\, dx \\
&= Q(d/\sqrt{2\eta_0}) \tag{7.46}
\end{aligned}
$$

All that remains is to relate the distance d to the signal energies.

   Let us consider the three specific binary geometries indicated in Figure 7.15. The actual signal shapes depend on which of the basis functions $\phi_r(t)$ and $\phi_s(t)$ refer to. If $\phi_r(t)$ refers to $\phi_0(t)$ (a constant), then the x set corresponds to baseband bipolar pulses (as in Figure 7.5). If $\phi_r(t)$ refers to a sinusoid, then the x set corresponds to the pulses used in phase shift keying. The 0 set refers to either a baseband pulse versus nothing or to a sinusoidal pulse versus nothing, and these are the pulses used in amplitude shift keying. If $\phi_r(t)$ and $\phi_s(t)$ refer to two different sinusoids, the Δ set corresponds to the pulses used in frequency shift keying. In any event, except for the absence of a pulse, all of the signals have the same energy E_s (i.e., are equidistant from the origin) and

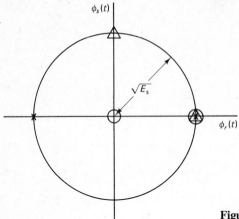

Figure 7.15 Three common binary geometries.

$$d_0 = \sqrt{E_s} \qquad\qquad (7.47)$$
$$d_x = 2\sqrt{E_s} \qquad\qquad (7.48)$$
$$d_\Delta = \sqrt{2E_s} \qquad\qquad (7.49)$$

If we consider the individual pulses of the three main modulated binary techniques, we have

$$P_{b(ASK)}(e) = Q\left(\sqrt{\frac{E_s}{2\eta_0}}\right) \qquad\qquad (7.50)$$

$$P_{b(FSK)}(e) = Q\left(\sqrt{\frac{E_s}{\eta_0}}\right) \qquad\qquad (7.51)$$

and

$$P_{b(PSK)}(e) = Q\left(\sqrt{\frac{2E_s}{\eta_0}}\right) \qquad\qquad (7.52)$$

We have verified the results of Chapter 6 that the PSK pulse requires 3 dB less energy than the FSK pulses, which require 3 dB less energy than the ASK pulses to achieve the same $P(e)$. These results are evident from Equations 7.47–7.49. We must be careful to note that there is a subtle distinction between this binary problem and PCM, a point that will be cleared up in Section 7.8.

The fact that the PSK pulses are the best is seen directly by observing the geometries in Figure 7.15. The x signals are clearly the furthest apart. Indeed we can imagine a hypersphere of as many dimensions as we like and cannot get two signal points further apart than the x points. The geometric picture gives us other benefits. An important problem when the communication link is not line of sight, but rather uses some sort of tropospheric scattering, is the phenomenon of *fading*. Fading can be thought of as a change in the signal energy with time. It can be visualized geometrically as all of the signal points (except the one at the origin) of Figure 7.15 drifting randomly along a line that extends from the origin to the point on the circumference. As the energy decreases, the probability of error increases according to Equations 7.50-7.52, provided the thresholds (or hyperplanes) remain halfway between the signal points. If the thresholds are no longer in the right place, the equations are no longer valid and $P(e)$ becomes quite large. Examination of Figure 7.15 makes it clear that fixed thresholds remain halfway between the signals in the presence of fading for all but the 0 or ASK signal set. It follows that ASK is much more sensitive to fading than are the other binary techniques. There seems to be little reason to use the ASK technique.

7.8 CODED SIGNALS

Many encoding techniques, such as PCM, encode each message into a sequence of N pulses of duration T/N s. If each pulse has L different levels or values, then one of M messages requires $N = \log_L M$ pulses. Binary techniques use binary pulses ($L = 2$), which require $\log_2 M$ pulses to represent M possible messages. Coded signals can be represented by a modification of the Fourier series basis functions, as shown in Figure 7.16. These basis functions are still orthogonal (i.e., $\int_0^T \phi_n^i(t) \, \phi_m^j(t) \, dt = 0$ for all n and m when $i \neq j$). It follows that all of the preceding discussion in this chapter applies to coded signals. We indicated in Chapters 3 and 6 that a feasible receiver for coded signals is one that makes a decision on each pulse and stores the result. The final message is determined from the N decisions. This may or may not be an optimum procedure. The development of Section 7.5 indicates that the optimum receiver includes a bank of M matched filters each matched to an entire signal.

Both baseband PCM using bipolar pulses as well as PSK can be visualized geometrically as in Figure 7.17. If the basis functions are the flat pulses indicated as $\phi_0^i(t)$, $\phi_0^j(t)$, and $\phi_0^k(t)$ in Figure 7.16, the geometry represents baseband PCM. If the basis functions are sinusoids, such as those indicated by $\phi_1^i(t)$, $\phi_1^j(t)$, and $\phi_1^k(t)$ in Figure 7.16, the geometry represents PSK. In both cases the signals are visualized as vertices of a hypercube that just fits in the hypersphere of radius $\sqrt{E_s}$. It follows that the $P(e)$ of bipolar PCM and PSK are identical if their energies are the same.

The distance from the origin to each of the signal points is by definition $\sqrt{E_s}$. If $\sqrt{E_b}$ is half the length d of one of the sides of the hypercube, as shown in Figure 7.17, it follows from the Pythagorean theorem that

$$E_s = NE_b \tag{7.53}$$

where N is the dimension of the hypercube. We identify E_b with the energy of each pulse, and Equation 7.53 indicates that the total signal energy is the sum

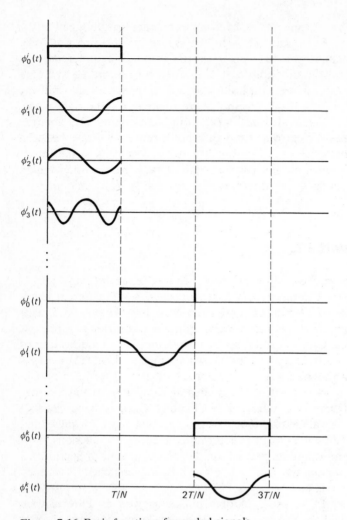

Figure 7.16 Basis functions for coded signals.

of the energies in each pulse. This geometry is clearly symmetric, and from Equation 7.44

$$P(e) = P(e/\mathbf{s}_1)$$

where $\mathbf{s}_1$ is arbitrarily picked in Figure 7.17. We know that

$$P(e/\mathbf{s}_1) = 1 - \text{Pr(correct decision/}\mathbf{s}_1) \tag{7.54}$$

and

$$\text{Pr(correct decision/}\mathbf{s}_1) = P(\text{all } n_j < d/2) \tag{7.55}$$

for $j = 1, 2, \ldots, N$. Thus, if any of the relevant noise components are greater than $d/2$, the received signal will be closer to a different vertex than

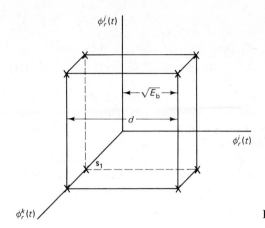

Figure 7.17 Vertices of a hypercube.

s_1, and the optimum receiver will be in error. Since the noise components are all statistically independent and are all zero-mean Gaussian with variance $\eta_0/2$,

$$\Pr\left(\text{all } n_j < \frac{d}{2}\right) = \left(\int_{-\infty}^{d/2} \frac{1}{\sqrt{\pi\eta_0}} e^{-n^2/\eta_0} \, dn\right)^N$$

$$= [1 - Q(d/\sqrt{2\eta_0})]^N \qquad (7.56)$$

From Equations 7.48 and 7.53, we have

$$d = 2\sqrt{E_b} = 2\sqrt{E_s/N} \qquad (7.57)$$

Combining Equations 7.54, 7.56, and 7.57 results in

$$P(e) = 1 - \left[1 - Q\left(\sqrt{\frac{2E_s}{N\eta_0}}\right)\right]^N \qquad (7.58)$$

In most cases the probability of pulse error is quite small, or $NQ(\sqrt{2E_s/N\eta_0})$ $\ll 1$. In this event Equation 7.58 can be closely approximated by

$$P(e) \approx NQ\left(\sqrt{\frac{2E_s}{N\eta_0}}\right) = NQ\left(\sqrt{\frac{2E_b}{\eta_0}}\right) \qquad (7.59)$$

Let us now reconsider the PCM receiver suggested in Chapter 6. This consists of an optimum binary receiver followed by an N-bit memory where the binary decisions are stored. We must modify the probability of error expression of Equation 7.52 to

$$P_{b(\text{PSK})}(e) = Q(\sqrt{2E_b/\eta_0}) \qquad (7.60)$$

where E_b is the energy per pulse or $E_b = E_s/N$. The probability that all N bits are correctly received is therefore

$$P(N \text{ bits correct}) = 1 - [1 - Q(\sqrt{2E_s/N\eta_0})]^N$$

$$\approx NQ(\sqrt{2E_s/N\eta_0})$$

Hence the probability of error for PSK when the receiver makes an optimum decision on each pulse is precisely the same as that of Equation 7.58, which is the probability of error for the optimum receiver. We have thus shown (for the first time) that an optimum binary receiver followed by a shift register memory is indeed the optimum receiver for PCM.

The question naturally arises whether, for the general class of coded signals, a receiver that makes an optimum decision on each pulse and then stores this result in memory is always optimum. Such a receiver is optimum if and only if all L^N possible combinations of decisions represent possible transmitted messages. We shall see examples where this is not the case and such a receiver is inferior to the optimum receiver, which consists of a bank of M matched filters followed by a decision device. Nevertheless, if we wish to calculate the performance of such a receiver, whether or not it is optimum, the following procedure is used.

1. Relate the energy of the individual pulses to the total signal energy E_s. Normally $E_p = E_s/N$, where N is the number of pulses ($N = \log_L M$).
2. Calculate the probability of error per pulse $P_p(e)$ in terms of E_s.
3. Calculate $P(e)$ as

$$P(e) = 1 - [1 - P_p(e)]^N$$
$$\approx N P_p(e) \tag{7.61}$$

where $P_p(e)$ is the pulse error for any one of the N pulses.

7.9 BANDWIDTH RELATIONSHIPS

The geometric representation can enable us to obtain bandwidth relationships for various signal geometries very quickly. The answers we obtain are theoretical minimum bandwidths which may be less than the practical bandwidths, as discussed in Chapter 3. This approach is very useful for comparing the bandwidth of signal sets, particularly those that have not been previously discussed in Chapter 3.

We have seen that the number of dimensions D required to represent signals of bandwidth B and duration T is given by $D = 2BT$, so that $B = (1/2T)D$. Because $1/T = f_s$ is the message rate, it follows that

$$B = \frac{D}{2} f_s \tag{7.62}$$

For coded signals, we can set $D = ND_p$, where N is the number of pulses, or $\log_L M$, and D_p is the number of dimensions per pulse. Hence for coded signals

$$B = N \frac{D_p}{2} f_s = \frac{D_p}{2} f_s \log_L M \tag{7.63}$$

Consider as an example baseband PCM. Each pulse requires one dimension [$\phi_0^i(t)$ of Figure 7.16]; hence $B = \frac{1}{2}(\log_2 M)f_s$. This is the same formula as considered in Chapter 3 with the factor $m = \frac{1}{2}$. We can achieve this small factor with pulse shaping.

Both ASK and PSK require two dimensions per pulse. We must always use two dimensions to characterize both the amplitude and phase of a sinusoidal burst; i.e., $\phi_k^i(t) = \sqrt{2/T} \cos 2\pi ft$, $\phi_{k+1}^i(t) = \sqrt{2/T} \sin 2\pi ft$, where k is odd. Thus the bandwidth for ASK or PSK is twice that of baseband PCM (i.e., it is a modulated PCM signal). FSK requires four dimensions per pulse (i.e., two different frequencies), and so the bandwidth is twice that of PSK. Note that although this relative result is correct, the formula $B_{FSK} = 2 \log_2 M f_s$ is different from the practical one of Chapter 3, $B_{FSK} = 2m \log_2 M f_s + (f_2 - f_1)$, where theoretically $f_2 - f_1$ can be made quite small. It should, however, be as large as the first term to be practical.

Because D_1 or D_p for coded signals is easily determined from the signal geometry, the theoretical minimum bandwidth is easily derived from Equation 7.62 or 7.63.

EXAMPLE 7.7 ■

The geometry for a QPSK pulse is shown in Figure 7.18. Thus QPSK requires two dimensions per pulse ($D_p = 2$). The number N of pulses per message is given by

$$N_{QPSK} = \log_4 M = \tfrac{1}{2}\log_2 M$$

Substituting this into Equation 7.63, we see that the theoretically minimum bandwidth for QPSK is given by

$$B_{QPSK} = \tfrac{1}{2}f_s \log_2 M$$

which is half the theoretically minimum bandwidth of PSK.　　　　　■ ■

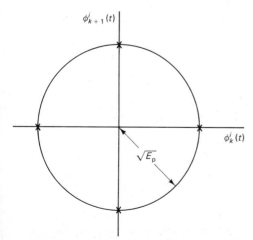

Figure 7.18 QPSK geometry.

7.10 UNION BOUND APPROXIMATIONS

Before we apply the techniques developed to other signal sets, let us develop
an easy way to approximate $P(e/s_i)$, which is needed to evaluate the
probability of error (see Equation 7.44). The approximation we shall develop
not only makes our analysis easy, but in addition the form of the answer is
much more suitable for comparison purposes than more accurate, but more
cumbersome, expressions.

Consider first the event $\mathcal{E}_{ki}$, where

$$\mathcal{E}_{ki} = \{|\mathbf{r} - \mathbf{s}_k|^2 < |\mathbf{r} - \mathbf{s}_i|^2 / \mathbf{s} = \mathbf{s}_i\}$$

and $k = 0, 1, \ldots, M - 1$ but $k \neq i$. If the event $\mathcal{E}_{ki}$ has occurred, that is, the
received signal $\mathbf{r}$ is closer to $\mathbf{s}_k$ than the transmitted signal $\mathbf{s}_i$, then clearly an
error has occurred. It does not, however, mean that the receiver selects $\mathbf{s}_k$
since it is possible that $\mathbf{r}$ is closer to $\mathbf{s}_m$ than $\mathbf{s}_k$. Thus these events are not
disjoint, as seen by Figure 7.19. If $\mathbf{r}$ is the received signal and $\mathbf{s}_1$ was
transmitted, then both the events $\mathcal{E}_{21}$ and $\mathcal{E}_{31}$ have occurred. If, on the other
hand, $\mathbf{s}_2$ was transmitted, then no such event has occurred (the receiver will
not make an error).

For a symmetric signal geometry, $P(e) = P(e/s_1)$ (see Equation 7.44),
where $\mathbf{s}_1$ is arbitrarily chosen. In terms of the events $\mathcal{E}_{k1}$,

$$P(e/s_1) = \Pr(\mathcal{E}_{21} \cup \mathcal{E}_{31} \cup \cdots \cup \mathcal{E}_{M1})$$

From the union bound (Equation 5.7),

$$P(e/s_1) \leq \sum_{k=2}^{M} \Pr(\mathcal{E}_{k1}) \tag{7.64}$$

$\Pr(\mathcal{E}_{ki})$, however, is the binary probability of error discussed in Section 7.7,
and from Equation 7.46

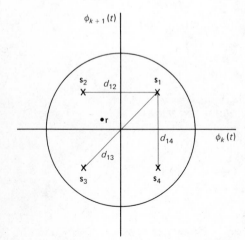

Figure 7.19 Union bound definitions.

$$\Pr(\mathscr{E}_{k1}) = Q(d_{1k}/\sqrt{2\eta_0}) \tag{7.65}$$

where d_{1k} is the distance between the assumed transmitted signal s_1 and the other signals s_k (see Figures 7.19 or 7.20). Combining the previous two equations yields the bound

$$P(e) = P(e/s_1) \le \sum_{k=2}^{M} Q\left(\frac{d_{1k}}{\sqrt{2\eta_0}}\right) \tag{7.66}$$

In Figure 7.20 we visualize the bound in connection with the vertices of a hypercube geometry. Let us assume that the probability that the received signal lies outside the outer circle is extremely small. The union bound involves adding the probabilities that the received signal lies in the hatched region. In Figure 7.20a this is the correct probability of error; in Figure 7.20b the three hatched regions overlap and the approximation overestimates the probability of error. We see that the approximation is quite good (or tight) for modest amounts of noise or small probabilities of error. This conclusion generalizes to all symmetric signal geometries.

Observe from Figure 7.20 that the contributions to the bound from nearest neighbors are more significant than the other contributions. This gives rise to another simple approximation,

$$P(e) = P(e/s_1) \simeq \sum_{\substack{\text{nearest} \\ \text{neighbors}}} Q\left(\frac{d_{1K}}{\sqrt{2\eta_0}}\right) = KQ\left(\frac{d_{\min}}{\sqrt{2\eta_0}}\right) \tag{7.67}$$

where K is the number of nearest neighbors (those signals whose distance to s_1 is a minimum). If in Figure 7.20b we eliminate the C hatched region because s_3 is not a near neighbor of s_1, the approximation is even better. We cannot in general argue that the approximation of Equation 7.67 is tighter than the bound of Equation 7.66. It is at least as tight, however, and easier to use, and therefore it will be used more often. If it is desirable to retain an upper bound, Equation 7.66 must be used.

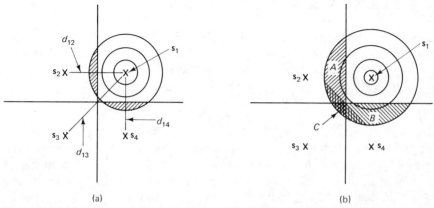

(a)   (b)

Figure 7.20 Union bound approximations: (a) slightly and (b) very noisy.

EXAMPLE 7.8 ■

Let us reconsider the PSK technique whose geometric visualization is the vertices of a hypercube as in Figure 7.17. We see that the nearest neighbors lie at the N other vertices of the N-dimensional hypercube edges that include the signal s_1. The distance to these N nearest neighbors is $2\sqrt{E_b}$ or $2\sqrt{E_s/N}$. From Equation 7.67 we have

$$P_{PSK}(e) \approx NQ\left(\sqrt{\frac{2E_s}{N\eta_0}}\right) \tag{7.68}$$

where

$$N = \log_2 M$$

which is precisely the approximation of Equation 7.59. As argued earlier, this approximation is tight if $P_b(e)$ is small. ■ ■

EXAMPLE 7.9 ■

Consider the QPSK signal set whose geometry for a signal pulse was shown in Figure 7.18. From Equation 7.61, $P_{QPSK}(e) = NP_p(e)$, where $N = \log_4 M = \frac{1}{2}\log_2 M$ and $P_p(e)$ is the probability of error for one pulse. From Figure 7.18 we see that there are two nearest neighbors, so from Equation 7.67

$$P_p(e) \approx 2Q\left(\frac{d_{min}}{\sqrt{2\eta_0}}\right)$$

where $d_{min} = \sqrt{2E_p} = \sqrt{2E_s/N} = \sqrt{4E_s/\log_2 M}$. It follows that

$$P_{QPSK}(e) \simeq \log_2 M\, Q\left(\sqrt{\frac{2E_s}{\eta_0 \log_2 M}}\right) \tag{7.69}$$

Comparing Equations 7.68 and 7.69, we see that the probability of error for QPSK is the same as that for PSK. ■ ■

7.11 *L*-LEVEL PHASE SHIFT KEYING

Let us extend the results of Examples 7.8 and 7.9 to a general L-level phase modulation technique. We wish to encode the M messages into $N = \log_L M$ pulses each of energy $E_p = E_s/N$, where the pulses are modeled by

$$s_n^j(t) = \sqrt{\frac{2E_s}{T}}\cos\left(2\pi ft + n\frac{2\pi}{L}\right)$$

where $n = 1, 2, \ldots, L$ and $jT/N \le t \le (j + 1)T/N$. This corresponds to PSK for $L = 2$ and QPSK for $L = 4$. Regardless of the value of L, each pulse is a sinusoidal pulse that requires two dimensions to be visualized as in Figure 7.21. We see that as far as the individual pulses are concerned there are two near neighbors for $L > 2$ (only one for $L = 2$) and $d_{min}/2\sqrt{E_p} = \sin \pi/L$.

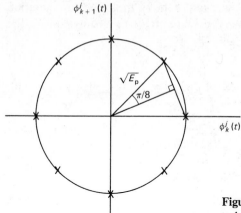

Figure 7.21 Eight-level phase modulated pulse.

Hence from Equation 7.67

$$P_p(e) \approx 2Q\left[\sin\left(\frac{\pi}{L}\right)\sqrt{\frac{2E_p}{\eta_0}}\right] \quad \text{for} \quad L > 2$$

where $E_p = E_s/N$. From Equation 7.61 for coded signal sets

$$P(e) \approx NP_p(e)$$

or

$$P(e) \approx 2NQ\left[\sin\left(\frac{\pi}{L}\right)\sqrt{\frac{2E_s}{N\eta_0}}\right] \tag{7.70}$$

where $N = \log_L M$. Since $\log_2 M = \log_2 M/\log_2 L$, Equation 7.70 can be written

$$P(e) \approx 2\frac{\log_2 M}{\log_2 L}Q\left[\sin\left(\frac{\pi}{L}\right)\sqrt{\frac{2E_s \log_2 L}{\eta_0 \log_2 M}}\right] \tag{7.71}$$

For $L = 4$, Equation 7.71 reduces to Equation 7.69, which is the probability of error for QPSK.

Let us consider why the form of the approximation is so convenient for comparison purposes. When we compare the performance of schemes, we are not concerned with evaluating the different probabilities of error for a given signal energy E_s, which can be misleading. Rather we wish to compare the amount of energy required by each scheme to achieve the same probability of error. This is desirable because the only alternative to a "better" encoding scheme is to increase the signal energy. We saw in Section 7.7 that for the binary case this simply involved comparing the argument of the function Q or the distance between signals. This remains the case, in general, to a good approximation because the function Q varies exponentially with its argument.

To a first-order approximation we can neglect the K multiplier of Equation 7.67 and compare the argument of the function Q or, equivalently, the distance between nearest neighbors directly.

EXAMPLE 7.10 ■

Let us compare the performance of eight-level phase modulation with PSK. Since $\log_2 8 = 3$, Equation 7.71 becomes

$$P_{8\,\text{level}}(e) \approx \tfrac{2}{3}\log_2 M\, Q\left[\sin\left(\frac{\pi}{8}\right)\sqrt{\frac{6E_s}{\log_2 M\, \eta_0}}\,\right]$$

$$= \tfrac{2}{3}\log_2 M\, Q\left(0.3827\sqrt{\frac{6E_s}{\log_2 M\, \eta_0}}\,\right)$$

For PSK (Equation 7.68),

$$P_{\text{PSK}}(e) \approx \log_2 M\, Q\left(\sqrt{\frac{2E_s}{\log_2 M\, \eta_0}}\,\right)$$

To compare, we equate

$$\tfrac{2}{3}\log_2 M\, Q\left(0.3827\sqrt{\frac{6E_{s8}}{\log_2 M\, \eta_0}}\,\right) = \log_2 M\, Q\left(\sqrt{\frac{2E_{s2}}{\log_2 M\, \eta_0}}\,\right)$$

To a high degree of accuracy (i.e., neglecting the multiplier $\tfrac{2}{3}$)

$$(0.3827)^2\, 6E_{s8} = 2E_{s2}$$

or

$$E_{s8} = 2.276 E_{s2}$$

Thus eight-level phase modulation requires roughly 2.276 times or 3.57 dB more energy than PSK to achieve the same probability of error.

■ ■

We must keep in mind that the bandwidth of eight-level phase modulation is one-third that of PSK; the multiple-threshold receiver, however, is more difficult to implement. It would appear that this technique should be considered whenever bandwidth is a premium, e.g., for high-speed modems operating over a telephone channel.

7.12 ORTHOGONAL SIGNAL SETS

If each of M signals is located on a separate dimension, as in Figure 7.22, they form an *orthogonal* signal set. We shall assume, as shown, that they all lie on the surface of the hypersphere and have energy E_s. The shape of the signals depends on the choice of basis functions.

All of the signals are equidistant from each other, $d_{ik} = \sqrt{2E_s}$ for any i and k; hence whether we use Equation 7.66 or 7.67 we have

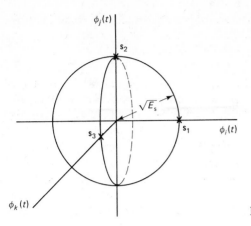

Figure 7.22 Orthogonal signal set.

$$P_{\text{orth}}(e) \leq (M - 1)\, Q\left(\sqrt{\frac{E_s}{\eta_0}}\right) \tag{7.72}$$

The minimum theoretical bandwidth is obtained by recognizing that the number of dimensions needed, D, is equal to the number of signals, M, and from Equation 7.62

$$B_{\text{orth}} \geq \tfrac{1}{2} M f_s \tag{7.73}$$

The actual bandwidth of specific orthogonal sets may differ and the receiver hardware problems could be very different.

Suppose that the basis functions are $\phi_0^i(t)$, $\phi_0^j(t)$, $\phi_0^k(t)$, ... from Figure 7.16. Thus the total time T alloted to each message is divided into M slots, and a single pulse of duration T/M is transmitted. The location of the pulse determines which message is intended. This scheme was referred to in Chapter 3 as PPM. These signals are useful for baseband or optical (laser) communications. They do not involve phase synchronization but do require accurate pulse synchronization. The assumption that each of the signals has energy E_s, regardless of the number of messages, can be questioned for PPM (but not the other orthogonal schemes considered). Usually transmitting hardware is more peak-power limited than energy limited. This means that the narrower the pulses, the harder it is to achieve a given energy. With present laser technology, however, the energy constraint assumption may not be unreasonable.

Assume now that the basis functions are sinusoidal bursts of length T each of a different frequency. This scheme was referred to in Chapter 3 as FPM. There is no conflict between peak power and energy constraints, and the assumption that E_s does not depend on M is valid. For optimum reception this signal set requires phase synchronization and accurate pulse synchronization. As stated earlier, each sinusoidal pulse requires two dimensions to characterize both its amplitude and phase. This means FPM requires twice as many dimensions as the number of messages and twice the bandwidth of other orthogonal sets such as PPM:

$$B_{FPM} > Mf_s \tag{7.74}$$

In Chapter 3 we noted the bandwidth differences between PPM and FPM and wondered whether FPM could be regarded as a modulated form of PPM. We now see that they are geometrically identical and, with an energy constraint and optimal receivers, they perform identically. Indeed the differences between these signal sets are almost the same as the difference between baseband PCM and a modulated form such as PSK.

Let us now compare the performance of these orthogonal signal sets with a binary signal set. Perhaps it is most fair to assume FSK as the binary signal set since FPM is a natural extension of FSK and has similar hardware problems. We have seen that

$$P_{FPM}(e) \le (M - 1) Q\left(\sqrt{\frac{E_s}{\eta_0}} \right)$$

and

$$P_{FSK}(e) \approx \log_2 M \, Q\left(\sqrt{\frac{E_s}{\log_2 M \, \eta_0}} \right)$$

To a first approximation, in order to achieve the same probability of error[3]

$$E_{s(FPM)} = E_{s(FSK)}/\log_2 M \tag{7.75}$$

For $M = 2$, FPM is FSK, and of course they require the same energy to achieve the same probability of error. For large M, however, FPM performs considerably better than FSK; hence the orthogonal set performs better than binary encoded schemes such as PCM. We must remember, however, that the bandwidth of orthogonal signal sets is much larger than those of binary signal sets for large M,

$$B_{FPM} \approx Mf_s > B_{FSK} \approx 2 \log_2 Mf_s \tag{7.76}$$

Also, the receiver of the orthogonal sets requires a bank of M matched filters and a series of comparators as opposed to one or two matched filters, a comparator, and an N-bit shift register memory. We must of course consider bandwidth and hardware complexities as well as performance in noise in order to determine which scheme to use.

EXAMPLE 7.11 ■

We have earlier considered the telephone dialing problem of communicating one of 12 messages. Using a binary scheme this would dictate four pulses. From Equation 7.75 we see that some orthogonal scheme would

---

[3]For large values of M, the multiplier $M - 1$ can be much larger than $\log_2 M$ and the approximation of Equation 7.75 may not be very accurate. Approximations for $Q(\alpha)$ are helpful for a more precise relationship (see Problem 7.16).

be on the order of 6 dB better. For the telephone company only the transmitter hardware is a major consideration. FPM looks attractive because it is easy to build an oscillator that can be tuned to 12 different frequencies with a series of switches or buttons. Furthermore, the message rate f_s (e.g., the number of buttons per second that can be pushed) is low, and bandwidth is not a problem. Thus if one can push buttons at a rate of 10/s, we see from Equation 7.76 that we need a bandwidth of at least 240 Hz for FPM. We saw in Chapter 4 that the typical telephone lines have a much larger bandwidth than this. Because of practical considerations, however, the scheme used is not really FPM. Problems 7.21 and 7.22 discuss the modifications and the reasons for them. ■ ■

Let us consider augmenting an orthogonal signal set by including the negatives of each of the signals. Such a set, shown geometrically in Figure 7.23, is called a *biorthogonal* set as it is no longer orthogonal. We see immediately that this set requires half the number of dimensions for M messages, and hence half the bandwidth. This is the primary motivation for considering this set. Hardware complexities may be the price that has to be paid, however, since if we augment FPM into a biorthogonal set, we no longer can remove the need for phase synchronization with suboptimal receivers. We also see immediately from Figure 7.23 that the distance between nearest neighbors is unaffected, and hence the performance is also unaffected. Only the negative of a signal is not a nearest neighbor, being $2\sqrt{E_s}$ away, and from Equation 7.66

$$P_{\text{biorth}}(e) \le (M - 2)Q\left(\sqrt{\frac{E_s}{\eta_0}}\right) + Q\left(\sqrt{\frac{2E_s}{\eta_0}}\right) \tag{7.77}$$

The second term is much smaller than the first; if it is left out, one obtains the approximation of Equation 7.67. For $M = 2$ the biorthogonal set, which can be implemented as PSK, is better than the orthogonal set, which can be

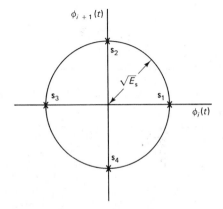

Figure 7.23 Biorthogonal signal set.

implemented as FSK. For $M > 2$, however, there is no substantial difference as far as probability of error is concerned.

EXAMPLE 7.12 ■

It is possible to have a set of orthogonal signals that do not appear to lie along the axes as in Figure 7.22. Presumably with a different set of basis functions than those of the Fourier series, they would appear as in Figure 7.22. We can still recognize them to be orthogonal from the relationship

$$\int_0^T s_i(t)s_j(t)\, dt = \delta_{ij}$$

The signal set shown in Figure 7.24, for $M = 8$, can be shown to be orthogonal (see Problem 7.16). This binary signal set along with FPM and PPM represents the three most common orthogonal sets. This set is not PCM in that only eight of the 256 possible combinations are messages. The optimum receiver is not a PCM receiver but contains a bank of eight MFs, each matched to one of the signals of Figure 7.24. From Equation 7.72,

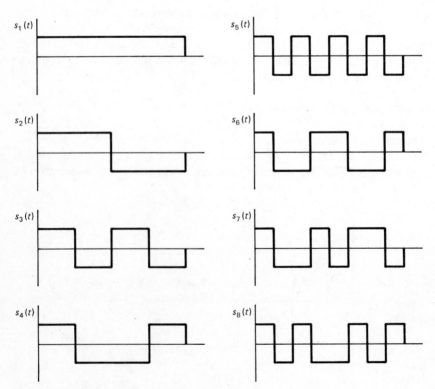

Figure 7.24 Binary orthogonal signal set.

$$P_{opt}(e) \approx 7Q(\sqrt{E_s/\eta_0})$$

Suppose, however, we use a suboptimum PCM receiver that makes a decision on every pulse and decodes if and only if one of the eight signals of Figure 7.24 is stored in the shift register. It follows from Equations 7.61 and 7.59 that

$$P_{subopt}(e) \approx 8Q(\sqrt{2E_s/8\eta_0})$$

This suboptimum receiver requires four times, or 6 dB, more energy than the optimum receiver to achieve the same probability of error.   ■ ■

EXAMPLE 7.13 ■

Let us now augment the signals of Figure 7.24 by including the negatives of the signals, i.e., an $M = 16$ biorthogonal set. From Equation 7.77,

$$P_{opt}(e) \approx 14Q(\sqrt{E_s/\eta_0})$$

and

$$P_{subopt}(e) \approx 8Q(\sqrt{2E_s/8\eta_0})$$

as before. If we consider a straight four-pulse bipolar PCM signal, then

$$P_{PCM}(e) \approx 4Q(\sqrt{2E_s/4\eta_0})$$

which requires 3 dB more energy than the biorthogonal signal that employs the optimum detector but 3 dB less energy than if the suboptimal receiver were used.   ■ ■

If we examine closely the biorthogonal signal set of the signals in Figure 7.24 and their negatives, another receiver suggests itself. Suppose we use a PCM receiver whose decoder attempts to correct for errors by determining the "closest" signal when the decisions stored in the shift register do not correspond to one of the 16 possible signals (i.e., an error has been detected). Such a receiver can correct all single errors (see Problem 7.19). Thus the probability of a correct decision, $P(e)$, now becomes the probability of no errors plus the probability of single errors, or

$$Pr(correct) = (1 - p)^8 + 8p(1 - p)^7 \approx 1 - 28p^2 + 112p^3$$

where $p = Q(\sqrt{E_s/4\eta_0})$. Now the probability of error is

$$P(e) = 1 - Pr(correct) \approx 28Q^2(\sqrt{E_s/4\eta_0})$$

assuming $8Q(\sqrt{E_s/4\eta_0}) \ll 1$. With the rough approximation

$$Q(\alpha) < \tfrac{1}{2} \exp(-\tfrac{1}{2}\alpha^2)$$

this becomes

$P(e) \leq 7 \exp(-E_s/4\eta_0)$

Using the same approximation for the optimum biorthogonal receiver, we obtain

$P_{opt}(e) \leq 7 \exp(-E_s/2\eta_0)$

and the four-pulse PCM signal

$P_{PCM}(e) \leq 2 \exp(-E_s/4\eta_0)$

Error correction has helped, but such a receiver performs no better than the PCM scheme.

We shall see in Chapter 9 that binary schemes that are not PCM (and not necessarily orthogonal either) but that employ PCM receivers followed by error correction schemes can perform very well. Procedures for doing this will be discussed in some detail in Chapters 9 and 10.

7.13 SIMPLEX SIGNALS

Let us approach the optimum signal set by fixing the number of messages and utilizing our geometric intuition. It is clear that the optimum set is one in which the distance between nearest neighbors (d_{min}) is as large as possible. Remember that all the signals are constrained to lie inside a hypersphere of radius $\sqrt{E_m}$, where E_m is the maximum energy the transmitter can generate per message. For $M = 2$, it is intuitively clear that, regardless of the number of dimensions, points on opposite ends of a major diameter $(d_{min} = 2\sqrt{E_m})$ are as far apart as possible. This set requires only one dimension, as shown in Figure 7.25a. This binary set was discussed in Section 7.7 and corresponds to either a PSK-type pulse or bipolar flat-topped pulses.

If we imagine a plane that goes through the origin and pick three points on the intersection of this plane and the surface of the hypersphere so as to form an equilateral triangle, we obtain the geometry of Figure 7.25b. We cannot find three points further apart; this set, which requires two dimensions

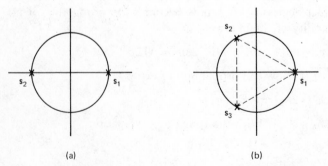

(a)   (b)

Figure 7.25 Simplex signals for (a) $M = 2$ and (b) $M = 3$.

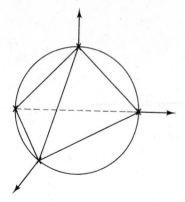

Figure 7.26 Simplex set for $M = 4$.

to characterize, must be optimum. From Section 7.11 we see that $d_{min} = 2 \sin(\pi/3)\sqrt{E_m}$ for this set.

We are tempted to argue by induction that the optimum signal set, called a *simplex*, are the vertices of a "completely symmetric" geometric form with M vertices, which requires $M - 1$ dimensions. Indeed, we cannot find four signals that are further apart than the vertices of a regular tetrahedron (a pyramid with all sides of equal length), which requires three dimensions to visualize, as in Figure 7.26. Unfortunately, while some geometrists may be capable of such abstractions, it is difficult for us to think in more than three dimensions and discuss the simplex for $M > 4$. Fortunately, we can relate the simplex to an orthogonal set through a translation and rotation. We shall argue that if an orthogonal set is first translated so that its center of gravity becomes the origin, then rotated from M into $M - 1$ dimensions, it forms the shape of a simplex. These simplexes do not lie on the surface of the hypersphere (i.e., the distances between signals can be increased by stretching the geometries) but have the same probability of error as the original orthogonal set. For $M = 2$ this transformation is rather obvious. If you form an orthogonal set with your thumb and first two fingers and then rotate into two dimensions (i.e., bring the tip of each finger into contact with this page), you will see the equilateral triangle. Geometrists assure us that four perpendicular vectors rotate into a regular tetrahedron, and so on.

Let us consider the translation first. If s_i represents an orthogonal signal,

$$s_i \cdot s_j = E_m \delta_{ij} \tag{7.78}$$

Thus the signals are all orthogonal and have maximum energy E_m. The vector **a** points to the center of gravity of the signal set if

$$a = \frac{1}{M} \sum_{i=1}^{M} s_i \tag{7.79}$$

It can be seen that

$$\mathbf{a} \cdot \mathbf{s}_i = \frac{E_m}{M} \quad \text{and} \quad |\mathbf{a}|^2 = \mathbf{a} \cdot \mathbf{a} = \frac{E_m}{M} \tag{7.80}$$

The translated signal geometry is determined from $\hat{\mathbf{s}}_i = \mathbf{s}_i - \mathbf{a}$. It follows that

$$\begin{aligned}
\hat{\mathbf{s}}_i \cdot \hat{\mathbf{s}}_j &= (\mathbf{s}_i - \mathbf{a}) \cdot (\mathbf{s}_j - \mathbf{a}) \\
&= \mathbf{s}_i \cdot \mathbf{s}_j - \mathbf{a} \cdot (\mathbf{s}_i + \mathbf{s}_j) + |\mathbf{a}|^2
\end{aligned} \tag{7.81}$$

Substituting Equations 7.78 and 7.80 yields

$$\hat{\mathbf{s}}_i \cdot \hat{\mathbf{s}}_j = E_m \delta_{ij} - 2\frac{E_m}{M} + \frac{E_m}{M}$$

or

$$\hat{\mathbf{s}}_i \cdot \hat{\mathbf{s}}_j = \begin{cases} E_m(1 - 1/M) & i = j \\ -E_m/M & i \neq j \end{cases} \tag{7.82}$$

The fact that $\hat{\mathbf{s}}_i \cdot \hat{\mathbf{s}}_j$ is not zero for $i \neq j$ simply means that the translated set is no longer orthogonal. On the other hand,

$$|\hat{\mathbf{s}}_i|^2 = E_m\left(1 - \frac{1}{M}\right) \tag{7.83}$$

means that the energy per message has been reduced by the factor $1 - 1/M$. Presumably stretching the simplex to the surface of the hypersphere means increasing the signal energies to E_m, thereby decreasing the probability of error.

Recall that our comparisons have been based on the energies required by different signal geometries to achieve the same probability of error. It follows that the simplex requires $1 - 1/M$ less energy than the orthogonal set to achieve the same probability of error. Thus for $M = 2$, the simplex visualized as the xs in Figure 7.15 is 3 dB better than orthogonal signals visualized as the Δs. On the other hand, for $M = 8$ the improvement is only 0.58 dB. We conclude that for large M (perhaps $M > 8$) the orthogonal set is nearly optimum. The bandwidth of the simplex is only trivially less than that of the orthogonal set since $M - 1$ dimensions instead of M are required. Thus

$$B_{\text{simplex}} > (M - 1)f_s \tag{7.84}$$

which is not much different than Equation 7.74 for $M \gg 2$.

Generating the simplex signals appears more difficult than generating the orthogonal set. Since many of the signals do not lie along the axes, the case analogous to FPM would require a series of tones with different amplitudes and phases. It would also be difficult to implement matched filters and simple suboptimum filters for such peculiarly shaped pulses. On the other hand, a binary signal set similar to the orthogonal set of Figure 7.24 should not require more hardware complexity.

We now are in position to conclude that the use of a modified FPM scheme for telephone dialing, as in Example 7.11, is more than reasonable.

TABLE 7.1   ENERGY E_s/η_0 REQUIRED TO ACHIEVE $P(e) = 10^{-4}$

M	2	4	8	16	$\cdots$	1024
$N = \log_2 M$	1	2	3	4	$\cdots$	10
Simplex	4.25	7.2	9.14	10.5	$\cdots$	16.78
Orthogonal	8.5	9.6	10.45	11.2	$\cdots$	16.8
	−3 dB	−1.25 dB	−0.58 dB	−0.28 dB		—
PSK or	4.25	9.2	14.4	19.8	$\cdots$	61
baseband	0 dB	−1.06 dB	−1.97 dB	−2.76 dB		−5.60 dB
FSK	8.5	18.4	28.8	39.6	$\cdots$	122
	−3 dB	−4.06 dB	−4.97 dB	−5.76 dB		−8.60 dB

7.14 CONCLUSIONS

It appears that unless we have severe bandwidth constraints or particular hardware constraints, the primary discrete signals that should be considered are the binary signals, the orthogonal signals, and perhaps, for $2 < M \leq 8$, simplex signals. We must consider not only their performance in additive noise but bandwidth differences and, most important, hardware differences. Using the formulas generated in this chapter, let us compute the signal energy needed for these most common signals to achieve an error probability of 10^{-4}. The results are tabulated in Table 7.1 for various values of M along with the loss in decibels relative to the simplex set. We must keep in mind that the orthogonal sets require M matched filters versus one or two, along with an N-bit memory, for the binary sets. Thus one pays a significant hardware price for the large improvement over PCM for large values of M. Also the bandwidth of the orthogonal sets increases much faster with M than the binary coded sets. For example, the bandwidths tabulated in Table 7.2 assume that one of M messages is transmitted in 1 ms (i.e., $f_s = 10^3$). For very large values of M we pay a large bandwidth price for the improvement of orthogonal sets over binary ones.

There are only two possible ways to decrease the probability of error for any of the techniques other than increasing the power available. We saw in Chapter 6 that, if one is not constrained to operate in real time, the energy per signal can be increased by time scaling. This is true for any discrete technique but it is perhaps easier to do for binary techniques which can be

TABLE 7.2   MINIMUM THEORETICAL BANDWIDTH FOR $f_s = 10^3$

B_{min}	M					
	2	4	8	16	$\cdots$	1024
Orthogonal (kHz)	1	2	4	8	$\cdots$	512
Binary (kHz)	0.5	1	1.5	2	$\cdots$	5

stored easily in computer memory. Earlier we saw that the communication technique used to transmit pictures of Mars from the Mariner satellite is PSK with time scaling. The cost or complexity of a PSK optimal receiver over an FSK suboptimal receiver is certainly not a significant factor. The choice of PSK is clearly logical.

In later chapters we shall discuss the other method of reducing the probability of error, known as block encoding. We shall learn that the only practical methods (both from hardware and bandwidth considerations) employ redundant binary signals and PCM-type receivers, followed by either logic hardware or computers that correct errors. Although these are not PCM signals, PCM signals and receivers can be modified to incorporate this type of encoding. This represents still another reason why PCM and modulated forms of PCM are the most common discrete techniques.

PROBLEMS

7.1. Consider the bipolar raised-cosine-shaped pulses visualized in Example 7.2 $\{s_i(t) = \pm\frac{1}{2}A[1 + \cos(2\pi/T)t]\}$. Assume that there is a peak-power constraint of A^2 V. (All terms are defined in Figure 7.6.)
(a) Compute the energy E_b in pulse $s_i(t)$ and prove that $E_b = \frac{3}{8}E_m$.
(b) Using the representation, verify that $a = \sqrt{2E_b/3}$ and $b = \sqrt{E_b/3}$.
(c) What is the energy difference between the two pulses in terms of E_m?

7.2. Consider the signal $x_1(t)$:

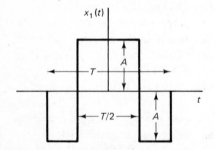

(a) Extend this signal periodically and expand in the normalized Fourier series, that is valid for $-T/2 < t < T/2$. Verify that

$$x_1(t) = \sum_{k=1}^{\infty} c_k \phi_k(t)$$

where

$$c_k = \begin{cases} \dfrac{2(-1)^{(k-1)/4}}{k+1} \dfrac{\sqrt{8}}{\pi}\sqrt{E} & k = 1, 5, 9, \ldots \\ 0 & \text{otherwise} \end{cases}$$

with E the energy of $x_1(t)$.

(b) Consider the bandlimited signal $\hat{x}_1(t) = \Sigma_{k=1}^{6} c_k \phi_k(t)$. Plot the vector representing $\hat{x}_1(t)$ as a point in a three-dimensional hypersphere. Identify the basis functions used and the radius of the hypersphere.

7.3. Consider the signal $x_2(t)$:

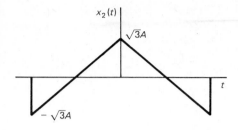

(a) Verify that this signal has the same energy E as $x_1(t)$ of Problem 7.2.

(b) Repeat Problem 7.2a for $x_2(t)$, where

$$
c_k = \begin{cases} \left(\dfrac{2}{k+1}\right)^2 \dfrac{8}{\pi^2} \sqrt{\dfrac{3E}{2}} & k = 1, 5, 9, \ldots \\ 0 & \text{otherwise} \end{cases}
$$

(c) Repeat Problem 7.2b for $x_2(t)$.

7.4. Indicate the vectors that represent the signals $\hat{x}_1(t)$ and $\hat{x}_2(t)$ (from Problems 7.2 and 7.3) in the same three-dimensional hypersphere.

(a) Identify the energy difference in terms of the length of a line segment in this hypersphere.

(b) Compute this energy difference.

7.5. Reconsider the geometric visualization of the signals $\hat{x}_1(t)$ and $\hat{x}_2(t)$ in Problem 7.4. Divide the sphere into two hemispheres with a plane that corresponds to the optimum ML decision rule.

7.6. Repeat Problem 7.5 for the geometric visualization of the QPSK pulse (see Figure 7.18).

7.7. Sketch an implementation of the optimum receiver for QPSK which includes the details of the MF (as in Figure 6.16). For QPSK only two MFs need be implemented (phase synchronization to 0° and 90°). Explain why, and indicate how, the "select largest" component could work.

7.8. (a) By rotating the geometry of a QPSK pulse, prove that the probability of error per pulse is identical to the probability of error for two PSK pulses.

(b) Identify the basis functions required for the PSK geometry of two pulses to correspond correctly to PSK.

7.9. Return again to the signals $\hat{x}_1(t)$ and $\hat{x}_2(t)$ of Problem 7.4. Assume that these are the binary pulses to be used for a baseband PCM signal where $A = 2$ V, $M = 128$, $f_s = 10^4$ samples/s, and $\eta_0/2 = 10^{-6}$ V^2/Hz.

(a) Calculate the minimum theoretical bandwidth for this signal.

(b) Determine the energy E of the signals and the energy difference between them.

(c) Calculate $P_b(e)$ and $P_w(e)$ assuming white Gaussian additive noise.

    (d) How much more energy (dB) does this signal set require than the NRZ signal set to achieve the same $P(e)$?

7.10. Consider the L-level phase techniques discussed in Problem 3.18 (where $L = 2$, 3, 4 and $M = 60$).

    (a) Sketch separately the signal geometries for a single pulse for each of these techniques. Label the radius in terms of E_s (the energy for an entire message).

    (b) Determine the distance between nearest neighbors for each scheme in terms of E_s.

    (c) Determine which scheme results in the smallest probability of error per message and how much more energy is required for the other schemes to achieve the same probability of error.

7.11. **(a)** Consider an eight-level ASK signal. Using the appropriate two dimensions, sketch the signal geometry for one of these pulses.

    (b) On this same figure, sketch the geometry of an eight-level PSK signal. By inspection, which results in a smaller $P(e)$?

    (c) By comparing the minimum distances directly, determine how much more energy (in dB) the eight-level ASK signal requires than eight-level PSK in order to achieve the same $P(e)$?

7.12. Consider an eight-level PSK signal that employs two different amplitudes, as shown below:

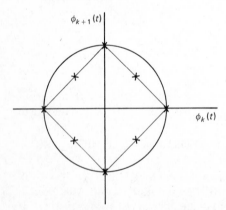

    (a) Compute the distance between nearest neighbors in terms of E_s.

    (b) Compare with the minimum distance for standard eight-level PSK. Which scheme results in the smallest $P(e)$?

7.13. Consider now the three-level ASK signal shown at the top of page 251, two pulses of which can be used to encode one of nine (or fewer) messages:

    (a) Identify E_s and E_p (for all but the zero pulse) on the figure.

    (b) Compare the $P(e)$ of this scheme for eight messages with that of the scheme in Problem 7.12. (*Hint:* Use method of rotation.)

    (c) Compare the theoretical minimum bandwidth of the two schemes.

    (d) Which receiver is easier to realize? Why?

7.14. Consider the three baseband communication schemes discussed in Problem 3.4.

    (a) Sketch the geometry of a single pulse for each scheme. Label the radius and

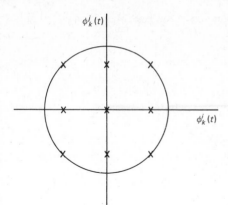

Problem 7.13

    minimum distance between signals in terms of E, (the energy of each message).

    (b) Compare E_s needed by each scheme to approximately achieve the same $P_w(e)$.

7.15. Consider the two signal sets that were compared in Problem 3.14 (PSK and four-level FPM).

    (a) Repeat Problem 7.14 for these signal sets.

    (b) Determine the minimum bandwidth required for each scheme.

7.16. **(a)** Using the approximation $Q(\alpha) < \frac{1}{2} \exp(-\frac{1}{2}\alpha^2)$, obtain a more precise comparison between FPM and FSK than that of Equation 7.75.

    (b) Using the tighter approximation $Q(\alpha) < (1/\sqrt{2\pi}\,\alpha)\exp(-\frac{1}{2}\alpha^2)$, repeat part (a). Show that this gives a transcendental equation that is hard to evaluate.

7.17. Your problem is to communicate one of 16 messages in 1 ms. The eight signals shown in Figure 7.24 and their negatives have been suggested. The eight signals shown are orthogonal, and the set can be considered to be a biorthogonal set.

    (a) Indicate the relationship that must be satisfied for the eight signals to be orthogonal. Verify this for one pair.

    (b) Sketch the maximum likelihood receiver in block diagram form.

    (c) Determine an approximate expression for the probability of error.

    (d) Compare with PCM, that is, show PCM requires twice the energy to achieve the same probability of error with half the bandwidth. What about receiver complexity?

7.18. We wish to communicate a sequence of signals, each one of which must represent one of eight possible messages, through a cable that introduces additive Gaussian white noise. We are considering one of three possible orthogonal signal sets FPM, PPM, or the eight-pulse binary code shown in Figure 7.24.

    (a) Select one of the signal sets and give a reasoned argument as to why you chose that particular one.

    (b) Compare the receiver hardware of your signal set with those of a PCM signal set.

    (c) Compare the bandwidths of all three sets as well as PCM.

    (d) How much energy has your choice saved relative to PCM?

7.19. Consider again the signal set corresponding to the eight signals of Figure 7.24 and their negatives. Assume a PCM receiver followed by error correction as discussed just after Example 7.13.

    (a) Show by example that if any single error is made, there is only one possible signal (of the 16 transmitted signals) that could have been sent.

    (b) Show by example that if any two errors are made, two possible signals could have been sent. Thus double errors are detected but cannot be corrected.

7.20. Consider the three signal sets considered in Problem 3.16.

    (a) Sketch the geometry of a single pulse for each scheme. Label the radius and minimum distance between signals in terms of E_s.

    (b) Suppose the received signal has a peak value of 10 V. Determine E_s.

    (c) Suppose the additive noise is Gaussian and white with spectral level $\eta_0/2 = 10^{-6}$ V²/Hz. Determine an expression for $P_w(e)$ for each of the three signal sets.

7.21. Let us assume (incorrectly) that each of the 12 buttons on a pushbutton telephone headset generates one of 12 different sinusoidal signals. The receiver would not be an FPM signal set because there is a 13th signal corresponding to no button being pushed. This is a necessary consequence of the loss of pulse synchronization.

    (a) Sketch this modified FPM geometry. (Show only two or three of the sinusoidal signals and the zero signal.) Argue from the sketch that it is no longer orthogonal.

    (b) Determine an approximate expression for the probability of error in terms of the probability P_0 of no button being pushed.

    (c) Calculate how much more energy is required to achieve the same $P(e)$ as in a 12-signal FPM set.

7.22. In reality, each button of a pushbutton telephone headset generates two of seven possible sinusoids, as indicated by the following matrix:

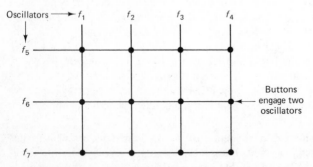

    (a) Show that while some of the signals remain $\sqrt{2E_s}$ apart, the distance between nearest neighbors is reduced to $\sqrt{E_s}$. [*Hint:* This is difficult to do geometrically. Use the fact that the square of the distance between signals corresponds to $\int_0^T [s_1(t) - s_2(t)]^2 \, dt$ (i.e., the energy difference).][4]

    (b) Recalculate an approximation expression for the probability of error and compare with that of Problem 7.21.

[4] Actually the principal source of noise is not stationary Gaussian noise but the result of impulses that cause ringing in the line. Thus the probability of mistaking no tone for one of the tones is high, but the error due to two simultaneous tones as a result of an impulse is unlikely. Thus for this nonstationary non-Gaussian noise this modified receiver works much better.

(c) Sketch a reasonable receiver for this signal set. Your receiver should not use matched filters because of the lack of pulse synchronization. Your receiver will probably cost significantly more ($\approx$6 dB) than the few dB expected from the use of envelope detectors. Thus, the modification is in fact costly even though the results of (b), which are based on unrealizable matched filters, indicate otherwise.

REFERENCES

1. J. M. Wozencraft and I. M. Jacobs, *Principles of Communication Engineering*, Wiley, New York, 1965.
2. W. B. Davenport and W. L. Root, *An Introduction to the Theory of Random Signals and Noise*, McGraw-Hill, New York, 1958.
3. J. B. Thomas, *An Introduction to Statistical Communication Theory*, Wiley, New York, 1969.

Chapter 8

Continuous Waveform and Other Suboptimal Techniques

8.1 INTRODUCTION

Continuous waveform (cw) techniques, such as amplitude and frequency modulation, are used primarily because of their small bandwidth, which makes them useful for commercial radio and TV broadcasting where bandwidth is at a premium. Another reason for their use is the simplicity and reliability of cw hardware. In this chapter we shall attempt to analyze and compare cw techniques with discrete methods.

Some of the preliminary analysis will also be useful for the evaluation of envelope detectors. The cost of replacing matched filters with narrow-band filters and envelope detectors in ASK and FPM receivers will be determined.

8.2 MODELING OF BANDPASS GAUSSIAN PROCESSES

Because cw signals are invariably narrow bandpass signals, the relevant noise (i.e., the output of the i.f. stage of a receiver) is a narrow bandpass process that can be assumed to be Gaussian. We seek a model for such a process of the form

$$n(t) = c(t) \cos 2\pi f_0 t + s(t) \sin 2\pi f_0 t \tag{8.1}$$

where f_0 is the center frequency of the noise spectrum and $c(t)$ and $s(t)$ are low-frequency baseband Gaussian processes called *qudrature processes*. If both quadrature processes are Gaussian, then a sample of $n(t)$ is the sum of

Gaussian random variables; hence it too is a Gaussian random variable. Thus, if the autocorrelation function of the model $R_n(\tau)$ is well defined, the model is a Gaussian process. Can we assign autocorrelation functions and a cross-correlation function to the quadrature processes such that the autocorrelation function $R_n(\tau)$ or power spectrum $S_n(f)$ is the same as that of the actual bandpass noise process? If so, the model will indeed represent the noise process (see property 5.1 of Gaussian processes). The autocorrelation function of the model is given by

$$R_n(\tau) = \tfrac{1}{2}[R_c(\tau) + R_s(\tau)]\cos 2\pi f_0\tau + \tfrac{1}{2}[R_{sc}(\tau) - R_{cs}(\tau)]\sin 2\pi f_0\tau \qquad (8.2)$$

where $R_c(\tau)$ and $R_s(\tau)$ are the autocorrelation functions of the quadrature processes and $R_{sc}(\tau)$ and $R_{cs}(\tau)$ are their cross-correlation functions (see problem 8.1). The autocorrelation functions of the quadrature processes must satisfy the properties of autocorrelation functions. The only properties that the cross-correlation functions must satisfy are

$$R_{cs}(\tau) = R_{sc}(-\tau) \qquad (8.3)$$

and they must be real.

Let us define the following baseband frequency functions:

$$S_q(f) = S_{n+}(f + f_0) + S_{n-}(f - f_0) \qquad (8.4)$$
$$C(f) = S_{n+}(f + f_0) - S_{n-}(f - f_0) \qquad (8.5)$$

where the subscripts $+$ and $-$ refer to the positive-frequency portion and the negative-frequency portion, respectively, of the noise spectrum $S_n(f)$. Since the spectrum $S_n(f)$ must have even symmetry about $f = 0$, it is easily argued that $S_q(f)$ is an even-baseband function and $C(f)$ is an odd-baseband function (see Figure 8.1).

We can now argue that the model of Equation 8.1 represents a narrow bandpass Gaussian process with power spectrum $S_n(f)$ provided

$$\mathcal{F}\{R_c(\tau)\} = \mathcal{F}\{R_s(\tau)\} = S_q(f) \qquad (8.6)$$

and

$$\mathcal{F}\{R_{sc}(\tau)\} = -\mathcal{F}\{R_{cs}(\tau)\} = jC(f) \qquad (8.7)$$

We first note that since $S_q(f)$ is an even-baseband function, the autocorrelation function of the quadrature processes [now denoted $R_q(\tau)$] must satisfy the correct properties. To verify that $R_{sc}(\tau)$ is real, the following must be true:

$$R_{sc}(\tau) = \mathcal{F}^{-1}\{jC(f)\} = \int_{-\infty}^{\infty} jC(f)\, e^{j2\pi f\tau}\, df$$

$$= j\int_{-\infty}^{\infty} C(f)\cos 2\pi f\tau\, df - \int_{-\infty}^{\infty} C(f)\sin 2\pi f\tau\, df$$

$$= -2\int_{0}^{\infty} C(f)\sin 2\pi f\tau\, df \qquad (8.8)$$

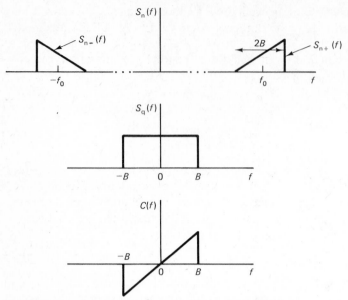

Figure 8.1 Relationship between $S_n(f)$, $S_q(f)$, and $C(f)$.

since $C(f)$ is an odd function. We also observe from Equation 8.8 that

$$R_{sc}(-\tau) = -R_{sc}(\tau) = R_{cs}(\tau) \tag{8.9}$$

and

$$R_{sc}(0) = 0 \tag{8.10}$$

Using Equations 8.3, 8.6, and 8.9, we can rewrite Equation 8.2

$$R_n(\tau) = R_q(\tau) \cos 2\pi f_0 \tau + R_{sc}(\tau) \sin 2\pi f_0 \tau \tag{8.11}$$

Case 1: Spectra Symmetric About f_0

If $S_n(f)$ is symmetric about f_0, it follows directly from Equations 8.4 and 8.5 that $S_q(f)$ has the same shape as $2S_{n+}(f)$ when shifted to the origin and $C(f) = 0$ for all f. Thus, the cross-correlation function is defined to be identically zero and the quadrature processes, being Gaussian, are statistically independent. For independent quadrature processes the autocorrelation function of the model becomes

$$R_n(\tau) = R_q(\tau) \cos 2\pi f_0 \tau \tag{8.12}$$

We have seen (Equation 2.68, for example) that this is the relationship between a bandpass process centered and symmetric about f_0 and its low-pass equivalent process $S_q(f)$. Thus, taking transforms of Equation 8.12, we obtain

$$S_n(f) = \tfrac{1}{2}[S_q(f + f_0) + S_q(f - f_0)] \tag{8.13}$$

which is the inverse of Equation 8.4.

Case 2: Nonsymmetric Spectra

For nonsymmetric spectra the quadrature processes are no longer statistically independent. They are uncorrelated, however $[R_{sc}(0) = 0]$, which means that a sample of $c(t)$ and a sample of $s(t)$ both taken at the same time are statistically independent. Taking Fourier transforms of Equation 8.11 yields

$$S_n(f) = \tfrac{1}{2}[S_q(f + f_0) + S_q(f - f_0)] + \mathscr{F}\{R_{sc}(\tau) \sin 2\pi f_0 \tau\} \tag{8.14}$$

But

$$\mathscr{F}\{R_{sc}(\tau) \sin 2\pi f_0 \tau\} = \int_{-\infty}^{\infty} R_{sc}(\tau)\left(\frac{e^{j2\pi f_0 \tau} - e^{-j2\pi f_0 \tau}}{2j}\right)e^{-j2\pi f \tau}\, d\tau$$

$$= \frac{1}{2j}\int_{-\infty}^{\infty} R_{sc}(\tau)e^{-j2\pi (f-f_0)\tau}\, d\tau$$

$$- \frac{1}{2j}\int_{-\infty}^{\infty} R_{sc}(\tau)e^{-j2\pi (f+f_0)\tau}\, d\tau$$

$$= \tfrac{1}{2}[C(f - f_0) - C(f + f_0)] \tag{8.15}$$

Thus,

$$S_n(f) = \tfrac{1}{2}[S_q(f + f_0) + S_q(f - f_0)] + \tfrac{1}{2}[C(f - f_0) - C(f + f_0)] \tag{8.16}$$

Using Equations 8.4 and 8.5, we see that

$$\tfrac{1}{2}[S_q(f - f_0) + C(f - f_0)] = S_{n+}(f) \qquad \tfrac{1}{2}[S_q(f + f_0) - C(f + f_0)] = S_{n-}(f)$$

Hence the spectrum of the model is the actual noise spectrum. We may verify the result graphically if we plot both parts of Equation 8.16 using the example of Figure 8.1, as shown in Figure 8.2.

Finally, if N_0 is the total average power both of the bandpass noise

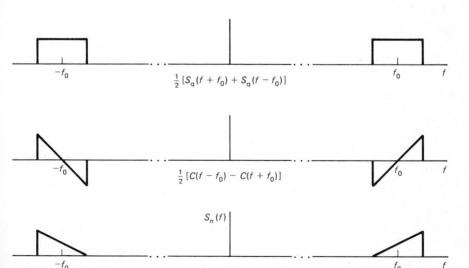

Figure 8.2 Verification of Equation 8.16.

process and of the quadrature process, then the joint PDF of a sample of $c(t)$ and one of $s(t)$ taken at the same time is always given by

$$f(c, s; 0) = f(c)f(s) = \frac{1}{2\pi N_0} \exp\left[-\frac{1}{2N_0}(c^2 + s^2)\right] \tag{8.17}$$

If the samples were taken at different times, Equation 8.17 would still hold provided the noise spectrum were symmetric about its center frequency f_0.

8.3 ENVELOPE AND PHASE OF BANDPASS PROCESSES

Using the trig identity

$$A \cos \theta + B \sin \theta = \sqrt{A^2 + B^2} \cos[\theta - \tan^{-1}(B/A)]$$

we can write the bandpass model

$$\begin{aligned} n(t) &= c(t) \cos 2\pi f_0 t + s(t) \sin 2\pi f_0 t \\ &= E(t) \cos[2\pi f_0 t - \theta(t)] \end{aligned} \tag{8.18}$$

where

$$E^2(t) = c^2(t) + s^2(t) \tag{8.19}$$

and

$$\theta(t) = \tan^{-1} \frac{s(t)}{c(t)} \pm 2\pi q \qquad q = 0, 1, \ldots \tag{8.20}$$

or, equivalently,

$$c(t) = E(t) \cos \theta(t)$$

and $\hspace{8cm}$ (8.21)

$$s(t) = E(t) \sin \theta(t)$$

The processes $E(t)$ and $\theta(t)$ are called the *envelope* and *phase*, respectively, of the bandpass process. A sketch of a typical narrow bandpass process (Figure 8.3) reveals that it is somewhat sinusoidal with amplitude and phase

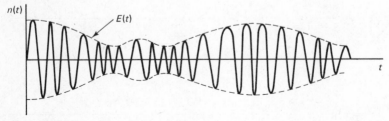

Figure 8.3 Narrow bandpass process.

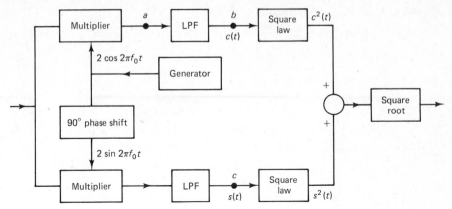

Figure 8.4 Ideal envelope detector.

varying in a random manner. The envelope shown dashed in Figure 8.3 is the function $E(t)$.

Both the envelope and phase processes can be generated from the bandpass process itself. The systems required to accomplish this are highly nonlinear, and we would hardly expect the envelope and phase to be Gaussian. The system of Figure 8.4 is often called an *ideal envelope detector*. Its output is very similar, but not identical, to the simple envelope detector considered earlier (Figure 3.39). If we model the noise by Equation 8.1, the signal at point a of Figure 8.4 is

$$2c(t) \cos^2 2\pi f_0 t + 2s(t) \sin 2\pi f_0 t \cos 2\pi f_0 t = c(t) + c(t) \cos 4\pi f_0 t$$
$$+ s(t) \sin 4\pi f_0 t$$

The LPF removes those components centered at $2f_0$, and the signal at point b is $c(t)$. Similarly, at point c, the signal is $s(t)$. The quadrature processes are physical processes in that they can be generated. The output of the system is $\sqrt{c^2(t) + s^2(t)}$, which is the formal definition of the envelope.

Let us assume that there is a narrow-band signal centered at some carrier frequency f_c. A general equation for such a signal is

$$s[t; m(t)] = A(t) \cos [2\pi f_c t - \psi(t)] \qquad (8.22)$$

where either $A(t)$ or $\psi(t)$ or both depend on the actual message $m(t)$ that is being transmitted. The received waveform $r(t)$ consists of signal plus noise,

$$r(t) = s(t) + n(t) \qquad (8.23)$$

where

$$s(t) = A(t) \cos \psi(t) \cos 2\pi f_c t + A(t) \sin \psi(t) \sin 2\pi f_c t$$

and where the noise is modeled by

$$n(t) = c(t) \cos 2\pi f_c t + s(t) \sin 2\pi f_c t$$

Thus

$$r(t) = [A(t) \cos \psi(t) + c(t)] \cos 2\pi f_c t$$
$$+ [A(t) \sin \psi(t) + s(t)] \sin 2\pi f_c t \qquad (8.24)$$

This can be put into the envelope-phase form:

$$r(t) = E(t) \cos[2\pi f_c t - \theta(t)]$$

where

$$E^2(t) = c^2(t) + s^2(t) + A^2(t) + 2A(t)[c(t) \cos \psi(t) + s(t) \sin \psi(t)] \qquad (8.25)$$

and

$$\theta(t) = \tan^{-1} \frac{A(t) \sin \psi(t) + s(t)}{A(t) \cos \psi(t) + c(t)} \pm 2\pi q \qquad q = 0, 1, \ldots \qquad (8.26)$$

Equivalently,

$$A(t) \cos \psi(t) + c(t) = E(t) \cos \theta(t)$$
$$A(t) \sin \psi(t) + s(t) = E(t) \sin \theta(t) \qquad (8.27)$$

As a check, observe that in the absence of noise

$$E(t) = A(t) \quad \text{and} \quad \theta(t) = \psi(t)$$

8.4 AMPLITUDE MODULATION ANALYSIS

We saw in Chapter 4 that double-sideband AM is achieved when

$$s(t) = A[1 + km(t)] \cos 2\pi f_c t$$

where k is the modulation index and it is assumed that $|m(t)| \leq 1$ and $f_c \gg B$, the bandwidth of $m(t)$. The signal $s(t)$ has a bandpass spectrum of bandwidth $2B$. If we substitute $A(t) = A[1 + km(t)]$ and $\psi(t) = 0$ into Equation 8.24, the received signal becomes

$$r(t) = \{A[1 + km(t)] + c(t)\} \cos 2\pi f_c t + s(t) \sin 2\pi f_c t \qquad (8.28)$$

The output of an ideal envelope detector with the dc component removed[1] becomes

$$y_i(t) = \sqrt{\{A[1 + km(t)] + c(t)\}^2 + s^2(t)} - A \qquad (8.29)$$

The PSK demodulator of Figure 3.40 has somewhat less noise distortion than the ideal envelope detector. This receiver, called a *synchronous AM detector*, is shown in simplified form in Figure 8.5. The output of the multiplier, from Equation 8.28, is

$$\{A[1 + km(t)] + c(t)\}[1 + \cos 4\pi f_c t] + s(t) \sin 4\pi f_c t$$

and the LPF output (with the dc removed) is

---

[1]The dc component, which cannot be heard, is filtered out to protect the speaker.

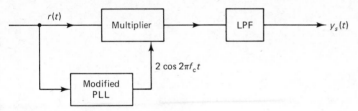

Figure 8.5 Synchronous AM detector.

$$y_s(t) = Akm(t) + c(t) \tag{8.30}$$

Comparing Equations 8.29 and 8.30, we observe that the ideal envelope detector has more noise distortion than the synchronous detector.

If the noise spectrum is white with level $\eta_0/2$, the input noise power N_i, assuming an input filter of bandwidth $2B$, is

$$N_i = 2\eta_0 B \tag{8.31}$$

The input signal power S_i is

$$S_i = \tfrac{1}{2}A^2 \langle [1 + km(t)]^2 \rangle = \tfrac{1}{2}A^2(1 + k^2\langle m^2(t) \rangle) \tag{8.32}$$

if $\langle m(t) \rangle = 0$, where $\langle m^2(t) \rangle < 1$ and is typically $\ll \tfrac{1}{2}$. The input signal-to-noise ratio SNR_i is therefore

$$SNR_i = \frac{S_i}{N_i} = \frac{\tfrac{1}{2}A^2[1 + k^2\langle m^2(t) \rangle]}{2\eta_0 B} \tag{8.33}$$

For the synchronous detector the output noise is $c(t)$. The output signal power $S_0 = A^2 k^2 \langle m^2(t) \rangle$, and

$$N_i = \langle n^2(t) \rangle = \langle c^2(t) \rangle = \langle s^2(t) \rangle \tag{8.34}$$

It therefore follows that, for this detector, the output SNR is

$$SNR_{AM} = \frac{A^2 k^2 \langle m^2(t) \rangle}{2\eta_0 B} = \alpha \, SNR_i \tag{8.35}$$

where $\alpha = 2k^2 \langle m^2(t) \rangle / [1 + k^2 \langle m^2(t) \rangle] < 1$.

If the SNR_i is large $[A^2 \gg N_i = \langle n^2(t) \rangle = \langle c^2(t) \rangle = \langle s^2(t) \rangle]$, then, for almost every t, $c(t)/A[1 + km(t)]$ or $s(t)/A[1 + km(t)]$ is much less than 1. Equation 8.29 can be approximated for large SNR_i, using $\sqrt{1 + \epsilon} \approx 1 + \tfrac{1}{2}\epsilon$:

$$y_i(t) \approx A[1 + km(t)]\left\{ 1 + \frac{c(t)}{A[1 + km(t)]} + \frac{1}{2}\frac{c^2(t) + s^2(t)}{A^2[1 + km(t)]^2} \right\} - A$$

$$= Akm(t) + c(t) + \frac{c(t)}{2}\frac{c(t)}{A[1 + km(t)]} + \frac{s(t)}{2}\frac{s(t)}{A[1 + km(t)]}$$

$$\approx Akm(t) + c(t) \tag{8.36}$$

Thus for large SNR_i there is no substantial difference in the performance of the receivers. This applies to the standard envelope detector of Figure 3.39 as

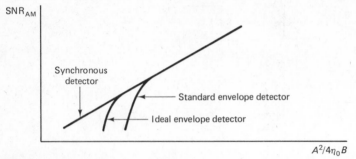

Figure 8.6 Envelope detection threshold effect.

well. When SNR_i is not large, the output SNR of the envelope detectors is less than that of the synchronous detector (see Figure 8.6). This degradation of the performance of envelope detectors below a particular SNR_i is called the AM *threshold effect*. Because commercial broadcasting is primarily concerned with the large-SNR_i case, Equation 8.35 is regarded as applying to DSB-AM modulation rather than exclusively to the synchronous detector.

Recall from Chapter 4 that the synchronous detector works for SSB-SC signals as well as for DSB. For SSB-SC the input signal power is $\frac{1}{2}A^2 k^2 \langle m^2(t) \rangle$, the output signal power remains $A^2 k^2 \langle m^2(t) \rangle$, and the input (as well as the output) noise power is $\frac{1}{2}(2\eta_0 B)$. Thus for SSB-SC the output SNR becomes

$$SNR_{SSB\text{-}SC} = \frac{A^2 k^2 \langle m^2(t) \rangle}{\eta_0 B} = 2\,SNR_{i(SSB)} = 4\,SNR_{i(DSB)} \qquad (8.37)$$

Comparing Equations 8.35 and 8.37, we see that SSB-SC has an output SNR that is $4/\alpha$ times larger (>6 dB) than standard DSB-AM for a fixed input signal power.

EXAMPLE 8.1 ■

We wish to communicate, using AM techniques, a signal whose bandwidth is 10 kHz and whose average power $k^2 \langle m^2(t) \rangle$ is 0.2 (-7 dB). The noise spectral level is $\frac{1}{2}\eta_0 = 10^{-6}$ V^2/Hz, and the maximum average power at the receiver is 500 V^2. Consider DSB techniques first:

$S_i \simeq \frac{1}{2}A^2(1 + 0.2) = 500 \quad$ or $\quad A \approx 28.9$ V

$N_i = 2\eta_0 B = 4 \times 10^{-6} \times 10^4 = 0.04$ V^2

$S_i/N_i = 500/0.04 = 1.25 \times 10^4 \qquad$ (41 dB)

$SNR_{DSB\text{-}AM} = \alpha\,SNR_i \quad$ where $\quad \alpha = 2(0.2)/(1 + 0.2) = \frac{1}{3}$

Therefore,

$SNR_{DSB\text{-}AM} = (12.5/3) \times 10^3 = 4.17 \times 10^3 \qquad$ (36.2 dB)

Consider now SSB-SC techniques:

$$S_i \approx \tfrac{1}{2}A^2(0.2) = 500 \quad \text{or} \quad A \approx 70.7 \text{ V}$$

$$N_i = \tfrac{1}{2}(2\eta_0 B) = 0.02 \text{ V}^2$$

$$S_i/N_i = 500/0.02 = 2.5 \times 10^4 \quad (44 \text{ dB})$$

$$\text{SNR}_{\text{SSB-SC}} = 2\,\text{SNR}_i = 5 \times 10^4 \quad (47 \text{ dB}) \qquad \blacksquare\ \blacksquare$$

8.5 FREQUENCY MODULATION ANALYSIS

We introduced FM modulation in Chapter 4, where

$$s(t) = A \sin\left\{2\pi\left[f_c + \Delta f \int m(t)\,dt\right] + \theta_0\right\} \tag{8.38}$$

where $|m(t)| \leq 1$ and $f_c \gg B$, the bandwidth of $m(t)$. We saw that the bandwidth depends on the modulation index $\beta = \Delta f/B$, and

$$B_{\text{FM}} \approx 2(\beta + 1)B \tag{8.39}$$

We shall see that, even though the bandwidth of FM is larger than that of AM, the performance of FM (SNR_{out}) is better for $\beta > 1$ because of this larger bandwidth. If the noise is white with level $\tfrac{1}{2}\eta_0$ V^2/Hz, the input noise power N_i, assuming an input filter of bandwidth $2(\beta + 1)B$, is

$$N_i = 2(\beta + 1)\eta_0 B \tag{8.40}$$

The input signal power S_i is independent of β and equal to $A^2/2$. Thus

$$\text{SNR}_i = \frac{S_i}{N_i} \approx \frac{A^2}{(\beta + 1)4\eta_0 B} \tag{8.41}$$

If we compare this with Equation 8.33, we see that it is smaller by a factor of at least $1/(\beta + 1)$.

We assume that an ideal frequency discriminator is a system whose response $y_i(t)$ to a narrow-band signal $A(t)\cos[2\pi f_c t + \psi(t)]$ is

$$y_i(t) = \frac{d\psi(t)}{dt} \tag{8.42}$$

From Equation 8.26,

$$\psi(t) = \tan^{-1}\frac{A \sin \psi_0(t) + s(t)}{A \cos \psi_0(t) + c(t)} \pm 2\pi q \qquad q = 0, 1, 2, \ldots \tag{8.43}$$

where from Equation 8.38

$$\psi_0(t) = 2\pi \Delta f \int m(t)\,dt + \theta_0 \tag{8.44}$$

Since $d(2\pi q)/dt = 0$, the absolute value of the phase is not relevant, and we shall drop the $\pm 2\pi q$ term from future consideration. The quadrature processes $s(t)$ and $c(t)$ are statistically independent Gaussian processes with spectra shown in Figure 8.7.

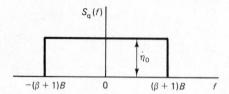

Figure 8.7 Power spectrum of quadrature processes.

We should point out that just as the ideal envelope detector was not the best AM demodulator, so the ideal frequency discriminator is not the best FM demodulator. As in the AM case, however, for large input SNR_i the performance of all FM receivers is about the same.

We assume first that the SNR_i is large. The significance of this assumption is seen by rewriting Equation 8.43 as

$$\psi(t) = \tan^{-1}\left\{\tan\psi_0(t)\left[\frac{1 + s(t)/A\ \sin\psi_0(t)}{1 + c(t)/A\ \cos\psi_0(t)}\right]\right\} \tag{8.45}$$

With large SNR_i, we can argue that terms like $c(t)/A\ \cos\psi_0(t)$ are usually quite small and we shall "linearize" Equation 8.45 by approximation techniques. Thus since

$$\frac{1 + \epsilon_1}{1 + \epsilon_2} \approx 1 + \epsilon_1 - \epsilon_2$$

and

$$\tan^{-1}[u(1 + \epsilon)] \approx \tan^{-1}u + \epsilon\frac{d}{du}\tan^{-1}u$$

$$= \tan^{-1}u + \epsilon\frac{u}{1 + u^2}$$

it follows that

$$\psi(t) \approx \psi_0(t) + \frac{\tan\psi_0(t)}{1 + \tan^2\psi_0(t)}\left[\frac{s(t)}{A\ \sin\psi_0(t)} - \frac{c(t)}{A\ \cos\psi_0(t)}\right]$$

or

$$\psi(t) \approx \psi_0(t) + \frac{s(t)}{A}\cos\psi_0(t) - \frac{c(t)}{A}\sin\psi_0(t) \tag{8.46}$$

We should note that Equation 8.46 is not always a good approximation to Equation 8.45 even when SNR_i is large. In the first place, even though $\langle c^2(t)\rangle = \langle n^2(t)\rangle = N_i \ll A^2$, it happens that $c(t) \geq A$ with a small, but finite, probability. Also, the $\sin\psi_0(t)$ or $\cos\psi_0(t)$ terms can be small. We call those rare times when Equation 8.46 does not adequately describe the output the *anomaly error*. This anomaly error will be discussed further.

It follows from Equations 8.42, 8.44, and 8.46 that for large SNR_i

$$y_i(t) = \frac{d\psi(t)}{dt} \approx 2\pi\ \Delta f m(t) + \frac{1}{A}\frac{d}{dt}N(t) \tag{8.47}$$

where

$$N(t) = s(t) \cos \psi_0(t) - c(t) \sin \psi_0(t) \tag{8.48}$$

The first term in Equation 8.47 is the undistorted message, and the second term is, to a first-order approximation, the effect of noise. The output signal power is therefore

$$S_o = (2\pi \Delta f)^2 \langle m^2(t) \rangle \tag{8.49}$$

and the output noise power is

$$N_o = \frac{1}{A^2} \int_{-B}^{B} (2\pi f)^2 S_N(f) \, df \tag{8.50}$$

This last expression comes from noting that the operation $(1/A) \, d/dt$ can be thought of as a linear filter with transfer function $(1/A)j2\pi f$; hence the spectrum of the noise is given by $(1/A^2)(2\pi f)^2 S_N(f)$. Furthermore, since the signal $m(t)$ is concentrated in a band of B Hz, we can use an LPF to remove any of the noise beyond B Hz without distorting the signal.

The problem remains to determine the spectrum of $N(t)$ as given by Equation 8.48. We first find the autocorrelation function $R_N(\tau) = \langle N(t)N(t - \tau) \rangle$:

$$R_N(\tau) = \langle s(t)s(t - \tau) \cos \psi_0(t) \cos \psi_0(t - \tau) \rangle$$
$$+ \langle c(t)c(t - \tau) \sin \psi_0(t) \sin \psi_0(t - \tau) \rangle$$
$$- \langle s(t)c(t - \tau) \cos \psi_0(t) \sin \psi_0(t - \tau) \rangle$$
$$- \langle c(t)s(t - \tau) \sin \psi_0(t) \cos \psi_0(t - \tau) \rangle$$

The terms $\langle s(t)s(t - \tau) \rangle = \langle c(t)c(t - \tau) \rangle = R_q(\tau)$ and $\langle s(t)c(t - \tau) \rangle = 0 = \langle c(t)s(t - \tau) \rangle$ since the quadrature processes are statistically independent, and

$$\langle s(t)s(t - \tau) \cos \psi_0(t) \cos \psi_0(t - \tau) \rangle = R_q(\tau) \langle \cos \psi_0(t) \cos \psi_0(t - \tau) \rangle$$

since $s(t)$ and $\psi_0(t)$ can be assumed to be statistically independent. Therefore,

$$R_N(\tau) = R_q(\tau) \langle \cos \psi_0(t) \cos \psi_0(t - \tau) + \sin \psi_0(t) \sin \psi_0(t - \tau) \rangle$$
$$= R_q(\tau) \langle \cos[\psi_0(t - \tau) - \psi_0(t)] \rangle \tag{8.51}$$

Because multiplication in the time domain becomes convolution in the transform domain, the Fourier transform of Equation 8.51 becomes

$$S_N(f) = S_q(f) * \xi(f) \tag{8.52}$$

where $*$ symbolizes convolution and $\xi(f)$ is the Fourier transform of $\langle \cos[\psi_0(t - \tau) - \psi_0(t)] \rangle$.

We now argue that the quadrature processes [with bandwidth $(\beta + 1)B$; see Figure 8.7] vary more rapidly than $m(t)$ (with bandwidth B), which in turn varies more rapidly than $\psi_0(t) = 2\pi \Delta f \int m(t) \, dt$. Equivalently, the spectrum of the quadrature process is wider than that of $m(t)$, which is wider than that of $\psi_0(t)$. It can be argued that the transform $\xi(f)$ is quite narrow compared to $S_q(f)$ as seen in Figure 8.8. Using graphical methods of evaluating

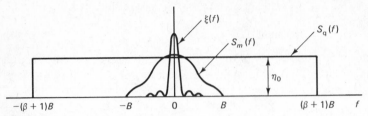

Figure 8.8 Comparison of spectrum.

a convolution integral and noting that $\int_{-\infty}^{\infty} \xi(f)\, df = \langle \cos 0 \rangle = 1$, we can sketch the spectrum $S_N(f)$ as in Figure 8.9a and the noise spectrum $(1/A^2)(2\pi f)^2 S_N(f)$ as in Figure 8.9b. The precise shape near $(\beta + 1)B$ Hz, and hence the precise shape of $\xi(f)$, is not significant since only the frequency range between 0 and B Hz is retained by the LPF. Thus, from Equation 8.50, the output noise power is

$$N_o = 2\left(\frac{2\pi}{A}\right)^2 \eta_0 \int_0^B f^2 \, df = \frac{2}{3A^2}(2\pi B)^2 \eta_0 B \tag{8.53}$$

From Equations 8.49 and 8.53 the output SNR for FM is

$$\text{SNR}_o = 3A^2 \beta^2 \frac{\langle m^2(t) \rangle}{2\eta_0 B} \tag{8.54}$$

since $\beta = \Delta f / B$. If we compare this result with AM (Equation 8.35) assuming 100% modulation ($k = 1$), we get

$$\text{SNR}_{\text{FM}} = 3\beta^2 \, \text{SNR}_{\text{DSB-AM}} \tag{8.55}$$

If follows that for large β FM performs much better than AM. Thus FM with

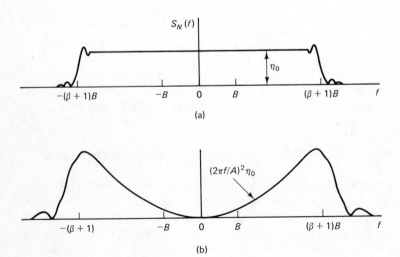

(a)

(b)

Figure 8.9 Output noise spectrum.

$\beta \leq 4$ performs as much as 48 times (16.8 dB) better than standard AM. Comparing the output SNR with the input SNR (Equation 8.41) gives

$$\text{SNR}_o = 3\beta^2(\beta + 1)\, 2\langle m^2(t)\rangle\, \text{SNR}_i \qquad\qquad (8.56)$$
$$\geq 3\beta^2(\beta + 1)\, \text{SNR}_i \qquad\qquad (8.57)$$

since typically $\langle m^2(t)\rangle < \frac{1}{2}$. Unlike for AM, the output SNR can be much larger than the input SNR. This difference and the reason FM performs so well can be seen with the help of Figure 8.9b. The bulk of the large input noise demodulates into a frequency range that is outside the range of the message. Thus the final low-pass filter of bandwidth B passes the message and removes most of the noise.

Equation 8.54 implies that the output SNR can be made as large as desired simply by increasing β. There is, of course, a limitation. The output signal power is a constraint of the transmitter, whereas the input noise power grows with β or the bandwidth (see Equation 8.40). Thus SNR_i decreases as β increases. The analysis is based on the assumption that SNR_i is large; when it drops below some threshold value, the equations are no longer valid and the output SNR drops sharply. These ideas are shown in Figure 8.10. Thus the SNR_o increases as β^2 increases (see Equation 8.54) until the FM threshold phenomena occurs, and then it decreases sharply.

Unlike the AM case, in which the threshold effect was a consequence of envelope detection, the *threshold effect* for FM occurs for all receivers. The location of the threshold, however, depends on the receiver. The threshold of an ideal frequency discriminator would occur for larger β (smaller SNR_i) than more typical FM receivers, as shown in Figure 8.10. FM receivers have been studied that can extend this threshold effect [1, 2]. The demodulator of Figure 4.17, which employs a PLL, is one such receiver [3]. Another way to view the difference between receivers is to plot (as in Figure 8.11) the output SNR as a function of the input power S_i for fixed β. Imagine reducing S_i by moving away from the transmitting antenna; there would be no noticeable difference between receivers until the threshold for the standard receiver is reached. Beyond this point, receivers with extended thresholds can pick up signals that are too weak for the standard FM receiver.

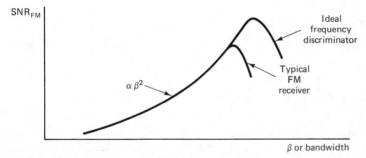

Figure 8.10 SNR$_o$ versus bandwidth.

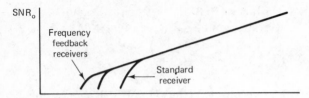

Figure 8.11 FM threshold effect.

EXAMPLE 8.2 ■

We wish to design a point-to-point FM communications link with the largest possible SNR_o. Our constraints are that the signal power ($A^2/2$) at the receiver is 6 V^2, the spectral level $\eta_0/2$ at the receiver is 10^{-6} V^2/Hz, the message bandwidth $B = 15$ kHz, the normalized message average power $\langle m^2(t) \rangle = 0.2$, and receiver threshold occurs at +10 dB. From Equation 8.41 and the receiver threshold

$$\frac{1}{\beta + 1} \frac{A^2/2}{2\eta_0 B} = \frac{1}{(\beta + 1)} \frac{6}{4 \times 10^{-6} \times 15 \times 10^3} \geq 10$$

Solving for β, we get $\beta \leq 9$. From Equation 8.56, using $\beta = 9$ to maximize SNR_o, we obtain

$$SNR_o = 3 \times 81 \times 10 \times 2 \times 0.2 \, SNR_i = 972 \, SNR_i$$
$$= 972 \times 10 = 9720 \quad (39.9 \text{ dB})$$

For comparison, had we used SSB-SC

$$S_i \simeq \tfrac{1}{2} A^2 \langle m^2(t) \rangle = 6 \times 0.2 = 1.2$$
$$N_i \simeq \tfrac{1}{2}(2\eta_0 B) = 2 \times 10^{-6} \times 15 \times 10^3 = 3 \times 10^{-2}$$
$$SNR_i = 1.2/(3 \times 10^{-2}) = 40$$
$$SNR_o = 2 \, SNR_i = 80 \quad (19 \text{ dB})$$

Note that the FM bandwidth $2(\beta + 1)B = 300$ kHz, is 20 times larger than that of SSB-SC AM (15 kHz). If we had a bandwidth constraint as well, our answer for FM would change. Thus, if the bandwidth were constrained to 150 kHz, then our design would require $\beta = 4$, and from Equation 8.54

$$SNR_o = (3 \times 12 \times 16 \times 0.2)/(4 \times 10^{-6} \times 15 \times 10^3)$$
$$= 1920 \quad (32.8 \text{ dB}) \qquad \blacksquare \blacksquare$$

8.6 PREEMPHASIS TECHNIQUES

The output SNR of any cw technique can usually be improved by taking advantage of our knowledge of typical spectra (speech or music) shapes. If the shape of the signal spectra is significantly different from that of the output

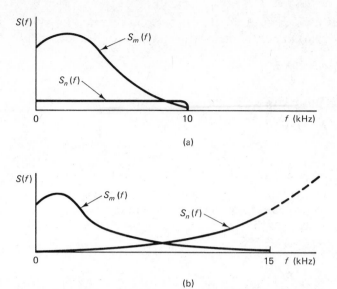

(a)

(b)

Figure 8.12 Signal and noise power spectra for (a) AM and (b) FM.

or demodulated noise, then an increase in the SNR can be achieved. Observe these spectra in Figure 8.12, where the bandwidth of broadcast FM is larger than that of broadcast AM, i.e., where there is higher fidelity. Note the striking difference between the signal and noise spectra for the FM case in particular. The ratio of signal power per hertz to noise power per hertz is much larger for low frequencies than for high frequencies. The SNR can be increased by first distorting the signal at the transmitter by a preemphasis filter that exaggerates the high frequencies and then correcting the distortion with an appropriate deemphasis filter at the receiver (see Figure 8.13). If the preemphasized signal spectrum still falls off with frequency much faster than the output noise spectrum, then the deemphasis filter removes significantly more noise power than signal power.

With the deemphasis filter $H^{-1}(f)$, the spectrum of the noise in the band of interest ($|f| < B$) is now proportional to $f^2|H^{-1}(f)|^2$ instead of to f^2. Thus the output noise power is reduced by

$$\int_0^B f^2|H^{-1}(f)|^2 \, df \bigg/ \int_0^B f^2 \, df$$

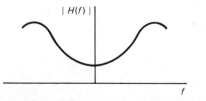

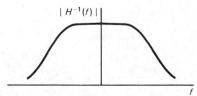

Figure 8.13 Preemphasis and deemphasis filters.

Since the deemphasis filter changes the signal spectrum from $S_m(f)|H(f)|^2$ (preemphasized spectrum) to $S_m(f)$, the output signal power is reduced by

$$\int_0^B S_m(f)\,df \Big/ \int_0^B S_m(f)|H(f)|^2\,df$$

It follows that the SNR is increased by the factor I, where

$$I = \left[\int_0^B f^2\,df \Big/ \int_0^B f^2|H^{-1}(f)|^2\,df \right]$$

$$\times \left[\int_0^B S_m(f)\,df \Big/ \int_0^B S_m(f)|H(f)|^2\,df \right] \tag{8.58}$$

We note that phase modulation (PM) can be regarded as a generalized FM technique with a preemphasis filter that is an integrator, $[|H(f)| = 2\pi f]$ and a deemphasis filter that is a differentiator $[|H^{-1}(f)| = 1/2\pi f]$. Thus for PM

$$I_{\text{PM}} = \left[\int_0^B f^2\,df \Big/ \left(\frac{1}{2\pi}\right)^2 \int_0^B df \right]$$

$$\times \left[\int_0^B S_m(f)\,df \Big/ (2\pi)^2 \int_0^B S_m(f) f^2\,df \right]$$

$$= \frac{B^2}{3} \int_0^B S_m(f)\,df \Big/ \int_0^B S_m(f) f^2\,df \tag{8.59}$$

Hence the SNR for PM is achieved by multiplying the SNR for FM (Equation 8.54) by I_{PM} (Equation 8.59).

EXAMPLE 8.3 ■

Let us model $S_m(f)$ by $K/[1 + (f/f_c)^2]^2$, where $f_c = 2.1$ kHz, and determine the SNR improvement of PM over FM. Substituting into Equation 8.59 and changing variables $x = f/f_c$, we get

$$I_{\text{PM}} = \frac{B^2}{3} f_c \int_0^{B/f_c} \frac{dx}{(1+x^2)^2} \Big/ f_c^3 \int_0^{B/f_c} \frac{x^2}{(1+x^2)^2}\,dx$$

From tables of integration

$$\int_0^{B'} \frac{dx}{(1+x^2)^2} = \frac{1}{2}\left(\tan^{-1} B' + \frac{B'}{1+B'^2} \right)$$

and

$$\int_0^{B'} \frac{x^2\,dx}{(1+x^2)^2} = \frac{1}{2}\left(\tan^{-1} B' - \frac{B'}{1+B'^2} \right)$$

Therefore,

$$I_{\text{PM}} = \frac{1}{3} B'^2 \frac{\tan^{-1} B' + B'/(1+B'^2)}{\tan^{-1} B' - B'/(1+B'^2)}$$

where $B' = 15/2.1 = 7.14$. Evaluating gives $I_{\text{PM}} = 26.13$, which corresponds to an improvement of 14.17 dB. ■ ■

For standard broadcast FM, the deemphasis filters are simple first-order LPFs with cutoff frequency of 2.1 kHz; $|H^{-1}(f)|^2 = K/[1 + (f/f_c)^2]$. Thus Equation 8.58 becomes

$$I_{FM} = \left[\int_0^B f^2\,df \bigg/ K^2 \int_0^B f^2 \frac{1}{1 + (f/f_c)^2}\,df\right]$$
$$\times \left\{\int_0^B S_m(f)\,df \bigg/ \frac{1}{K^2}\int_0^B S_m(f)[1 + (f/f_c)^2]\,df\right\}$$
$$= \frac{1}{3}\frac{B'^3}{B' - \tan^{-1} B'}\int_0^B S_m(f)\,df \bigg/ \int_0^B S_m(f)[1 + (f/f_c)^2]\,df \qquad (8.60)$$

where $B = 15$ kHz, $f_c = 2.1$ kHz, and $B' = B/f_c = 7.14$. Using the same model for $S_m(f)$, we can determine (see Problem 8.12) that the improvement is 10.66 dB. Standard FM with preemphasis is on the order of 28 dB better than standard AM.

8.7 ANOMALY ERROR AND COMPARISON OF FM WITH PCM

The output of an ideal frequency discriminator is given by $d\psi(t)/dt$, where $\psi(t)$ is the phase of the received signal, given by

$$\psi(t) = \tan^{-1}\left[\frac{A\sin\psi_0(t) + s(t)}{A\cos\psi_0(t) + c(t)}\right] \pm 2\pi q \qquad (8.61)$$

where $\psi_0(t) = 2\pi\,\Delta f \int m(t)\,dt + \theta_0$. When the phase differs from the linearized approximation of Equation 8.37, the difference is called the anomaly error. Consider the plot of $\psi(t)$ in Figure 8.14a. Equation 8.46 is valid as long as $\psi(t)$ is close to $\psi_0(t)$. An occasional discrepancy, as indicated by the burst, is an anomaly. If $\psi(t)$ lies between $0°$ and $90°$ (both numerator and denominator are positive) and if there is a negative burst of $c(t)$ that is sufficiently negative to make the denominator negative, the phase would

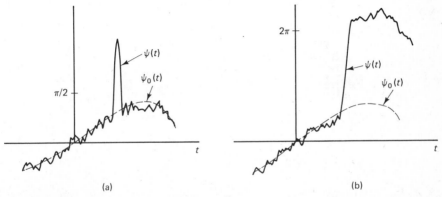

Figure 8.14 Phase versus time with anomalies.

temporarily increase above $90°$ (or $\pi/2$ rad). This type of anomaly, pictured in Figure 8.14a, occurs fairly regularly. Suppose now that, almost simultaneously, a negative burst of $s(t)$ occurs that makes the numerator negative before the denominator returns to its normally positive value. Such an occurrence is extremely rare for large SNR_i since $c(t)$ and $s(t)$ are statistically independent. Nevertheless, when this happens the phase will shift to the second quadrant (between $90°$ and $180°$) when $c(t)$ goes negative, to the third quadrant (between $180°$ and $270°$) when $s(t)$ goes negative, then to the fourth quadrant when $c(t)$ returns to a small value, and then back to the first quadrant when $s(t)$ returns to a small value. The phase $\psi(t)$ will have suddenly increased by $360°$, or 2π rad, when this rare anomaly occurs, as shown in Figure 8.14b. After the anomaly the phase is close to $\psi_0(t) + 2\pi$, but the significant point is the jump itself. The output of the ideal frequency discriminator is not $\psi(t)$ but $d\psi(t)/dt$, as shown in Figure 8.15. The more typical anomalies result in the doublet pulse of Figure 8.15a. These doublet pulses have a relatively small percentage of their energy in the signal band (see Problem 8.16) and are therefore largely removed by the final filters. On the other hand, the less-typical anomalies result in the pulses of Figure 8.15b. A significant portion of the energy of these pulses is in the very low-frequency band; hence they will introduce a "click" in the output. Usually these clicks occur so seldom that we do not notice them. The clicks become quite noticeable, however, just beyond the FM threshold.

Rice [4] has assumed a simple model in which the noise consists of two additive parts. One part is the noise distortion in the approximation of Equation 8.47, whose spectrum was shown in Figure 8.9b, and the other part is an impulse noise which models the clicks. If the average number of clicks per second is inversely related to the input SNR, the overall performance is very close to the true measured performance, as indicated in Figure 8.10. It follows that, conceptually at least, the anomaly effect can be regarded separately from the way in which FM utilizes the channel (i.e., $SNR \propto \beta^2$).

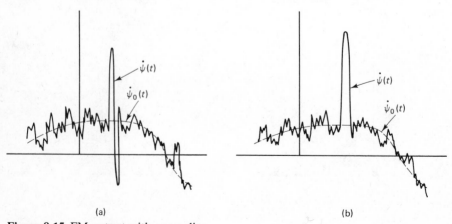

(a)   (b)

Figure 8.15 FM output with anomalies.

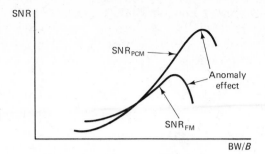

Figure 8.16 Comparison of FM and PCM.

For small bandwidths (or large input SNR) the anomaly effect can be ignored. However, for large bandwidths it becomes the dominant factor and causes the threshold effect. With this insight we can compare the performance of FM with discrete techniques such as PCM.

In Section 6.6, we discussed the output SNR for PCM, which is plotted versus the bandwidth in Figure 6.8. For discrete techniques there are two very distinct noise components. For a small number of quantization intervals, and hence a small bandwidth, the quantization error is the dominant effect. The actual noise causes a rare probability of error (or anomaly error) when the input SNR is large. If the bandwidth is increased by increasing the number of quantization intervals, the input SNR decreases and the anomaly error (probability of error) becomes the dominant factor. It seems reasonable, therefore, to compare the linearized analysis of FM in Section 8.5 with the quantization error of PCM (or any discrete technique).

We saw in Section 6.6 that for PCM

$$\text{SNR}_q = cM^2 = c \, \exp\!\left(\ln 2 \, \frac{B_{\text{PCM}}}{B}\right) \tag{8.62}$$

where B is the bandwidth of $m(t)$. This should be compared with Equation 8.55, which when combined with Equation 8.39 can be rewritten

$$\text{SNR}_{\text{FM}} = 3\!\left(\frac{B_{\text{FM}}}{2B} - 1\right)^{2} \text{SNR}_{\text{AM}} \tag{8.63}$$

Exponential growth far exceeds square-law growth, as seen in Figure 8.16. When the input SNR is small enough that the error probabilities for discrete techniques are significant, then the cw techniques are well beyond threshold and perform quite poorly.

8.8 EFFECT OF NONLINEAR OPERATIONS ON PDF MODELS

The rest of this chapter is concerned with evaluating the cost of using envelope detectors instead of matched filters when it is possible to make this replacement. In order to carry out this analysis, it is necessary to determine

PDF models for the envelope as given by Equations 8.19 and 8.25. These equations represent nonlinear operations, and we cannot presume the envelope to be Gaussian. In this section we consider some techniques for finding PDF models after nonlinear operations.

Single-Variable Case

Let us consider a random variable Y that is uniquely related to another random variable X by the relationship

$$y = h(x) \tag{8.64}$$

where $h(\ )$ is any absolutely continuous function. The function $h(\ )$ is said to be a one-to-one (1-1) mapping if there is a unique point x specified by every point y. An n-1 mapping is such that n points are specified by every point y. Figure 8.17a shows a 1-1 mapping, Figure 8.17b shows a 2-1 mapping, Figure 8.17c represents a more complicated mapping.

For an n-1 mapping, where x_{0i} is one of n points that map into y_0 [i.e., $y_0 = h(x_{0i})$ for $i = 1, \ldots, n$], it follows that

$$\Pr(y_0 < Y \le y_0 + \Delta y) = \sum_{i=1}^{n} \Pr(x_{0i} < X \le x_{0i} + \Delta x_i) \tag{8.65}$$

or

$$\int_{y_0}^{y_0+\Delta y} f_Y(y)\, dy = \sum_{i=1}^{n} \int_{x_{0i}}^{x_{0i}+\Delta x_i} f_X(x)\, dx$$

In the limit as $\Delta y \to 0$,

$$f_Y(y_0)\, \Delta y = \sum_{i=1}^{n} f_X(x_{0i})\, \Delta x_i \tag{8.66}$$

Thus for all possible output values y we can write

$$f_Y(y) = \sum_{i=1}^{n} \frac{f_X(x_i)}{\lim(\Delta y/\Delta x_i)} = \sum_{i=1}^{n} \frac{f_X(x_i)}{|h'(x_i)|}\Bigg|_{x_i = h_i^{-1}(y)} \tag{8.67}$$

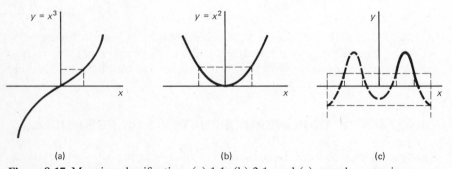

(a) (b) (c)

Figure 8.17 Mapping classification: (a) 1-1, (b) 2-1, and (c) complex mappings.

where $h'(x)$ is the derivative of $h(x)$ with respect to x and $h_i^{-1}(y)$ is the inverse function. For a 1-1 mapping Equation 8.67 becomes

$$f_Y(y) = \frac{f_X(x)}{|h'(x)|}\bigg|_{x=h^{-1}(y)} \tag{8.68}$$

EXAMPLE 8.4 ■

$Y = aX$ is a linear 1-1 mapping. From Equation 8.68,

$$f_Y(y) = \frac{f_X(x)}{|a|}\bigg|_{x=y/a} = \frac{f_X(y/a)}{|a|} \tag{8.69}$$

If X is a Gaussian random variable $N(m, \sigma^2)$,

$$f_Y(y) = \frac{1}{|a|\sqrt{2\pi\sigma^2}} \exp\left[-\frac{1}{2}\frac{(y/a-m)^2}{\sigma^2}\right]$$

$$= \frac{1}{\sqrt{2\pi a^2\sigma^2}} \exp\left[-\frac{1}{2}\frac{(y-am)^2}{a^2\sigma^2}\right]$$

Thus Y is Gaussian with mean am and variance $a^2\sigma^2$. We already know that a linear operation on a Gaussian random variable yields another Gaussian random variable, a result that could have been derived without the help of Equation 8.68.   ■ ■

EXAMPLE 8.5 ■

$Y = X^3$ is a nonlinear 1-1 mapping. Since $h'(x) = dy/dx = 3x^2$ and $x = \sqrt[3]{y}$, it follows from Equation 8.68 that

$$f_Y(y) = \frac{f_X(\sqrt[3]{y})}{3|y^{2/3}|} \tag{8.70}$$

Assuming X is uniformly distributed in $(0, a]$, $f(x) = 1/a$ for $0 < x \le a$ and zero elsewhere, it follows that

$$f_Y(y) = \begin{cases} \dfrac{1/a}{3|y^{2/3}|} & 0 < y \le a^3 \\ 0 & \text{elsewhere} \end{cases}$$

or

$$f_Y(y) = \begin{cases} \frac{1}{3}y^{-2/3} & 0 < y \le a^3 \\ 0 & \text{elsewhere} \end{cases} \tag{8.71}$$

■ ■

EXAMPLE 8.6 ■

Let $Y = X^2$, a nonlinear 2-1 mapping. Since $h'(x) = dy/dx = 2x$ and $x_i = \pm\sqrt{y}$, it follows from Equation 8.67 that

$$f_Y(y) = \frac{f_X(+\sqrt{y}) + f_X(-\sqrt{y})}{2|\sqrt{y}|} \tag{8.72}$$

If $f_X(x)$ is Gaussian with zero mean and variance σ^2,

$$f_X(+\sqrt{y}) = f_X(-\sqrt{y}) = \frac{1}{\sqrt{2\pi\sigma^2}} \exp\left(-\frac{1}{2}\frac{y}{\sigma^2}\right)$$

$$f_Y(y) = \begin{cases} \frac{1}{\sqrt{2\pi\sigma^2 y}} \exp\left(-\frac{1}{2}\frac{y}{\sigma^2}\right) & y > 0 \\ 0 & y \leq 0 \end{cases} \tag{8.73}$$

This is the χ^2 PDF with one degree of freedom.   ■ ■

Multiple-Variable Case

Let us now consider a random variable Y_1 that is uniquely related to m other random variables X_i by the relationship

$$y_1 = h_1(x_1, x_2, \ldots, x_m) \tag{8.74}$$

where $h_1(\ )$ is any continuous function of m variables that corresponds to a 1-1 mapping. Quite often Equation 8.74 is one of a set of such equations

$$y_i = h_i(x_1, x_2, \ldots, x_m) \tag{8.75}$$

$i = 1, 2, \ldots, m$, that defines a transformation of a vector $\mathbf{X}$ into a vector $\mathbf{Y}$. When this is not the case, we can always augment Equation 8.74 by arbitrarily defining the remaining equations to be $Y_j = X_j$ for $j > 1$. Since these are all 1-1 mappings, it follows that

$$\Pr(\mathbf{y} < \mathbf{Y} \leq \mathbf{y} + \Delta\mathbf{y}) = \Pr(\mathbf{x} < \mathbf{X} \leq \mathbf{x} + \Delta\mathbf{x}) \tag{8.76}$$

where $\mathbf{x}$ is the value of the $\mathbf{X}$ vector that maps into the value $\mathbf{y}$. Thus in the limit as $\Delta\mathbf{y} \to 0$

$$f_Y(\mathbf{y})\, \Delta\mathbf{y} = f_X(\mathbf{x})\, \Delta\mathbf{x} \tag{8.77}$$

which is a vector version of Equation 8.66 for 1-1 mappings. Equation 8.68 can be extended to the vector case. It has been proven [5] that the $\Delta\mathbf{x}$ incremental volume is $|\mathcal{J}(\mathbf{x/y})|$ times larger than the $\Delta\mathbf{y}$ incremental volume where $\mathcal{J}(\mathbf{x/y})$ is called the Jacobian and defined by the determinant

$$\mathcal{J}(\mathbf{x/y}) = \begin{vmatrix} \dfrac{\partial x_1}{\partial y_1} & \dfrac{\partial x_1}{\partial y_2} & \cdots & \dfrac{\partial x_1}{\partial y_m} \\ \dfrac{\partial x_2}{\partial y_1} & \dfrac{\partial x_2}{\partial y_2} & \cdots & \dfrac{\partial x_2}{\partial y_m} \\ \vdots & \vdots & & \vdots \\ \dfrac{\partial x_m}{\partial y_1} & \dfrac{\partial x_m}{\partial y_2} & \cdots & \dfrac{\partial x_m}{\partial y_m} \end{vmatrix} \tag{8.78}$$

where $|\mathbf{A}|$ represents the determinant of the matrix $\mathbf{A}$. It follows that

$$f_{\mathbf{Y}}(\mathbf{y}) = f_{\mathbf{X}}(\mathbf{x})|\mathcal{J}(\mathbf{x}/\mathbf{y})|\bigg|_{\mathbf{x}=h^{-1}(\mathbf{y})} \tag{8.79}$$

If we seek only the PDF of Y_1, this is determined from

$$f_{Y_1}(y_1) = \int_{-\infty}^{\infty}\int_{-\infty}^{\infty}\cdots\int_{-\infty}^{\infty} f_{\mathbf{Y}}(\mathbf{y})\, dy_2\, dy_3 \cdots dy_m \tag{8.80}$$

EXAMPLE 8.7 ■

Consider the transformation $Z = X + Y$. By arbitrarily defining $V = Y$ we have a mapping of the vector (X, Y) into the vector (Z, V). The Jacobian of this transformation is

$$\mathcal{J} = \begin{vmatrix} \dfrac{\partial z}{\partial x} & \dfrac{\partial z}{\partial y} \\[2mm] \dfrac{\partial v}{\partial x} & \dfrac{\partial v}{\partial y} \end{vmatrix} = \begin{vmatrix} 1 & 1 \\ 0 & 1 \end{vmatrix} = 1$$

which means that the differential volumes are equal. It follows that

$$f_{Z,V}(z, v) = f_{X,Y}(x, y)\bigg|_{\substack{x=z-v \\ y=v}}$$

$$= f_{X,Y}(z - v, v) \tag{8.81}$$

To find the PDF of Z, if X and Y are independent,

$$f_Z(z) = \int_{-\infty}^{\infty} f_X(z - v)f_Y(v)\, dv \tag{8.82}$$

which we previously argued in an alternative way (Equation 5.74).

■ ■

8.9 ENVELOPE AND PHASE STATISTICS

We have seen in Section 8.3 that if $E(t)$ and $\theta(t)$ are the envelope and phase of narrow-band Gaussian noise with average power N_0, they are related to the quadrature processes by the equations

$$\begin{aligned} c(t) &= E(t)\cos\theta(t) \\ s(t) &= E(t)\sin\theta(t) \end{aligned} \tag{8.83}$$

where

$$f(c, s; 0) = \frac{1}{2\pi N_0} \exp\left[-\frac{1}{2N_0}(c^2 + s^2)\right] \tag{8.84}$$

Furthermore, we have argued that $E(t)$ is the output of the ideal envelope detector of Figure 8.4. If we regard Equations 8.83 as defining a mapping from the vector (c, s) to the vector (E, θ), it follows that

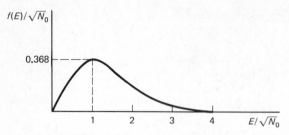

Figure 8.18 PDF of the envelope of narrowband noise.

$$f(E, \theta) = f(c, s)|\mathcal{J}| \Big|_{\substack{c=E\cos\theta \\ s=E\sin\theta}} \tag{8.85}$$

where

$$\mathcal{J} = \begin{vmatrix} \dfrac{\partial c}{\partial E} & \dfrac{\partial c}{\partial \theta} \\ \dfrac{\partial s}{\partial E} & \dfrac{\partial s}{\partial \theta} \end{vmatrix} = \begin{vmatrix} \cos\theta & -E\sin\theta \\ \sin\theta & E\cos\theta \end{vmatrix}$$

$$= E\cos^2\theta + E\sin^2\theta = E \tag{8.86}$$

Since $c^2 + s^2 = E^2$,

$$f(E, \theta) = \begin{cases} \dfrac{E}{2\pi N_0} \exp\left(-\dfrac{1}{2N_0}E^2\right) & E \geq 0 \quad 0 < \theta \leq \pi \\ 0 & \text{elsewhere} \end{cases} \tag{8.87}$$

In order to determine the envelope and phase statistics, we use Equation 8.80:

$$f(E) = \int_0^{2\pi} f(E, \theta)\, d\theta = \frac{E}{N_0} \exp\left(-\frac{1}{2}\frac{E^2}{N_0}\right) \qquad E \geq 0 \tag{8.88}$$

Similarly

$$f(\theta) = \int_0^{\infty} f(E, \theta)\, dE = \frac{1}{2\pi} \qquad 0 < \theta \leq 2\pi \tag{8.89}$$

The envelope has a Rayleigh PDF, Equation 8.88, which is shown in Figure 8.18, and the phase is uniformly distributed. It is easily seen that the phase and envelope are uncorrelated; that is, $f(E, \theta) = f(E)f(\theta)$.[2]

EXAMPLE 8.8 ∎

If one omits the final square root device of the ideal envelope detector of Figure 8.4, then the output is $E^2(t)$. Following Example 8.6 but with X positive only, i.e., a 1-1 mapping,

---

[2]This means that a sample of $E(t)$ and one of $\theta(t)$ taken at the same time are statistically independent. However, $E(t)$ and $\theta(t)$ are not statistically independent processes [6].

$$f_Z(z) = \frac{1}{2\sqrt{z}} f_E(\sqrt{z})$$

where $Z = E^2$. From Equation 8.88,

$$f_Z(z) = \frac{1}{2\sqrt{z}} \frac{\sqrt{z}}{N_0} \exp\left(-\frac{z}{2N_0}\right) = \frac{1}{2N_0} \exp\left(-\frac{z}{2N_0}\right) \qquad z > 0 \qquad (8.90)$$

We see that the square of the envelope of narrow-band Gaussian noise is exponentially distributed. ■ ■

We have also seen in Section 8.3 that if, in addition to narrow-band noise, there is a narrow-band signal defined by

$$s(t) = A(t) \cos[2\pi f_c t - \psi(t)] \qquad (8.91)$$

then the envelope and phase processes are defined by

$$\begin{aligned}A(t) \cos \psi(t) + c(t) &= E(t) \cos \theta(t) \\ A(t) \sin \psi(t) + s(t) &= E(t) \sin \theta(t)\end{aligned} \qquad (8.92)$$

For this signal-plus-noise case, Equation 8.85 still holds where $\mathcal{J} = E$ as before (i.e., $\partial c/\partial E$ and $\partial c/\partial \theta$ are the same for Equations 8.92 as they are for Equations 8.83). The difference is

$$\begin{aligned}c^2 + s^2 &= (E \cos \theta - A \cos \psi)^2 + (E \sin \theta - A \sin \psi)^2 \\ &= E^2 + A^2 - 2AE(\cos \theta \sin \psi + \sin \theta \cos \psi) \\ &= E^2 + A^2 - 2AE \cos(\theta - \psi)\end{aligned} \qquad (8.93)$$

Combining Equations 8.84, 8.85, and 8.93 yields

$$f(E, \theta) = \frac{E}{2\pi N_0} \exp\left\{-\frac{1}{2\pi N_0}[E^2 + A^2 - 2AE \cos(\theta - \psi)]\right\} \qquad (8.94)$$

The PDF for the envelope of narrow-band signal plus noise becomes

$$\begin{aligned}f(E) &= \int_0^{2\pi} f(E, \theta)d\theta \\ &= \frac{E}{N_0} \exp\left[-\frac{1}{2N_0}(E^2 + A^2)\right] \frac{1}{2\pi} \int_0^{2\pi} \exp\left[\frac{1}{N_0} EA \cos(\theta - \psi)\right] d\theta\end{aligned}$$

Since

$$\frac{1}{2\pi} \int_a^{2\pi + a} \exp(\rho \cos \alpha)\, d\alpha = I_0(\rho)$$

where $I_0(\rho)$ is the modified Bessel function of the first kind and order zero, it follows that

$$f(E) = \frac{E}{N_0} \exp\left[-\frac{1}{2N_0}(E^2 + A^2)\right] I_0\left(\frac{EA}{N_0}\right) \qquad (8.95)$$

which is called the Ricean PDF [7] and is sketched in Figure 8.19. The statistics of the phase is somewhat complicated and $f(E, \theta) \neq f(E)f(\theta)$.

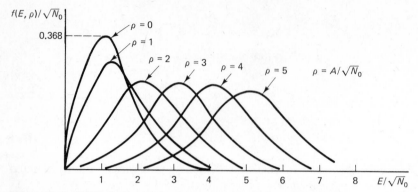

Figure 8.19 PDF of the envelope of narrowband signal and noise.

EXAMPLE 8.9 ■

Let us reconsider Example 8.8, where the final square root device of the ideal envelope detector is omitted and the output is $Z = E^2$. As before

$$f_Z(z) = \frac{1}{2\sqrt{z}} f_E(\sqrt{z})$$

Thus

$$f_Z(z) = \frac{1}{2N_0} \exp\left[-\frac{1}{2N_0}(z + A^2)\right] I_0\left(\frac{\sqrt{z}A}{N_0}\right) \tag{8.96}$$

■ ■

8.10 COMPARISON OF ENVELOPE DETECTORS AND MATCHED FILTERS

It has been mentioned a number of times that for certain modulated forms of discrete communication techniques (notably ASK, FSK, and FPM) the optimum matched filters can be replaced with envelope detectors at some cost in performance. We can now evaluate the performance of these suboptimal detectors.

Let us consider the decision making part of an ASK receiver as shown in Figure 8.20. Our goal is to compute an expression for the probability of error using this receiver and compare it with that based on the optimal receiver of Figure 6.16, which uses a matched filter. The input to the comparator is either

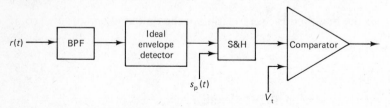

Figure 8.20 Subopimal binary decision device.

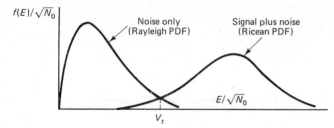

Figure 8.21 A posteriori PDFs for samples of envelope.

a sample of the envelope of narrow-band noise only or a sample of the envelope of narrow-band signal plus noise. The PDF models are Rayleigh and Ricean, respectively, as seen in Figure 8.21. Determining whether a signal pulse or no pulse is more likely is seen to be equivalent to comparing the sample with some threshold voltage V_t. We shall assume that the two signals have equal a priori probabilities and the threshold is set accordingly.

If one performed a monotonic operation on the output of the envelope detector and adjusted the comparator threshold accordingly, the probability of error would not be affected. Since the envelope is always positive, a square law device is strictly monotonic. Squaring the envelope is advantageous, of course, since that omits the square root device from the end of the ideal envelope detector. With this modification we have from Examples 8.8 and 8.9

$$f_n(x) = \frac{1}{2N_0} \exp\left(-\frac{x}{2N_0}\right) \qquad x \geq 0 \tag{8.97}$$

and

$$f_{s+n}(x) = \frac{1}{2N_0} \exp\left[-\frac{1}{2N_0}(x + A^2)\right] I_0\left(\frac{\sqrt{x}A}{N_0}\right) \tag{8.98}$$

The overall probability of error assuming both signals are equally likely is

$$P_b(e) = \frac{1}{2}\int_{V_t}^{\infty} f_n(x)\, dx + \frac{1}{2}\int_{0}^{V_t} f_{s+n}(x)\, dx \tag{8.99}$$

where V_t is determined from $f_n(V_t) = f_{s+n}(V_t)$. This expression could be evaluated with a computer as a function of the parameter $A/\sqrt{N_0}$ [8].

It is easier to evaluate the probability of error for an FSK signal set with a receiver employing envelope detectors instead of matched filters. Such a receiver (see Figure 3.44, for example) bases its decision on the sign of the difference between the output of the two envelope detectors. Thus, if x is the output of the envelope detector that contains a signal pulse and y is the output of the other detector that contains noise only, the probability of error becomes

$$P_b(e) = \Pr(y - x > 0) \tag{8.100}$$

where y is distributed according to Equation 8.97 and x according to Equation 8.98. We evaluate Equation 8.100 by

$$P_b(e) = \int_0^\infty \int_x^\infty f(x, y)\, dy\, dx$$

$$= \int_0^\infty f_{s+n}(x) \int_x^\infty f_n(y)\, dy\, dx \tag{8.101}$$

since the outputs of the two detectors are independent (see Example 5.15). Performing the inside integral first, we obtain

$$\int_x^\infty f_n(y)\, dy = \int_x^\infty \frac{1}{2N_0} \exp\left(-\frac{y}{2N_0}\right) dy = \exp\left(-\frac{x}{2N_0}\right)$$

It follows that

$$P_b(e) = \frac{1}{2N_0} \int_0^\infty \exp\left(-\frac{A^2}{2N_0}\right) \exp\left(-\frac{2x}{2N_0}\right) I_0\left(\frac{\sqrt{x}A}{N_0}\right) dx$$

If we set $x = \xi/2$ and rearrange terms slightly, this becomes

$$P_b(e) = \frac{1}{2} \exp\left(-\frac{A^2}{4N_0}\right) \int_0^\infty \frac{1}{2N_0} \exp\left(-\frac{A^2}{4N_0}\right) \exp\left(-\frac{\xi}{2N_0}\right) I_0\left(\frac{\sqrt{\xi}A}{\sqrt{2}N_0}\right) d\xi$$

The integrand is the PDF of Equation 8.98 with A^2 replaced by $A^2/2$ and as such must integrate out to unity. It follows that for the suboptimum FSK receiver

$$P_b(e) = \frac{1}{2} \exp\left(-\frac{A^2}{4N_0}\right) \tag{8.102}$$

The value of N_0 depends on the type of LPF used in the ideal envelope detector (see Figure 8.4). The "narrowest" LPF that can be implemented is an integrator that integrates over an entire pulse, i.e., a filter "matched" to the envelope. For such a filter $h(\tau) = 1/T$ for $t - T \le \tau \le t$ and zero elsewhere, or $|H(f)| = \operatorname{sinc} fT$. Thus the spectrum of the noise after the filter (the input is a quadrature process of level η_0) is $S(f) = \eta_0 \operatorname{sinc}^2 fT$ and the minimum power is

$$N_0 = \int_{-\infty}^\infty S(f)\, df = \frac{\eta_0}{T} \tag{8.103}$$

Thus for the "best" envelope detectors that use integrators for LPFs,

$$P_b(e) = \frac{1}{2} \exp(-A^2 T/4\eta_0)$$
$$= \frac{1}{2} \exp(-E_s/2\eta_0) \tag{8.104}$$

since E_s is the energy in a pulse or $E_s = \frac{1}{2}A^2 T$.

We saw (Equation 7.51) that for the optimum FSK receiver that uses matched filters

$$P_{b(opt)}(e) = Q\left(\sqrt{\frac{E_s}{\eta_0}}\right) \approx \frac{1}{\sqrt{2\pi E_s/\eta_0}} \exp\left(-\frac{E_s}{2\eta_0}\right) \tag{8.105}$$

The latter inequality is the tight bound discussed in Section 5.7.

Since, to a first approximation, it is sufficient to compare the arguments of the exponential only, it follows that the best envelope detector causes no appreciable degradation in performance. There is, of course, a trivial difference due to the different multipliers of the exponential. Thus to achieve a probability of error per pulse of 10^{-4} requires 0.5 dB more energy with the envelope detectors than the optimum matched filters. For smaller error probabilities this difference decreases toward zero.

We briefly discussed in Chapter 6 the fact that integrators can be approximated by simple first-order LPFs with a degradation of 0.9 or 2.5 dB.[3] Thus, if the LPFs of the ideal envelope detector were first-order filters instead of integrators, there would be a degradation of between 1 and 3 dB relative to optimum matched filters. Finally, if the simple envelope detectors used in standard AM demodulation (see Figure 2.11) were used, the degradation would be even greater, on the order of 3 dB, compared with matched filters. It is with these ideas in mind that we conclude that the use of envelope detectors instead of matched filters causes a degradation somewhere between 0 and 3 dB depending on the details of the envelope detector.

PROBLEMS

8.1. Following the procedure of Chapter 2, prove Equation 8.2.

8.2. Determine and completely specify a bandpass model for a zero-mean Gaussian process whose spectrum is given below:

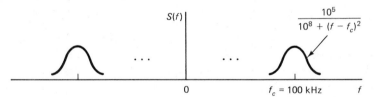

What are the autocorrelation functions and cross-correlation functions of the quadrature processes?

8.3. In some cases, such as single-sideband transmission, it is desirable to have a bandpass model for a signal in terms of a frequency f_0 that is not the center frequency of the spectrum. Determine and completely specify a bandpass model for a Gaussian process with the spectrum shown in terms of the indicated f_0 rather than the center frequency.

---

[3] Depending on whether the LPF is discharged suddenly, as the integrator is, to remove intersymbol interference.

8.4. You have the following constraints for a cw communications link:

  i. $\langle m_n^2(t) \rangle = 0.4$ V^2.
  ii. The bandwidth of $m(t)$ is 10 kHz.
  iii. The noise at the receiver is white with spectral level $\eta_0/2 = 10^{-5}$ V^2/Hz.
  iv. The received signal power S_i is given by $S_i = 10^8/(10^4 + r^2)$ V^2, where r is the distance (in meters) between the receiver and the transmitter.

  (a) Compute SNR$_i$ for $r = 1000$ m.
  (b) Compute SNR$_o$ for DSB-AM assuming a modulation index $k = 0.9$.
  (c) Compute SNR$_o$ for SSB-SC-AM.

8.5. You have the same constraints as in Problem 8.4.
  (a) Assume that your receiver is an envelope detector with a threshold at SNR$_i$ = 13 dB. What is the distance between your receiver and the transmitter at the threshold? Estimate the output SNR at threshold. (Assume the large SNR$_i$ still holds.)
  (b) Assume now that SSB-SC modulation was employed and your receiver is coherent. How far can your receiver be from the transmitter and have the same SNR$_o$ as in part (a)?

8.6. An audio signal that is bandlimited to 10 kHz is to be transmitted by either AM or FM. The amplitude is such as to provide 100% carrier modulation in the AM case and a frequency deviation $\Delta f = 75$ kHz in the FM case. The input SNR is 30 dB for both the AM and FM systems.
  (a) Caluclate the transmission bandwidth required in both cases.
  (b) Calculate the output SNR for each system.

8.7. You are to design a point-to-point FM modulation system that has the largest possible output SNR. You have the following constraints:

  i. The received signal (carrier) amplitude is 10 V.
  ii. The received two-sided noise spectral level $\eta_0/2 = 10^{-5}$ V^2/Hz.
  iii. The analog message bandwidth is 20 kHz.
  iv. The receiver threshold occurs when the input SNR is equal to 10 (i.e., for SNR$_i$ < 10, the receiver will be beyond the threshold and the output will be very noisy).

  (a) Determine the maximum amount of modulation Δf that you can use.
  (b) What is the output SNR obtained?

8.8. Suppose in the previous problem there was an additional available channel bandwidth constraint of 200 kHz. Would this have changed the results? If so, repeat parts (a) and (b).

8.9. You wish to multiplex M messages, each having a bandwidth of 5 kHz, through a cable whose bandwidth is 360 kHz and in which there is white additive noise of level $\eta_0/2$ V^2/Hz. Your system must include guard bands of at least 20% of the message bandwidth (i.e., leave at least 1.2 times the message bandwidth for each channel). The total average input power is P V^2, that is, P/M V^2 for each message. The message variance $\langle m^2(t) \rangle = 0.5$. You are considering using either SSB-SC modulation for multiplexing or frequency modulation.

(a) For each scheme, determine the maximum number of messages (M) that can be multiplexed.

(b) For each scheme calculate an expression for the largest possible output SNR for each message as a function of M.

(c) Show that for $M < 14$ the performance of the FM scheme is better than that of the SSB-SC scheme.

8.10. You wish to design a point-to-point communication system using FM. You have the same constraints as those of Problem 8.4. In addition the signal bandwidth can be no larger than 140 kHz and your receiver's threshold occurs at $\text{SNR}_i = 20$. Assume that SNR_o decreases to zero almost instantaneously beyond the threshold.

(a) Sketch and label the output SNR versus r with the best choice for a fixed β.

(b) Assume now that the bandwidth can be optimized by varying β subject to the constraint $\text{BW} \leq 140$ kHz. Now sketch the output SNR versus r.

8.11. You wish to design a point-to-point communication system with the largest possible output SNR. Your constraints are as follows:

    i. The receiver signal (carrier) amplitude is 10 V.
    ii. The received two-sided noise spectral level $\eta_0/2 = 5 \times 10^{-6}$ V²/Hz.
    iii. The analog message bandwidth is 10 kHz, and $\langle m_n^2(t) \rangle = 0.4$.
    iv. There is an available channel bandwidth of 400 kHz centered at 3.6 MHz.
    v. The receiver FM threshold occurs when $\text{SNR}_i \approx 50$.

(a) Design an FSK system with the two frequencies far enough apart for suboptimal noncoherent reception (assume $m = 1$). Specify the frequencies, number of quantization levels, and bandwidth.

(b) Design an FM system. Specify Δf and the bandwidth.

(c) Calculate the output SNR for each system and compare your results. (Recall that for a discrete system $\text{SNR} = cM^2$ for large input SNRs. Assume $c = 0.8$.)

8.12. Assume that $S_m(f) = K/[1 + (f/f_c)^2]^2$ and $f_c = 2.1$ kHz. Evaluate, in decibels, the improvement of standard FM due to deemphasis by a standard first-order LPF.

8.13. A PM signal is one whose phase varies proportionally with the normalized signal $m_n(t)$:

$$s(t) = A \cos[2\pi f_c t + 2\pi \Delta f \, m_n(t) + \theta_0]$$

(a) Determine the preemphasis filter that converts FM to PM.

(b) Sketch the PM receiver hardware, assuming a standard FM receiver with the appropriate deemphasis filter. Show that the carrier must be filtered out after the AM demodulator.

8.14. Reevaluate Problem 8.7 assuming the addition of the standard first-order deemphasis filter at the receiver. Assume the signal spectrum of Problem 8.12.

8.15. Reevaluate Problem 8.9 assuming the deemphasis filter and signal spectrum of Problem 8.12. Show now that for $M < 18$ the performance of the FM scheme is better than that of the SSB-SC scheme.

8.16. Suppose the doublet pulses and the more rare pulses caused by the anomaly error are modeled as below:

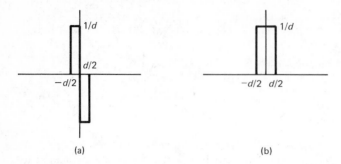

(a)   (b)

Compute and sketch the energy densities of both pulses. Argue from the sketch that a more significant portion of the single pulse than the doublet lies in the frequency range $[0, 1/2d]$.

8.17. Reconsider Problem 8.11. Assume that the SNR prior to the decision mechanism of the FSK receiver is 3 dB worse than if matched filters were used $(\frac{1}{2}E_b/\eta_0)$.

(a) Calculate SNR_d for the 400-kHz bandwidth and show that it is much larger than SNR_q.

(b) Suppose the allotted bandwidth was increased to 650 kHz. Calculate SNR_q and SNR_d for this larger bandwidth and argue that this FSK system is performing better than FM.

8.18. Assume the random variable X has a uniform distribution in the range $|x| \le A$. Determine the PDF of Y if $Y = X^2$.

8.19. It is intuitively clear that if X is Gaussian and $Y = X^2$, then $P(y > A^2) = 2P(x > A) = 2Q(A/\sigma_x)$. Prove this result by finding the PDF model of Y and integrating it from A^2 to ∞.

8.20. In Problem 5.20 we determined the mean and variance of the magnitude of a vector $\mathbf{X}$ whose components X_1 and X_2 are independent zero-mean Gaussian random variables with $\sigma_1^2 = \sigma_2^2 = 1$. Repeat this problem by first finding the PDF of the magnitude of $\mathbf{X}$ and then determining the mean and variance from this PDF model.

8.21. Assume that X is a uniformly distributed random variable where $0 \le x \le 1$. We define the random variable Y by $y = \ln x$. Determine and sketch the PDF of Y.

8.22. Consider the PDFs of Equations 8.97 and 8.98. Verify that if the random variables with these PDFs are inputs to a square root device $(y = +\sqrt{x})$, the resulting PDFs are the Rayleigh and Ricean PDFs.

REFERENCES

1. L. H. Enloe, "Decreasing the Threshold in FM by Frequency Feedback," *Proc. IRE* 50, 18–30 (1962).
2. D. L. Schilling and J. Billig, "On the Threshold Extension Capability of PLL and FDMFB," *Proc. IEEE* 52, 621–622 (1964).

3. M. Schwartz, W. R. Bennett, and S. Stein, *Communication Systems and Techniques*, McGraw-Hill, New York, 1966.

4. S. O. Rice, "Noise in FM Receivers," in *Proceedings Symposium of Time Series Analysis*, M. Rosenblatt, ed., Wiley, New York, 1963, Chapter 25, pp. 395–424.

5. H. Cramer, *Mathematical Methods of Statistics*, Princeton University Press, Princeton, NJ, 1946.

6. W. B. Davenport, Jr., and W. L. Root, *An Introduction to the Theory of Random Signals and Noise*, McGraw-Hill, New York, 1958.

7. S. O. Rice, "Statistical Properties of a Sine-Wave Plus Random Noise," *Bell System Tech. J.* 27, 109–157 (1948).

8. M. Schwartz, W. R. Bennett, and S. Stein, *Communication Systems and Techniques*, McGraw-Hill, New York, 1966.

Chapter 9

Introduction to
Error-Correcting Codes

9.1 INTRODUCTION

We saw in Chapter 7 that there are two obvious ways to decrease the probability of error for any of the discrete communication techniques. One, of course, involves increasing the available power. Another method involves increasing the signal energy by time scaling, which usually can be employed only when we are not constrained to operate in real time. Of course, if we wish to communicate continuously for a long time, this could result in considerable time delay.

A third method, which can be achieved while communicating in real time, is known as *block encoding*. Surprisingly, if we combine a series of messages or symbols together into "blocks" that are then regarded as individual messages, it is possible to reduce the probability of error. The definition of what constitutes a message is often not very clear. For teletype signals, we tend to think of a letter as a message, but why not a word or a sentence? For pictures, we might consider a sample as a message, but why not a line—or the entire picture? For the pushbutton telephone we regard a decimal number (or button) as a message, but why not an entire ten-digit telephone number? Redefining a message as a block of symbols should cause us no philosophical problems. Unfortunately, achieving better performance using block encoding is not an easy task.

9.2 BLOCK DEFINITION

If we wish to consider long sequences of symbols of duration T s, where T is not uniquely specified, we must redefine our terms slightly. For example, it is .

no longer useful to consider the energy per message E_s as a constraint; rather, one considers the average transmitter power P_s, defined by

$$P_s = \frac{1}{T} \int_0^T s^2(t) \, dt = \frac{E_s}{T} \tag{9.1}$$

It follows that if the transmitted power is a constraint, then the longer the block of symbols, the more energy per message block ($E_s = TP_s$). This alone gives some insight into why block encoding can be helpful.

On the other hand, the longer the block of symbols, the more messages we have to decide among. Thus, it is no longer helpful to use the number of messages M as a constraint. Let us define the *source rate* of any source which generates M messages in a T-s interval as

$$R = \frac{1}{T} \log_2 M \quad \text{bits/s} \tag{9.2}$$

This definition corresponds to the number of pulses per second that would be required to encode the source into PCM form. This definition is useful whether or not PCM is used. If we multiplex two channels that can transmit M_1 and M_2 messages, respectively, in time T, the multiplexed channel must be able to transmit $M = M_1 M_2$ messages in time T, and thus

$$R = \frac{1}{T} \log_2 M = \frac{1}{T} \log_2 M_1 M_2$$

$$= \frac{1}{T} \log_2 M_1 + \frac{1}{T} \log_2 M_2$$

$$= R_1 + R_2 \quad \text{bits/s}$$

Thus multiplexing n sources with rates of R bits/s requires a channel that can accommodate nR bits/s.

We see that if a source transmits at a rate of R bits/s, the number of messages we must distinguish among in T s is given by $M = 2^{RT}$. This exponential growth in the number of messages with block length is troublesome. These problems will be explored in Examples 9.2 and 9.3.

EXAMPLE 9.1: PCM Signaling ■

We have seen that for a system that transmits a sequence of N bipolar binary pulses in time T ($N = RT$) when the additive noise is white Gaussian, the probability of error $P(e)$ is given by

$$P(e) = 1 - (1 - p)^N = 1 - (1 - p)^{RT}$$

where $p = Q(\sqrt{2E_p/\eta_0})$. The energy per pulse E_p is given by

$$E_p = \frac{E_s}{N} = \frac{P_s}{R} \quad \text{V}^2/\text{bit} \tag{9.3}$$

For fixed T, we can make p, and hence $P(e)$, small only by making P_s large or R small. Note, however, that, for small p,[1]

$$1 - (1 - p)^{RT} = 1 - (1 - p)^{(1/p)pRT}$$
$$\approx 1 - e^{-pRT} \to 1 \qquad \text{as} \quad T \to \infty$$

We can see that block encoding with PCM makes things worse. ■ ■

9.3 BLOCK ORTHOGONAL SIGNALING

Suppose we consider orthogonal signaling such as FPM, which is optimum for large M. We saw in Chapter 7 that

$$P(e) < (M - 1) Q(\sqrt{E_s/\eta_0})$$
$$< MQ(\sqrt{P_s T/\eta_0})$$
$$\approx 2^{RT} Q(\sqrt{P_s T/\eta_0}) \qquad (9.4)$$

The energy increases with T, which causes $P(e)$ to decrease, but, because M increases with T, it is not clear whether $P(e)$ increases or decreases. In order to see which effect is dominant, let us use the approximation $Q(\alpha) < \frac{1}{2}e^{-\alpha^2/2}$. Equation 9.4 becomes

$$P(e) < e^{\ln 2 RT} \frac{1}{2} e^{-P_s T/2\eta_0}$$

$$= \frac{1}{2} \exp\left[-T\left(\frac{P_s}{2\eta_0} - R \ln 2 \right) \right]$$

It follows that as long as $R < P_s/2 \ln 2 \, \eta_0$, the probability of an error decreases exponentially with increasing T. Alternatively, if the power is above some threshold, we can make the probability of error as small as we please by using sufficiently large blocks.

EXAMPLE 9.2: Telephone Dialing ■

Suppose we assume for simplicity that pushbutton dialing uses an orthogonal FPM technique. Suppose we consider an entire ten-digit number as a message. We have a different button (tone) for each telephone in the country, and this tone maintained for a period that is ten times longer than each tone in the present system. Since there are approximately 10^7 phones in this country (not counting extensions), the transmitter requires 10^7 buttons, and the receiver requires 10^7 matched filters. ■ ■

The problems of block orthogonal signaling are obvious from this example. Because the number of messages grows exponentially with block size T, it follows that hardware complexity (i.e., the number of matched

[1]Since $\lim(1 - \Delta)^{1/\Delta} \to e^{-1}$ as $\Delta \to 0$, $(1 - \Delta)^{1/\Delta}$ can be approximated by e^{-1} for small Δ.

filters in the receiver) also grows exponentially. We do not propose a telephone with 10^7 buttons. Such a telephone would require a bandwidth for dialing of nearly 1 GHz (10^9 Hz), which is larger than the entire cable bandwidth. We saw in Section 7.10 that for orthogonal signals the number of dimensions needed corresponds to the number of messages. Because $D = M = 2BT$, where B is the bandwidth, it follows that

$$B = \frac{1}{2T} 2^{RT} \quad \text{Hz} \tag{9.5}$$

which grows exponentially with block size. In order to emphasize the absurdity of exponentially growing bandwidth, let us consider another example.

EXAMPLE 9.3: Mariner IV Pictures ■

The pictures from Mariner IV were scanned (200 lines), sampled (200 samples per line), quantized (64 levels per sample), and stored on a magnetic tape to be transmitted at a slow rate. The scanned signal was approximately 1 min in duration; thus the source rate (if communicated in real time) was 4000 bits/s (i.e., 240,000 pulses/min). The true source rate was considerably less than this because of the redundancy or correlation between samples. For the sake of argument, suppose the actual source rate is 1000 bits/s (i.e., there are only $2^{60,000}$ possible pictures that we need distinguish). Assume that we can muster sufficient average power so that if we encode an entire line ($T = 0.3$ s) in an orthogonal block, the error probability will be greatly reduced. The rate will have to be reduced by time scaling for this method to work. The required bandwidth, for communication in real time, is

$$B = \frac{2^{300}}{0.6} \simeq 3.4 \times 10^{90} \text{ Hz}$$

which is absurd. If we reduce this bandwidth to 7000 MHz by time scaling (needed anyway), the required transmission time for each picture will be 10^{75} years. ■ ■

It appears that block orthogonal signaling is not practical. There is no point considering any block encoding technique if either the bandwidth or the hardware complexity grows exponentially with block size.

9.4 BINARY CODES

The binary code of Example 7.6 gives us some insight into the only practical way of achieving block encoding. We saw that by using eight binary pulses to represent 16 messages (instead of the four required for PCM) we could generate biorthogonal sets and achieve improved performance (3 dB) relative

TABLE 9.1 A BIORTHOGONAL
             BINARY CODE

11111111	00000000
11110000	00001111
11001100	00110011
11000011	00111100
10101010	01010101
10011001	01100110
10010110	01101001
10100101	01011010

to PCM. These signals are reproduced in Table 9.1, where a 1 stands for a positive pulse and a 0 stands for negative pulse. The left column represents the orthogonal signals shown in Figure 7.21 and the right column their complements. The important fact is not that these messages are biorthogonal but that they are further apart than PCM signals.

Let us define the *Hamming distance* d_H between code words (i.e., binary messages) as the number of pulses that are different. This distance is monotonically related to our notion of distance in Chapter 7. Previously, the distance between two messages d_{12} was the square root of the energy difference ΔE_{12}. For each pulse that is different, however, the energy difference is $4E_p([A - (-A)]^2 T = 4A^2 T = 4E_p)$ where E_p is the energy per pulse:

$$E_p = \frac{P_s T}{N} = \frac{P_s T}{RT} = \frac{P_s}{R} \tag{9.6}$$

Thus

$$\Delta E_{12} = d_{H12} 4E_p \tag{9.7}$$

and it follows that

$$d_{12} = \sqrt{\Delta E_{12}} = \sqrt{4d_{H12}\frac{P_s T}{N}} = \sqrt{4d_{H12}\frac{P_s}{R}} \tag{9.8}$$

The bigger the Hamming distance between nearest neighbors, the better the performance of the signal set or code. From Table 9.1 we see that the minimum Hamming distance between code words or messages is 4. One must change at least four pulses to change one code word to another code word. Thus the minimum distance between nearest neighbors is

$$d_{min} = \sqrt{4 \times 4 \times \frac{P_s T}{8}} = \sqrt{2P_s T}$$

If one were to use a four-pulse PCM code, the minimum Hamming distance would be 1 and the minimum distance would be

$$d_{min}(\text{PCM}) = \sqrt{4 \times 1 \times P_s T/4} = \sqrt{P_s T}$$

We see that the encoded signals are further apart than PCM and, therefore, better.

Can we use block binary codes, however, that are not restricted to orthogonality, are better than PCM, and yet require reasonable bandwidths and hardware? Suppose, as an example, we combined a sequence of 13 messages into a block that would require 52 pulses if we used PCM. Instead of PCM, however, we use a 63-bit code and have the problem of selecting 2^{52} 63-bit sequences to represent the "messages" out of a possible 2^{63} sequences. Note that there are $2^{63}/2^{52} = 2048$ times as many possible sequences as there are desired code words which suggests the possibility of discovering a good code.[2] We shall see in the next few sections that it might be possible to find such a code with a minimum Hamming distance of 5. Thus, for the 63-bit code,

$$d_{min} = \sqrt{\frac{20P_sT}{63}}$$

as opposed to

$$d_{min} = \sqrt{\frac{4P_sT}{52}} \quad \text{(for PCM)}$$

Thus PCM requires 4.1 times the average power (6 dB) to achieve the same probability of error. Even though this 63-bit code is not biorthogonal, the messages are further apart than the 8-bit code. It appears that block binary encoding is possible. This code requires a bandwidth of only 63/52 times (21%) larger than PCM (remember the bandwidth is proportional to the pulse rate). Thus block binary encoding does not give us the exponentially growing bandwidth of block orthogonal coding.

Unfortunately, this 6-dB improvement is based on an optimum receiver that consists of 2^{52} filters matched to 63-bit sequences. If this discussion is to be of any value, we must find a suboptimum receiver that is reasonable to implement and that results in an improvement relative to PCM.

9.5 QUANTIZED RECEIVERS

Let us consider a receiver that makes an optimum decision on each binary pulse and on the basis of this received pulse train determines the most likely message. This suboptimum receiver is called a *quantized receiver*. The first part of this receiver is a standard PCM receiver. This is followed by a computer or by hardware logic to determine the correct message. If each of the legitimate messages were stored in the receiver computer, it would be easy to determine the Hamming distances between the received pulse train and each of the possible messages. An optimum decision would be to select the closest legitimate message.

[2] A good code is defined as one whose probability of error is less than that of PCM and whose bandwidth is reasonable.

Consider first the code of Table 9.1. If any of the 16 messages is transmitted and a single error is introduced, the received signal will lie closer to the correct message. It follows that, for this code, the suboptimum receiver will correct all single errors. If two errors occur, there will be a number of possible signals whose distance[3] from the received pulse train is 2. Thus, for example, if the first message (11111111) is sent and the first two bits are in error, the received pulse train is 00111111. This pulse train is equally close to the legitimate code words 11111111, 00001111, 00110011, and 00111100. It follows that while the receiver can detect double errors, it cannot correct double errors. The probability of error for this receiver is given by[4]

$$P(e) = 1 - \Pr(\text{no errors}) - \Pr(\text{single errors})$$
$$\approx \text{probability of double errors}$$
$$= C_2^8 P_b^2(e)[1 - P_b(e)]^6$$
$$\approx C_2^8 P_b^2(e) \qquad (9.9)$$

where

$$P_b(e) = Q\left(\sqrt{\frac{2E_p}{\eta_0}}\right) = Q\left(\sqrt{\frac{2P_sT}{8\eta_0}}\right) < \tfrac{1}{2}e^{-P_sT/8\eta_0}$$

Therefore,

$$P(e) < 28(\tfrac{1}{2}e^{-P_sT/8\eta_0})^2 = 7e^{-P_sT/4\eta_0}$$

For a PCM signal,

$$P(e) \approx 4P_b(e) = 4Q\left(\sqrt{\frac{2P_sT}{4\eta_0}}\right) < 2e^{-P_sT/4\eta_0}$$

We see that the eight-bit biorthogonal code performs about the same with the quantized receiver as a four-bit PCM signal.

There are circumstances in which this code is particularly useful. Suppose the communication link were a two-way channel where the receiver could request a retransmission. In this case, double errors would also be corrected, and the net probability of error would be

$$P(e) \approx C_3^8 Q^3\left(\sqrt{\frac{2P_sT}{8\eta_0}}\right) < 56(\tfrac{1}{2}e^{-P_sT/8\eta_0})^3 = 7e^{-3P_sT/8\eta_0}$$

This technique would require only two-thirds the average power of PCM. Consider still another situation where an eight-bit PCM communication link is available but the messages we wish to transmit have only 16 values. We could reduce the probability of error considerably by time scaling, but this requires a change in most of the hardware and all of the timing circuits. An equivalent technique, which is easier to implement, is to use the eight-bit

---

[3]We use Hamming distance for the remainder of the text.

[4]C_k^n refers to the number of ways of choosing k out of n events, which is equal to $n!/(n-k)!k!$.

biorthogonal code and error correction, perhaps with a microprocessor, after the receiver.

It is possible with larger block codes and suboptimal receivers to do better than PCM with time scaling. Suppose we can find the code of 2^{52} 63-bit words with a minimum Hamming distance of 5, as mentioned in Section 9.4. We shall see that such a code could correct single and double errors if it exists. It follows that

$$P(e) \simeq C_3^{63} Q^3 \left(\sqrt{\frac{2P_s T}{63\eta_0}} \right) = 39{,}711 Q^3 \left(\sqrt{\frac{2P_s T}{63\eta_0}} \right)$$
$$< 39{,}711 \left(\tfrac{1}{2} e^{-P_s T/63\eta_0} \right)^3 = 4964 e^{-3P_s T/63\eta_0}$$

To compare with PCM, we should consider the probability of error in any of the 13 messages in the block or equivalently any of the 52 pulses:

$$P_{PCM}(e) \simeq 52 Q \left(\sqrt{\frac{2P_s T}{52\eta_0}} \right) < 26 e^{-P_s T/52\eta_0}$$

To achieve the same probability of error, PCM requires roughly twice (or 3 dB) more power than the coded system using quantized receivers.

Unfortunately, the procedure of storing each of the 2^{52} 63-bit messages and comparing each of them with the received pulse train is still not practical. Replacing an exponentially growing number of matched filters with an exponentially growing memory and computation is marginally better and still not acceptable. A more efficient error-correcting algorithm is essential. We also need a method of finding good, large block codes. If both these problems can be solved, then we are on the right track.

9.6 CHANNEL CAPACITY

Shannon has determined a variety of coding bounds [1]. Let us consider one of these bounds, although the development is beyond the scope of this text. Assume a channel with additive Gaussian noise through which we can transmit a binary pulse train at a rate of f_b pulses/s with acceptable intersymbol interference ($f_b \approx$ channel bandwidth). Let us further assume that the suboptimal quantized receiver is used. If the source rate R is less than the channel capacity C, defined by

$$C = f_b [1 + p \log_2 p + (1 - p) \log_2 (1 - p)] \quad \text{bits/s} \tag{9.10}$$

where

$$p = Q \left(\sqrt{\frac{2P_s T}{\eta_0 N}} \right) = Q \left(\sqrt{\frac{2P_s}{\eta_0 f_b}} \right)$$

then we can achieve binary block encoding successfully. This capacity is plotted as the solid curve of Figure 9.1 [1]. If we communicate at a rate below

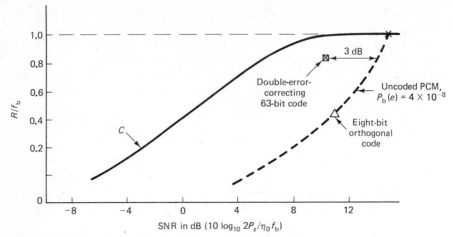

Figure 9.1 Channel capacity based on a quantized receiver.

this curve, we should be able to find a block binary code that will achieve as small an error probability as desired.

In order to understand the idea of channel capacity, let us try to achieve a bit error probability of 4×10^{-8} with a PCM code while transmitting at the maximum pulse rate $R = f_b$. Suppose this procedure (indicated by the $\times$ in Figure 9.1) requires too much average power P_s. To reduce P_s while achieving the desired error probability, we must reduce the source rate. One method is to slow down the pulse rate (a form of time scaling), as indicated by the dashed curve. Thus, for example, we can halve the pulse rate and decrease the P_s required by 3 dB (this is indicated by the $\triangle$ in Figure 9.1). The other method involves transmitting pulses at the maximum rate while using error-correcting codes. The eight-bit biorthogonal code reduces the rate by half (i.e., 2^4 messages instead of 2^8) and requires 3 dB less power (also indicated by the $\triangle$ in Figure 9.1). The 63-bit code that we have speculated about reduces the rate to $\frac{52}{63}$ of the maximum pulse rate and requires 3 dB less power than PCM with a reduced pulse rate (indicated by the $\boxtimes$ in Figure 9.1). In theory, we can operate as close to the channel capacity curve as we like, but this may require extremely long blocks. The location of the dashed curve (time scaling) depends on the desired bit error probability.

Although this channel capacity theorem does not help us find good codes, Shannon is able to argue that there are many good codes. This means that if one randomly selects 2^{52} 63-bit sequences, there will be a number of possible codes that work almost as well as the best selection. Although such a search is not feasible for blocks of any substantial size, we are encouraged in our quest.

Good codes have code words that are far apart. We can relate the error-correcting capability of a code to the minimum Hamming distance between code words. If the minimum distance between any two is 2, then a single error will result in a sequence that is not a code word (i.e., a detectable error), but may be equally close to two or more code words (i.e., not correctable). On the other hand, if the minimum distance is 3 and a single error occurs, the resulting sequence will be closer to the correct code (i.e., is correctable). These ideas can be extended using geometric arguments. Let us construct spheres around every code word that contains all sequences whose distance to the code word is e or less. If these spheres do not intersect or touch (i.e., given a minimum distance between code words of at least $2e + 1$), then all errors of e or less can be corrected. This leads to Theorem 9.1:

Theorem 9.1. If for any positive integer e the minimum Hamming distance is $2e + 1$, then all single, double, and so on up through e-tuple errors can be corrected. If the minimum distance is $2e$, then all errors up to $(e - 1)$-tuple errors can be corrected and e-tuple errors can be detected but not in general corrected.

Another version of this theorem can be seen by recognizing that, for N-bit sequences, the number of sequences inside the sphere of radius e is

$$1 + N + C_2^N + \cdots + C_e^N = \sum_{i=0}^{e} C_i^N$$

Corollary 9.1. A bound on the ratio of the total number of N-bit sequences (2^N) to the number of code words (S) that can correct up to e-tuple errors is given by

$$\frac{2^N}{S} \geq \sum_{i=0}^{e} C_i^N$$

As an example, for the hypothetical 63-bit code we considered, $2^{63}/2^{52} = 2^{11} = 2048$. In order to correct up to double errors, this ratio must be at least as large as $1 + 63 + C_2^{63} = 2017$. It follows that such a code may be possible.

9.7 INTRODUCTION TO PARITY CHECK CODES

A block encoder converts an M-bit word (denoted by the vector $\mathbf{x}$, where $x_i = 0$ or 1 for each i) into an N-bit code word $\mathbf{y}$, where $N > M$. If each of the y_i bits is the modulo-2 sum of some arbitrary set of bits x_i, the code is called a *parity check code* and can be implemented as in Figure 9.2. Modulo-2 addition is defined by the equations

$$0 \oplus 0 = 1 \oplus 1 = 0$$
$$0 \oplus 1 = 1 \oplus 0 = 1 \tag{9.11}$$

and is easily implemented with exclusive-or logic components. Parity check codes are sometimes called *linear codes* because the relationship between the output vector $\mathbf{y}$ and the input vector $\mathbf{x}$ is given by

$$\mathbf{y} = x_1 \mathbf{f}_1 \oplus x_2 \mathbf{f}_2 \oplus \cdots \oplus x_M \mathbf{f}_M \tag{9.12}$$

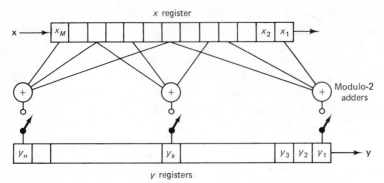

Figure 9.2 Parity check coder. The switches close after the entire input sequence x is fed into the x register.

where the $\mathbf{f}_h$, called connection vectors, are N-dimensional binary vectors defined by

$$f_{hj} = \begin{cases} 1 & \text{if } x_h \text{ affects } y_j \\ 0 & \text{otherwise} \end{cases}$$

and f_{hj} is the jth component of the vector $\mathbf{f}_h$. Equation 9.12 can be written in matrix form:

$$\mathbf{y} = \mathbf{H}\mathbf{x}^{\mathrm{T}} \tag{9.13}$$

where $\mathbf{H}$ is an $N \times M$ matrix of 1s and 0s called the *generating matrix* and $\mathbf{x}^{\mathrm{T}}$ is the transposed $\mathbf{x}$ vector, which is a column vector with x_M at the top. For the encoder of Figure 9.3 we have

$$
\begin{aligned}
y_8 &= x_4 \\
y_7 &= x_3 \\
y_6 &= x_2 \\
y_5 &= x_4 \oplus x_3 \oplus x_2 \\
y_4 &= x_1 \\
y_3 &= x_4 \oplus x_3 \oplus x_1 \\
y_2 &= x_4 \oplus x_2 \oplus x_1 \\
y_1 &= x_3 \oplus x_2 \oplus x_1
\end{aligned}
\tag{9.14}
$$

or

$$
\mathbf{y} = \begin{bmatrix}
1 & 0 & 0 & 0 \\
0 & 1 & 0 & 0 \\
0 & 0 & 1 & 0 \\
1 & 1 & 1 & 0 \\
0 & 0 & 0 & 1 \\
1 & 1 & 0 & 1 \\
1 & 0 & 1 & 1 \\
0 & 1 & 1 & 1
\end{bmatrix} \mathbf{x}^{\mathrm{T}}
\tag{9.15}
$$

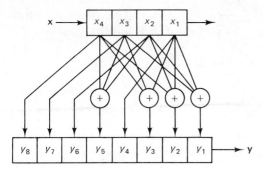

Figure 9.3 [8, 4] parity check encoder.

Thus if $\mathbf{x} = 1101$, we can determine from Equation 9.14 or 9.15 that $\mathbf{y} = 11001100$. This code word is the third word of Table 9.1. It is easily verified (see Problem 9.1) that the [8,4] parity check encoder of Figure 9.3 generates the biorthogonal code of Table 9.1.

We observe that four of the $\mathbf{y}$ bits (y_8, y_7, y_6, y_4) correspond to the information bits $\mathbf{x}$. The remaining four extra bits are called *parity checks*. Even if we cannot explicitly identify M information bits, we nevertheless refer to M information bits and $N - M$ parity check bits. Whenever M explicit information bits occur in the first or last M places, the code is called a *systematic code*.

For Equations 9.14, if the $\mathbf{x}$ bits are solved in terms of the $\mathbf{y}$ bits, then by substitution methods $N - M$ equations in $\mathbf{y}$ bits only can be obtained. Thus if we rewrite the four nontrivial equations while substituting $y_8 = x_4$, $y_7 = x_3$, etc., we obtain

$$y_5 = y_8 \oplus y_7 \oplus y_6$$
$$y_3 = y_8 \oplus y_7 \oplus y_4$$
$$y_2 = y_8 \oplus y_6 \oplus y_4$$
$$y_1 = y_7 \oplus y_6 \oplus y_4$$

With modulo-2 addition these equations can be rewritten

$$y_8 \oplus y_7 \oplus y_6 \oplus y_5 = 0$$
$$y_8 \oplus y_7 \oplus y_4 \oplus y_3 = 0$$
$$y_8 \oplus y_6 \oplus y_4 \oplus y_2 = 0 \tag{9.16}$$
$$y_7 \oplus y_6 \oplus y_4 \oplus y_1 = 0$$

Similar $N - M$ equations can be found for any parity check code. It is particularly easy to accomplish if M of the bits are explicitly information bits as in this example. Equation 9.16 can be written in matrix form as in Equation 9.17:

$$\mathbf{A}\mathbf{y}^T = 0 \tag{9.17}$$

The $N \times M$ matrix **A** is called the *parity code matrix*. The solutions of this equation are the code words themselves. For the [8,4] biorthogonal code,

$$
\mathbf{A}_1 = \begin{bmatrix}
1 & 1 & 1 & 1 & 0 & 0 & 0 & 0 \\
1 & 1 & 0 & 0 & 1 & 1 & 0 & 0 \\
1 & 0 & 1 & 0 & 1 & 0 & 1 & 0 \\
0 & 1 & 1 & 0 & 1 & 0 & 0 & 1
\end{bmatrix}
$$

The parity equations (Equation 9.16) are not unique. Depending on our solution technique, Equation 9.16 could be manipulated into a different set of equations (or rows in the matrix). If any of the equations (or rows) are replaced by the modulo-2 sum of two or more equations (or rows), the resulting parity code matrix corresponds to the same code. The equivalent matrices can be found by standard matrix row operations using modulo-2 addition. While it is by no means obvious, we shall argue in subsequent sections that column additions or interchanges result in an equivalent code. Equivalent codes have the same error-correcting capability but somewhat different encoders (i.e., connections of Figure 9.2) and decoders. We leave it to the reader to determine that, by simple row operations and column operations, A_1 can be converted to the following matrices, which we claim will result in equivalent codes (see Problem 9.4):

$$
\mathbf{A}_2 = \begin{bmatrix}
1 & 1 & 1 & 1 & 1 & 1 & 1 & 1 \\
0 & 1 & 1 & 1 & 1 & 0 & 0 & 0 \\
0 & 1 & 1 & 0 & 0 & 1 & 1 & 0 \\
0 & 1 & 0 & 1 & 0 & 1 & 0 & 1
\end{bmatrix}
$$

$$
\mathbf{A}_3 = \begin{bmatrix}
1 & 1 & 1 & 0 & 1 & 0 & 0 & 0 \\
1 & 1 & 0 & 1 & 0 & 1 & 0 & 0 \\
1 & 0 & 1 & 1 & 0 & 0 & 1 & 0 \\
0 & 1 & 1 & 1 & 0 & 0 & 0 & 1
\end{bmatrix}
$$

The last matrix ($\mathbf{A}_3$) corresponds to a systematic code. Any matrix that can be put into either of the forms

$$
N - M \left\{ \begin{bmatrix} \overbrace{}^{M} & \vdots & \overbrace{\mathbf{I}}^{N-M} \end{bmatrix} \right. \quad \text{or} \quad \begin{bmatrix} \overbrace{\mathbf{I}}^{N-M} & \vdots & \overbrace{}^{M} \end{bmatrix} \tag{9.18}
$$

where **I** is the identity matrix, corresponds to a systematic code. If one is given a parity code matrix, rather than the generating matrix or encoder, the easiest way to determine the encoder implementation and/or the code words themselves is to manipulate the matrix into this form and allow the first or last M bits to be information bits. If this can be achieved by row manipulations only, the code words will not be altered.

EXAMPLE 9.4 ■

Find the parity check code words that are the solutions to Equation 9.17 where the parity code matrix is given by

$$\mathbf{A} = \begin{bmatrix} 1 & 0 & 1 & 1 & 1 & 0 \\ 1 & 1 & 0 & 0 & 1 & 1 \\ 0 & 0 & 1 & 0 & 1 & 1 \end{bmatrix}$$

Using row additions only, we obtain

$$\begin{bmatrix} 1 & 0 & 1 & 1 & 1 & 0 \\ 1 & 1 & 0 & 0 & 1 & 1 \\ 0 & 0 & 1 & 0 & 1 & 1 \end{bmatrix} \Rightarrow \begin{bmatrix} 1 & 0 & 0 & 1 & 0 & 1 \\ 1 & 1 & 0 & 0 & 1 & 1 \\ 0 & 0 & 1 & 0 & 1 & 1 \end{bmatrix} \Rightarrow \begin{bmatrix} 1 & 0 & 0 & 1 & 0 & 1 \\ 0 & 1 & 0 & 1 & 1 & 0 \\ 0 & 0 & 1 & 0 & 1 & 1 \end{bmatrix}$$

This represents a [6, 3] systematic parity check code. If we allow y_3, y_2, and y_1 to correspond to the three information bits, then the parity check equations (Equation 9.16 or 9.17) can be written

$$y_4 = y_2 \oplus y_1$$
$$y_5 = y_3 \oplus y_2$$
$$y_6 = y_3 \oplus y_1$$

If we list all $8 = 2^3$ input words, we can use these equations to determine the code words:

y_6	y_5	y_4	y_3	y_2	y_1
0	0	0	0	0	0
1	0	1	0	0	1
0	1	1	0	1	0
1	1	0	0	1	1
1	1	0	1	0	0
0	1	1	1	0	1
1	0	1	1	1	0
0	0	0	1	1	1

The systematic encoder for this [6,3] parity check code is shown in Figure 9.4. We see that parity check codes are specified by their parity code matrix. ■ ■

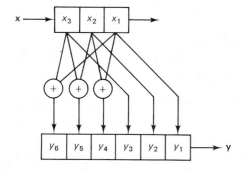

Figure 9.4 Example 9.4.

The parity code matrix $\mathbf{A}_2$ will give us some insight as to how we can correct errors without storing the entire code. The parity check equations specified by $\mathbf{A}_2$ are

$$\sum_{i=1}^{8} y_i = 0$$

$$y_4 \oplus y_5 \oplus y_6 \oplus y_7 = 0$$
$$y_2 \oplus y_3 \oplus y_6 \oplus y_7 = 0$$
$$y_1 \oplus y_3 \oplus y_5 \oplus y_7 = 0$$

If we perform the four operations (parity checks) indicated by these equations at the receiver (i.e., $\mathbf{Aw}^{\mathrm{T}}$, where $\mathbf{w}$ is the received eight-bit signal), the output will be 0000 if there are no errors. If any single bit is in error, the output of the first parity check becomes a 1, indicating a single error. Surprisingly, the output of the other three parity checks becomes the binary number corresponding to the location of the error. Thus, for example, if bit y_5 is in error, the response to the parity checks, called the *corrector*, becomes 1101. The first 1 indicates an error, and 101 is the standard binary code for 5. The reader can verify that this is true for all single errors. For any double error, the first parity check will be a 0, but the other three will not. Such an output corrector indicates a detected but not correctable double error. For this particular code, the receiver need only store the 4×8 parity code matrix rather than the 16 eight-bit code words.

We shall prove in subsequent sections that determining the corrector from the parity checks (which requires storing the $N - M$ by N parity code matrix) will enable us to implement the optimum quantized receiver. In most cases, the corrector will not point directly to the error pattern, and a table of correctors versus error patterns must also be stored. Nevertheless, the memory required is not growing exponentially with block size, and the receivers can be implemented.

EXAMPLE 9.5 ■

We have previously considered a hypothetical [63, 52] code that can correct single and double errors. We were concerned with the apparent need to store 2^{52} 63-bit messages, corresponding to 2.837×10^{17} bits. If such a code can be implemented as a parity check code and if the receiver just discussed works, the numbers are reasonable. We must store an 11×63 parity code matrix to determine the 11-bit correctors. There are 63 possible single errors and $C_2^{63} = 1953$ possible double errors. Thus we also need a table of 2016 11-bit correctors and 63-bit error patterns. The total number of bits required is

$$11 \times 63 + 2106(11 + 63) = 1.499 \times 10^5$$

This is still a large number of bits but this can be implemented.   ■ ■

Another version or corollary of Theorem 9.1 can be seen by examining the minimum number of parity checks needed to correct up to e-tuple errors.

TABLE 9.2 BOUNDS ON THE REQUIRED NUMBER OF PARITY CHECK BITS

N	Single-error correctors		Double-error correctors	
	K_{min}	M_{max}	K_{min}	M_{max}
1	1	—	—	—
2	2	—	2	—
3	2	1	3	—
4	3	1	4	—
5	3	2	4	1
6	3	3	5	1
7	3	4	5	2
8	4	4	6	2
9	4	5	6	3
10	4	6	6	4
⋮	⋮	⋮	⋮	⋮
15	4	11	7	8
⋮	⋮	⋮	⋮	⋮
63	6	57	11	52

The $K = N - M$ parity check bits (or corrector bits) must point uniquely to all of the correctable error patterns as well as the correct code word. Since there are $\sum_{i=0}^{e} C_i^N$ such patterns and since a K-bit word has 2^K unique combinations, we can conclude the following:

Corollary 9.2. The minimum number of parity checks K to correct single, double, and so on up to e-tuple errors is given by

$$2^K \geq \sum_{i=0}^{e} C_i^N \qquad (9.19)$$

Table 9.2 displays the minimum number of parity checks K for a given code size N. We see that, for small blocks, error correction is expensive in terms of source rate. For large blocks, however, most of the bits can be used as information bits.

It is often possible to find parity check codes that require the minimum number of parity checks. Thus, for example, we can find single-error-correcting codes such as [7, 4] and [63, 57] codes. On the other hand, a [63,52] double-error-correcting code has not been found, although a number of [63, 51] double-error-correcting codes have.

9.8 GROUP CODES

One tool to help us find good codes and develop efficient error-correcting algorithms is to associate algebraic properties with codes. A code is said to be a *group code* if it satisfies the properties of a group. A group is a set G of elements, together with some operation (modulo-2 addition, for example) that satisfies the following four properties:

1. There exists an identity element 0 such that $\forall a \in G$ (for any element a that is in G)

$$a \oplus 0 = 0 \oplus a = a$$

2. $\forall a \in G$, there exists an inverse element $-a \in G$ such that

$$(-a) \oplus a = a \oplus (-a) = 0$$

3. Associativity holds; that is,

$$a \oplus (b \oplus c) = (a \oplus b) \oplus c$$

4. The property of closure holds; $\forall a \in G, \forall b \in G,$

$$a \oplus b \in G$$

The main advantages of the group code structure are based on the closure property. If we define an error pattern or noise vector $\mathbf{z}$ as an N-bit binary word with 1s indicating the error locations and 0s otherwise, then the received vector $\mathbf{r}$ is related to the transmitted vector $\mathbf{y}$ by $\mathbf{r} = \mathbf{y} \oplus \mathbf{z}$. Any noise pattern $\mathbf{z}_i$ that is different from a code word of a group code must be detectable; if not

$$\mathbf{y}_1 \oplus \mathbf{z}_i = \mathbf{y}_2 \quad \text{or} \quad \mathbf{y}_1 \oplus \mathbf{y}_2 = \mathbf{z}_i$$

and, therefore, $\mathbf{z}_i$ must be in the group G. Because a few errors are more likely than a large number of errors, we wish to detect or correct noise vectors with a small number of 1s (the number of 1s is called the *weight* of the vector). It follows that a good group code is one whose code words (with the exception of the **0** word) have the largest possible weight.

Theorem 9.2. Group codes with minimum weight w can always detect ($w - 1$)-tuple or less errors. If w is odd, ($w - 1$)/2-tuple or fewer errors can be corrected. If w is even, ($w - 2$)/2-tuple or fewer errors can be corrected, and, in addition, $w/2$-tuple errors can be detected.

Proof: Since any error pattern that is different from a code word is detectable, it follows that all error patterns of weight $w - 1$ or less are detectable. Therefore, the minimum Hamming distance between code words is at least w. From Theorem 9.1, if w is odd (if $w = 2e + 1$, then $e = (w - 1)/2$), then all single, double, . . . , ($w - 1$)/2-tuple errors can be corrected. If w is even (if $w = 2e$, then $e = w/2$), then ($w/2 - 1$)-tuple errors can be corrected and $w/2$-tuple errors detected.

It is much easier to determine the minimum weight of a code than the minimum Hamming distance between code words. For example, it is easy to see by inspection that the minimum weight of the [8, 4] code of Table 9.1 is 4. If this code is a group code, it follows that the minimum Hamming distance is

4. Previously, we saw this by computing the distance between each of the 15 possible pairs of code words.

Theorem 9.3. All parity check codes [i.e., the 2^M set of N-bit solutions to $A\mathbf{y}^T = 0$, where A is an $(N - M) \times N$ matrix] form a group under modulo-2 addition.

Proof: We must verify that the four properties of a group are satisfied. We note that the **0** code word is the identity element under modulo-2 addition and is always a solution of $A\mathbf{y}^T = 0$. Therefore, property 1 is satisfied. Under modulo-2 addition any vector $\mathbf{y}$ satisfies $\mathbf{y} \oplus \mathbf{y} = 0$, and each solution is its own inverse. Associativity holds for modulo-2 addition. A standard property of matrix multiplication is $A(\mathbf{x}_1 \oplus \mathbf{x}_2) = A\mathbf{x}_1 \oplus A\mathbf{x}_2$, which also holds for modulo-2 addition. It follows that if $\mathbf{y}_1$ and $\mathbf{y}_2$ are code words, then

$$A(\mathbf{y}_1^T \oplus \mathbf{y}_2^T) = A\mathbf{y}_1^T \oplus A\mathbf{y}_2^T = 0 \oplus 0 = 0$$

Hence $\mathbf{y}_1 \oplus \mathbf{y}_2$ is also a code word. Thus all parity check codes are closed.

It can be proven that all group codes can be implemented as parity check codes, but this is not needed for our discussion.

EXAMPLE 9.6 ■

In Example 9.4 we considered a [6, 3] parity check code. According to Table 9.2, it might be possible to find a [6, 3] single-error-correcting code. An examination of the code words indicates that the minimum weight of the code in Example 9.4 is 3. It follows from Theorem 9.2 that this code is single-error correcting. ■ ■

We see that there are advantages in identifying algebraic properties of codes. We shall rely on this procedure more in Chapter 10.

9.9 ERROR-CORRECTING CAPABILITY

We stated in Section 9.7 that a receiver that stores the parity code matrix and computes the $N - M$ parity checks or corrector implied by

$$\mathbf{c} = A\mathbf{r}^T \tag{9.20}$$

where $\mathbf{r}$ is the received N-bit vector and $\mathbf{c}$ is the $(N - M)$-bit corrector, can be used to implement the optimum, minimum-distance, quantized receiver. Such a receiver would also require storage of a table of correctors and their associated error patterns. The proof of this important result will be delayed until Section 9.10. We further note that if a minimum-distance decoder were used with the [8, 4] and [6, 3] codes considered previously, some (but not all) of the double-error patterns would be corrected. How these patterns and their correctors are chosen must also be discussed.

In this section we assume that the correctors point to the appropriate error pattern to be corrected. Our purpose is to relate the error-correcting capability of a parity check code directly to the parity code matrix rather than to the code itself. This will give us some tools with which to find good parity check codes.

Let us define the error pattern or noise vector $\mathbf{z}$ as we did in the previous section. If $\mathbf{y}$ is the transmitted code word and $\mathbf{r}$ the received vector, then

$$\mathbf{r} = \mathbf{y} \oplus \mathbf{z} \tag{9.21}$$

This $\mathbf{z}$ is an N-bit vector with 1s at the error locations and whose weight, therefore, corresponds to the number of errors. For any code word $\mathbf{y}_1$ and any error pattern $\mathbf{z}$, the corrector formed by Equation 9.20 becomes

$$\mathbf{c} = \mathbf{A}(\mathbf{y}_i^T \oplus \mathbf{z}^T) = \mathbf{A}\mathbf{y}_i^T \oplus \mathbf{A}\mathbf{z}^T = \mathbf{A}\mathbf{z}^T \tag{9.22}$$

We see that the corrector is determined entirely by the error pattern and does not depend on the transmitted code word. Furthermore, we see that if $\mathbf{z}$ has 1s in positions $J_1, J_2, \ldots, J_e$, then the ith component of $\mathbf{c}$ is the modulo-2 sum of these components in the ith row of $\mathbf{A}$. More generally, the corrector $\mathbf{c}$ is the modulo-2 sum of the $J_1, J_2, \ldots, J_e$ columns of the matrix $\mathbf{A}$. We know that e-tuple and smaller errors can be corrected if and only if all error patterns of weight e or less yield distinct correctors.

Theorem 9.4. e-tuple and smaller errors are correctable if and only if no linear combination of e or less columns of $\mathbf{A}$ equals another such combination or, equivalently, every set of $2e$ columns of $\mathbf{A}$ are linearly independent.

It follows directly from Theorem 9.4 that column operations, such as addition or interchange, on the matrix $\mathbf{A}$ yield a code that is equivalent as far as error-correcting capability is concerned. Thus, as stated earlier, by using row and column operations, one can find a number of equivalent codes. Such methods can be used to find a systematic implementation of any parity check code.

We observe from Theorem 9.4 that any matrix whose rows are linearly independent and, in addition, whose columns are all distinct must be at least single-error correcting. We can easily check that the previous [8, 4] and [6, 3] codes have distinct columns and can correct single errors. Any parity code matrix whose columns are the standard binary representation of their location generates a Hamming weight-3 code that can correct single errors. Thus, for example, Hamming weight-3 [7, 4] and [15, 11] codes are generated by the matrices $\mathbf{A}_{[7,4]}$ and $\mathbf{A}_{[15,11]}$, where

$$\mathbf{A}_{[7,4]} = \begin{bmatrix} 1 & 1 & 1 & 1 & 0 & 0 & 0 \\ 1 & 1 & 0 & 0 & 1 & 1 & 0 \\ 1 & 0 & 1 & 0 & 1 & 0 & 1 \end{bmatrix} \tag{9.23}$$

$$\mathbf{A}_{[15,11]} = \begin{bmatrix} 1 & 1 & 1 & 1 & 1 & 1 & 1 & 1 & 0 & 0 & 0 & 0 & 0 & 0 & 0 \\ 1 & 1 & 1 & 1 & 0 & 0 & 0 & 0 & 1 & 1 & 1 & 1 & 0 & 0 & 0 \\ 1 & 1 & 0 & 0 & 1 & 1 & 0 & 0 & 1 & 1 & 0 & 0 & 1 & 1 & 0 \\ 1 & 0 & 1 & 0 & 1 & 0 & 1 & 0 & 1 & 0 & 1 & 0 & 1 & 0 & 1 \end{bmatrix} \qquad (9.24)$$

Recall that the $\mathbf{A}$ matrix for an (N, M) code is an $(N - M) \times N$ matrix. The correctors for these codes point directly to the location of a single error. Thus, it is not necessary to store the correctors and the associated error patterns for these single-error-correcting codes. Presumably for the Hamming [15, 11] code, for example, we need only store the 4×15 matrix instead of all 2^{11} 15-bit code words (or 30,720 bits).

It is possible to convert any weight-3 code to a weight-4 code (which corrects single errors and at the same time detects double errors) by adding an extra parity check that performs the modulo-2 sum of all the bits of a code word. This extra parity check forces each code word to have an even number of 1s and, therefore, increases the words of weight 3 to weight 4. Thus the Hamming [7, 4] weight-3 code becomes the Hamming weight-4 [8, 4] code if generated by the matrix

$$\mathbf{A}_{[8,4]} = \begin{bmatrix} 1 & 1 & 1 & 1 & 1 & 1 & 1 & 1 \\ 0 & & & & & & & \\ 0 & & & \mathbf{A}[7, 4] & & & & \\ 0 & & & & & & & \end{bmatrix}$$

$$= \begin{bmatrix} 1 & 1 & 1 & 1 & 1 & 1 & 1 & 1 \\ 0 & 1 & 1 & 1 & 1 & 0 & 0 & 0 \\ 0 & 1 & 1 & 0 & 0 & 1 & 1 & 0 \\ 0 & 1 & 0 & 1 & 0 & 1 & 0 & 1 \end{bmatrix} \qquad (9.25)$$

where y_1 is the added parity check and the additional parity check equation is

$$y_1 \oplus y_2 \oplus y_3 \oplus y_4 \oplus y_5 \oplus y_6 \oplus y_7 \oplus y_8 = 0$$

The error-correcting capability of this code is given by Theorem 9.2. This procedure of increasing the weight of a code by 1 or equivalently increasing the number of linearly independent columns by 1 can be generalized (see Problem 9.11).

While Theorem 9.4 gives considerable insight into finding multiple-error-correcting codes (see Problem 9.13), it is still a difficult task to find large-block-size matrices that generate such codes.

The matrix $\mathbf{A}_{[8,4]}$ of Equation 9.25 is, in fact, the matrix $\mathbf{A}_2$ considered earlier. It follows that the biorthogonal code we have discussed is equivalent to a Hamming weight-4 code. The fact that it is biorthogonal is irrelevant.

9.10 OPTIMUM PARITY CHECK DECODERS

We shall now prove that the optimum, minimum-distance, decoder can be implemented by computing the corrector (which requires storing the parity code matrix) and then determining the error pattern from a table.

Given a group code s that is a subgroup of G (all possible 2^N N-bit binary sequences), the *coset* associated with an element $\mathbf{z} \in G$ is the set of all elements $\mathbf{z} \oplus \mathbf{y}, \mathbf{y} \in s$. The coset associated with $\mathbf{z}$ is referred to as the $\mathbf{z} + s$ coset.

EXAMPLE 9.7a ■

Consider the [4, 2] systematic code generated by the matrix

$$\mathbf{A} = \begin{bmatrix} 1 & 0 & 1 & 0 \\ 0 & 1 & 1 & 1 \end{bmatrix}$$

The four code words (s) are shown below. We see from the matrix (second and fourth columns are the same) and the code s (the minimum weight is only 2) that this code is not single-error correcting and, therefore, not very good. Under the code s in Table 9.3 we list a series of cosets.   ■ ■

Let us now consider two key properties of cosets.

Property 9.1.   The binary patterns of any two cosets are either identical or completely disjoint.

Proof: (a)   If $\mathbf{z}_2$ is a pattern in the $\mathbf{z}_1 + s$ coset, then there exist $\mathbf{y}_i$ such that

$$\mathbf{z}_2 = \mathbf{z}_1 \oplus \mathbf{y}_i$$

and

$$\mathbf{z}_2 \oplus \mathbf{y}_j = \mathbf{z}_1 \oplus \mathbf{y}_i \oplus \mathbf{y}_j = \mathbf{z}_1 \oplus \mathbf{y}_k$$

since s is closed. Thus the patterns in the $\mathbf{z}_2 + s$ coset are also in the $\mathbf{z}_1 + s$ coset.

(b)   If $\mathbf{z}_2$ is not in the $\mathbf{z}_1 + s$ coset, then there cannot be k and m such that

$$\mathbf{z}_2 \oplus \mathbf{y}_k = \mathbf{z}_1 \oplus \mathbf{y}_m$$

for otherwise

TABLE 9.3   SOME COSETS FOR s OF EXAMPLE 9.7a

	\multicolumn{4}{c}{s}			
	0000	0101	1110	1011
$0110 + s$	0110	0011	1000	1101
$1111 + s$	1111	1010	0001	0100
$0001 + s$	0001	0100	1111	1010
$0010 + s$	0010	0111	1100	1001

$$\mathbf{z}_2 = \mathbf{z}_1 \oplus (\mathbf{y}_m \oplus \mathbf{y}_k) = \mathbf{z}_1 \oplus \mathbf{y}_j$$

and $\mathbf{z}_2$ would be in the $\mathbf{z}_1$ coset, which is a contradiction.

As a result of this property, we always obtain $2^N/2^M$ distinct cosets (counting s).

EXAMPLE 9.7b ■

We observe in the cosets listed in Example 9.7a, that the $1111 + s$ coset and the $0001 + s$ coset have identical patterns. The patterns in the other three cosets shown are completely distinct. We further note that the four distinct cosets shown ($2^4/2^2 = 4$) include all 16 possible patterns. ■ ■

The second key property of cosets involves considering all of the binary sequences as possible error patterns.

Property 9.2. All error patterns corresponding to sequences in the same coset of a group code s result in the same corrector. All sequences in different cosets result in different correctors.

Proof: If $\mathbf{z}_1$ and $\mathbf{z}_2$ are in the same coset, then $\mathbf{z}_2 = \mathbf{z}_1 \oplus \mathbf{y}_i$ for some $\mathbf{y}_i \in s$. Therefore, $\mathbf{Az}_2^T = \mathbf{A}(\mathbf{z}_1^T \oplus \mathbf{y}_i^T) = \mathbf{Az}_1^T = \mathbf{c}$. On the other hand, if $\mathbf{z}_1$ and $\mathbf{z}_2$ have the same corrector $\mathbf{c}$, then

$$\mathbf{Az}_1^T = \mathbf{Az}_2^T \Rightarrow \mathbf{A}(\mathbf{z}_1^T \oplus \mathbf{z}_2^T) = 0$$

Therefore, $\mathbf{z}_1 = \mathbf{z}_2 \oplus \mathbf{y}_i$ and $\mathbf{z}_1$ and $\mathbf{z}_2$ are in the same coset.

EXAMPLE 9.7c ■

In Table 9.4 we rewrite the distinct cosets of Example 9.7a along with the corrector of the first pattern (called the *coset leader*) in each coset. If the patterns in a coset are error patterns, they result in the same corrector. ■ ■

It appears that if we base our decoder on the correctors, then we must choose one of the error patterns in each coset to correct. For Example 9.7,

TABLE 9.4 EXAMPLE 9.7c

	s				
	0000	0101	1110	1011	c^a
$0110 + s$	0110	0011	1000	1101	10
$0001 + s$	0001	0100	1111	1010	01
$1101 + s$	0010	0111	1100	1001	11

[a] $c = 00$ for s.

the error patterns 0001 and 0100 have the same corrector, and only one of them can be corrected at the expense of the other being incorrectly decoded. As we shall now see, this is not a restriction.

Theorem 9.5. If, for a particular error pattern z, $y_i \oplus z$ is corrected to y_i by the optimum minimum-distance decoder then (a) $y \oplus z$ is corrected to y for every y and (b) no other error pattern in the $z + s$ coset can be corrected.

Proof: Let us use the notation $d(y_1, y_2)$ to mean the Hamming distance between y_1 and y_2. An obvious property of Hamming distance is $d(y_1 + y_3, y_2 + y_3) = d(y_1, y_2)$. Part (a) of the theorem states that if

$$d(y_i + z, y_i) \le d(y_i + z, y) \qquad \forall y_i \in s$$

then

$$d(y + z, y) \le d(y + z, y') \qquad \forall y, y' \in s$$

Now,

$$
\begin{aligned}
d(z + y, y') &= d(z + y + y_i - y, y' + y_i - y) \\
&= d(z + y_i, y_i + y' - y) \ge d(z + y_i, y_i) \\
&= d(z + y_i + y - y_i, y_i + y - y_i) \\
&= d(z + y, y)
\end{aligned}
$$

This means that a given error pattern z is always corrected or never corrected by a minimum-distance decoder.

To prove part (b), let z_1 be an error pattern that is always corrected. If z_2 is a pattern in the $z_1 + s$ coset, $z_2 = z_1 \oplus y$ for some $y \in s$. If z_2 is received ($r = z_2$), it could have come from either $z_1 \oplus y$ or $z_2 \oplus 0$. Since z_1 is always corrected, z_2 will not be decoded as 0, and therefore z_2 will never be corrected.

The optimum receiver corrects one error pattern in each coset, and the only question that remains is which one. It is intuitively clear that the minimum-weight sequence should be corrected.

Theorem 9.6. The minimum-distance decoder corrects the sequence of minimum weight in each coset.

Proof: If z is of minimum weight, then $d(z, y) = d(z - y, y - y) = d(z - y, 0) > d(z, 0)$ by the definition of minimum weight. Hence, z is at least as close to 0 as any other code word and, therefore, should be decoded as 0. By Theorem 9.5, $z \oplus y$ should always be decoded as y.

The procedure for determining the error patterns to be stored in a table versus correctors can now be developed. We choose the cosets by picking as a coset leader an error pattern of minimum weight that has not previously appeared in a coset. We continue this procedure until all 2^{N-M} cosets are found. The coset leaders are the patterns to be corrected.

EXAMPLE 9.8 ■

Determine the optimum decoder for the code generated by

$$A = \begin{bmatrix} 1 & 0 & 0 & 0 & 1 & 1 \\ 1 & 1 & 0 & 0 & 0 & 0 \\ 1 & 0 & 1 & 0 & 0 & 1 \\ 0 & 1 & 1 & 1 & 0 & 0 \end{bmatrix}$$

Convert this to systematic form by row operations

$$A_1 = \begin{bmatrix} 1 & 0 & 0 & 0 & 1 & 1 \\ 0 & 1 & 0 & 0 & 1 & 1 \\ 0 & 0 & 1 & 0 & 1 & 0 \\ 0 & 0 & 0 & 1 & 0 & 1 \end{bmatrix}$$

With bits y_1, y_2 as information bits, the four code words of the [6, 2] code are listed in Table 9.5. Since the minimum weight of the code is 4, we know this code can correct single errors and detect double errors. We begin to form the cosets by using single-error patterns as coset leaders. We first observe that all of the single-error patterns have unique correctors. We also note that there are no double-error patterns in the cosets considered so far; therefore, all double errors have correctors different from single errors and are at least detectable. The total number of correctors or cosets is $2^{6-2} = 16$, and we have already used up seven of them. Since there are $C_2^6 = 15$ double-error patterns, all these errors cannot be corrected. At best $16 - 7 = 9$ of the double-error patterns can be corrected. Continuing with weight-2 coset leaders, we get the cosets listed in Table 9.6. The seven cosets shown contain all 15 double-error patterns. Thus we correct seven out of 15 double-error patterns. Presumably, we can find two triple errors with unique correctors. These are

011100   101001   100110   010011   0111
101100   011001   010110   100011   1011

We conclude that the optimum decoder stores the parity code matrix and a table of the 15 correctors and error patterns corresponding to the

TABLE 9.5 EXAMPLE 9.8: THE FOUR CODE WORDS AND WEIGHT-1 COSETS

				c
000000	110101	111010	001111	0000
000001	110100	111011	001110	1101
000010	110111	111000	001101	1110
000100	110001	111110	001011	0001
001000	111101	110010	000111	0010
010000	100101	101010	011111	0100
100000	010101	010010	101111	1000

TABLE 9.6   EXAMPLE 9.8: WEIGHT-2 COSETS

				c
000011	110110	111001	001100	0011
000110	110011	111100	001001	1111
011000	101101	100010	010111	0110
110000	000101	001010	111111	1100
010100	100001	101110	011011	0101
101000	011101	010010	100111	1010
100100	010001	011110	101011	1001

coset leaders. This code corrects all single errors, seven out of 15 double errors and two triple errors.   ■ ■

9.11 CONCLUSIONS

We have shown that binary block codes can be implemented which will reduce the probability of error relative to standard PCM. In order to obtain significant improvement, the block codes have to be large. Because of this need, significant practical problems remain:

1. Theorem 9.4 is not adequate to find large block codes capable of correcting multiple errors. Trial-and-error approaches exhaust even the largest computers for sizable block codes.
2. The optimal decoding techniques of Section 9.10 become cumbersome and require considerable storage capabilities for large block codes.
3. In many applications it is sufficient to correct multiple errors that occur in a burst rather than randomly scattered over the block. The theory developed in this chapter gives us no insights into or techniques for finding special codes with such properties.

In Chapter 10 we shall see that, by considering more restricted classes of codes, we can attack many of these problems.

PROBLEMS

9.1. Verify that [8, 4] encoder of Figure 9.3 generates the biorthogonal code of Table 9.1. Do this by determining the output of Figure 9.3 or Equation 9.15 for various four-bit input vectors.

9.2. Show by substituting a few code words from Table 9.1 into the equation $A_1 y^T = 0$ that the solutions to this equation are indeed the [8, 4] biorthogonal code.

9.3. A five-symbol parity check code $(y_5 y_4 y_3 y_2 y_1)$ is generated from a three-symbol input code $(x_3 x_2 x_1)$ by

$$y_j = \sum_{i=1}^{3} a_{ji} x_i$$

where $a_{ji} = 0, 1$ where the summation is modulo 2. Three message words and the corresponding parity check codes $x_3 x_2 x_1$ and $y_5 y_4 y_3 y_2 y_1$ are

```
011   10101
101   01010
010   00110
```

(a) Find the generating matrix.
(b) Find the parity check code words for the remaining five input code words.
(c) Find the parity check matrix.

9.4. Prove that by simple row and column operations the matrix A_1 of Section 9.7 can be converted to the matrices A_2 and A_3.

9.5. Consider the following code, which might be useful for four messages: 000000, 110101, 111010, and 001111:
(a) Show that this code is a systematic parity check code by finding a matrix **A** with the form of Equation 9.18 such that the code words are solutions to $\mathbf{A y}^{\mathrm{T}} = 0$.
(b) Find the generating matrix for this code and sketch a block diagram of the encoder.

9.6. In Examples 9.4 and 9.6 we considered a single-error correcting [6, 3] code. Determine for each of the six single-error patterns the parity check solutions (i.e., the correctors). Verify that the six correctors are distinct, that is, can be used to point to the error location.

9.7. What is the error-correcting capability of the code in Problem 9.5? Is this consistent with the bounds of Table 9.2?

9.8. Consider the [5, 2] encoder shown below:

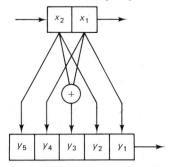

(a) List the four code words.
(b) Determine the error-correcting capability.
(c) Determine the parity check matrix.

9.9. By standard row and column operations convert the Hamming [7, 4] parity code matrix (Equation 9.23) into systematic form. Determine the code words from this new matrix. Verify that the minimum weight, and hence the minimum Hamming distance is 3.

9.10. Write a 4 × 12 matrix that will generate a [12, 8] single-error-correcting parity check matrix.

9.11. Given an e-tuple error-correcting parity check code with parity code matrix $\mathbf{A}_1$, show that the code generated by the matrix $\mathbf{A}_0$ will correct e-tuple errors and detect $(e + 1)$-tuple errors:

$$\mathbf{A}_0 = \begin{bmatrix} 1 & 1 & 1 & 1 & \cdots & 1 & 1 & 1 \\ 0 & & & & & & & \\ 0 & & & \mathbf{A}_1 & & & & \\ \vdots & & & & & & & \\ 0 & & & & & & & \end{bmatrix}$$

9.12. Find a parity code matrix of a single-error-correcting and double-error-detecting code with five check digits and eight information digits.

9.13. (a) Prove that the sum of any two columns of the matrix shown yields a corrector that is unique from all columns and all other pairs of columns:

$$\mathbf{A}_{[8,2]} = \begin{bmatrix} 0 & 0 & 0 & 0 & 0 & 1 & 1 & 0 \\ 0 & 0 & 0 & 0 & 1 & 0 & 1 & 0 \\ 0 & 0 & 0 & 1 & 0 & 0 & 1 & 1 \\ 0 & 0 & 1 & 0 & 0 & 0 & 1 & 1 \\ 0 & 1 & 0 & 0 & 0 & 0 & 1 & 1 \\ 1 & 0 & 0 & 0 & 0 & 0 & 0 & 1 \end{bmatrix}$$

(b) Determine the code words and verify that the minimum weight is indeed 5 and that this is a double-error-correcting code.

9.14. In Examples 9.4 and 9.6 we considered a $[6, 3]$ single-error-correcting code.

(a) By considering the number of cosets determine the maximum number of double-error patterns that can be corrected.

(b) Determine all the cosets and indicate the table of correctors versus error patterns that the receiver would need.

(c) What percentage of double-error patterns are corrected?

9.15. Repeat Problem 9.14 for the $[5, 2]$ code of Problem 9.8.

9.16. Consider the $[8, 2]$ code of Problem 9.13. Based on the number of cosets, determine the maximum number of triple-error patterns that can be corrected. Using Theorem 9.4 only, find at least one triple error whose corrector is the same as that of a double error and one triple error with a corrector different from that of any single or double error.

9.17. Show that changing any one element in column 6 of the parity code matrix of Example 9.8 will destroy the double-error-detecting capability of the code.

REFERENCE

1. C. E. Shannon, "Mathematical Theory of Communication," *Bell Syst. Tech. J.* 27, 379–423, 623–656 (1948).

Chapter 10

Cyclic and Convolution Codes

10.1 INTRODUCTION

It was indicated in Chapter 9 that some important practical problems need to be resolved. The most important are finding good, large, block-size codes and special codes with efficient decoding algorithms. In order to resolve such problems, it is necessary to restrict the class of codes to a small subset of parity check codes. These restrictions will not be so severe that we eliminate all or most of the good codes.

First, we shall restrict ourselves to those parity check codes that satisfy certain algebraic properties (other than those of a group). The idea is to use known algebraic theorems to help find good codes and efficient decoding techniques. The second approach is to restrict ourselves to a limited hardware structure that makes it feasible to conduct trial-and-error searches for good codes. The advantages and applications of, and the concepts underlying, these two approaches are very different.

10.2 DEFINITION OF CYCLIC CODES

Consider the operation of converting a vector $\mathbf{y} = (y_N, y_{N-1}, \ldots, y_2, y_1)$ into a vector $\mathbf{y}' = (y_1, y_N, \ldots, y_3, y_2)$.[1] This operation corresponds to shifting the components of $\mathbf{y}$ one unit to the right with the rightmost bit that is forced out of the shift register fed back in at the left (called a rotate right). Any parity check or group code that is closed under this operation (i.e., the resulting vector is another code word) is said to be a *cyclic code*.

[1]The following development is that of W. W. Peterson [1].

EXAMPLE 10.1 ■

Consider the [7, 4] Hamming weight-3 code whose parity code matrix is given by

$$\mathbf{A} = \begin{bmatrix} 1 & 1 & 1 & 0 & 1 & 0 & 0 \\ 0 & 1 & 1 & 1 & 0 & 1 & 0 \\ 1 & 1 & 0 & 1 & 0 & 0 & 1 \end{bmatrix}$$

The resultant code words are listed in Table 10.1, but not in the usual systematic order. The order is chosen to show explicity that the resultant code is closed under the rotate right operation and is, therefore, a cyclic code.   ■ ■

Many of the good parity check codes that have been found are equivalent to cyclic codes. It follows that restricting the class of binary codes to cyclic codes does not affect the search for good codes seriously.

It is advantageous to identify an N-dimensional vector with a polynomial of degree $N - 1$. Thus a vector $\mathbf{y} = (a_0, a_1, \ldots, a_{N-1})$ is identified with the polynomial $f(x) = a_0 + a_1 x + a_2 x^2 + \cdots + a_{N-1} x^{N-1}$. For example, the first nonzero vector in Table 10.1 is identified with the polynomial $g(x) = 1 + x^2 + x^3$. Any uncoded vector $\mathbf{x}$ can be identified with a polynomial of degree $M - 1$. Multiplication of any polynomial of degree $N - 1$ by x can be identified as a one-unit cyclic shift or rotation to the right provided one reduces the resultant polynomial to degree $N - 1$ by invoking the constraint $x^N = 1$, which corresponds to feeding the rightmost bit back at the left. This type of polynomial multiplication is called *multiplication modulo $x^N - 1$*. Multiplication modulo $x^N - 1$ of a polynomial $g(x)$ that corresponds to a code word in a cyclic group by any other polynomial $h(x) = a_0 + a_1 x + a_2 x^2 + \cdots$ can be written in the form

$$g(x)h(x) = a_0 g(x) \oplus a_1 x g(x) \oplus a_2 x^2 g(x) \oplus \cdots \tag{10.1}$$

Each of the polynomials $a_i x^i g(x)$ are code words (by the definition of cyclic codes), and the modulo-2 sum of code words is another code word (by the definition of a group code). Thus multiplication of a code word polynomial by

TABLE 10.1 A [7, 4] CYCLIC CODE

0000000	1110100
1011000	0111010
0101100	0011101
0010110	1001110
0001011	0100111
1000101	1010011
1100010	1101001
0110001	1111111

any other polynomial, modulo $x^N - 1$, results in another code word polynomial.

EXAMPLE 10.2 ∎

Let us multiply the first nonzero vector identified with $g(x)$ by $1 + x$:

$$
\begin{array}{ll}
\quad 1011000 & \quad 1 + x^2 + x^3 \\
\times 11 & \quad 1 + x \\
\hline
\quad 1011000 \\
\llap{\swarrow}\;1011000 \\
\hline
\quad 1110100 & \quad 1 + x + x^2 + x^4
\end{array}
$$

Note that this corresponds to the first word in the second column of Table 10.1. ∎ ∎

Cyclic codes form a group (called an *ideal*) that is closed under multiplication modulo $x^N - 1$, in addition to being closed under modulo-2 addition. A code word polynomial that generates the entire code when multiplied by the 2^M uncoded polynomials is called a *generating polynomial*. The polynomial corresponding to the first nonzero vector, $g(x) = 1 + x^2 + x^3$, will generate the [7, 4] cyclic code of Table 10.1. As we shall see, the implementation of the encoder can be visualized as multiplication modulo $x^N - 1$. Finding good codes corresponds to finding good generator polynomials.

A polynomial $g(x)$ is said to divide another polynomial $f(x)$ if there exists a polynomial $h(x)$ such that $g(x)h(x) = f(x)$. Every equivalent code word polynomial of the Hamming [7, 4] code can be divided by the generator polynomial, and the quotient corresponds to the uncoded message polynomial. This suggests an implementation for an optimum error-detecting decoder. The receiver divides the received polynomial by the generator polynomial (provided that division using modulo-2 arithmetic can be implemented). If the remainder is zero, then the quotient is assumed to be the uncoded message polynomial. If the remainder is nonzero, then an error has been detected. Thus the remainder serves the same function as the parity checks.

EXAMPLE 10.3 ∎

Assume, for example, that the received vector from a Hamming [7, 4] code generated by $g(x) = 1 + x^2 + x^3$ is 1001100. The receiver performs the following operation:

$$
\begin{array}{r}
0 + \;x\; + 0x^2 + 0x^3 \\
1 + 0x + \;x^2\; + \;x^3 \overline{)\,1 + 0x + 0x^2 + \;x^3\; + \;x^4 + 0x^5 + 0x^6} \\
\underline{x + 0x^2 + \;x^3\; + \;x^4} \\
1 - \;x
\end{array}
$$

The remainder is not zero; hence an error has been detected. Like the parity checks, the remainder cannot distinguish between single-bit errors and double-bit errors. ∎ ∎

10.3 IMPLEMENTATION OF CYCLIC CODES

Cyclic codes can be implemented with switching circuits composed of memory devices and modulo-2 adders. The memory devices are either delay devices or single stages of ordinary binary shift registers, and the modulo-2 adders are exclusive-or logical blocks. Input or output polynomials are transmitted higher-order coefficients first.

The circuit shown in Figure 10.1 is used to multiply any input polynomial

$$a(x) = a_0 + a_1 x + \cdots + a_{k-1} x^{k-1} + a_k x^k$$

by the fixed polynomial

$$h(x) = h_0 + h_1 x + \cdots + h_{r-1} x^{r-1} + h_r x^r$$

Since the polynomial coefficients are either 1 or 0, the multipliers in the circuit correspond to connections or absence of connections. The storage devices initially contain all 0s. When the first element a_k enters the circuit, the polynomial $a_k h(x)$ is entered in the shift register (highest-order coefficient actually at the output). At the next time unit, the polynomial $a_k h(x)$ is shifted to the right $[x a_k h(x)]$ and added modulo 2 to the polynomial $a_{k-1} h(x)$ formed by the second input element. It follows by induction that the output is

$$(\{[a_k h(x) x + a_{k-1} h(x)] x + a_{k-2} h(x)\} x + a_{k-3} h(x)) x + \cdots$$

or

$$a_k x^k h(x) + a_{k-1} x^{k-1} h(x) + a_{k-2} x^{k-2} h(x) + \cdots = a(x) h(x)$$

The circuit shown in Figure 10.2 is used to divide an input polynomial $a(x)$ by the fixed polynomial

$$g(x) = g_0 + g_1 x + \cdots + g_r x^r$$

The storage devices are initially set to 0. The output is 0 for the first r shifts, after which it becomes a 1. (It is assumed that a_k and g_r are both 1.) At this point the polynomial $g(x)$ is subtracted from the portion of the dividend in the shift register. The division continues whenever a 1 is shifted out of the last storage device, and the polynomial $g(x)$ is subtracted from the remainder in the shift register. This operation can be viewed as division exactly like long division performed by hand. After a total of k shifts the entire quotient has appeared at the output, and the remainder is in the shift register.

EXAMPLE 10.4 ■

Consider again the [7, 4] Hamming code generated by $g(x) = 1 + x^2 + x^3$. An implementation for this code, based on the circuits of Figures 10.1 and 10.2, is given in Figure 10.3. Consider now the same input as that of Example 10.3 (1001100), which was not a valid code word. Let us step through the decoder output as well as the values stored in the shift register to show that it behaves exactly as the division of Example 10.3.

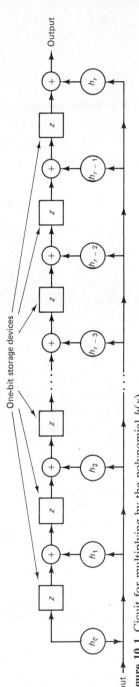

Figure 10.1 Cicuit for multiplying by the polynomial $h(x)$.

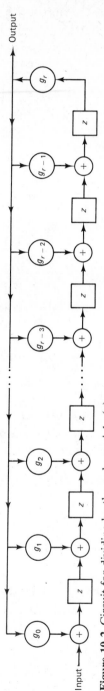

Figure 10.2 Circuit for dividing by the polynomial $g(x)$.

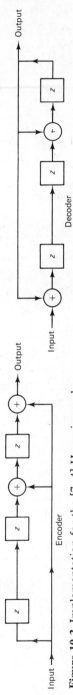

Figure 10.3 Implementation for the [7, 4] Hamming code.

We begin after the first three zeros have shifted out and the first three input bits appear inside the shift register of the circuit in Figure 10.3b.

$$\overset{\downarrow\ \downarrow\uparrow}{1001100}\rightarrow$$

$$\overset{\downarrow\ \downarrow\uparrow}{100110}\rightarrow \qquad\qquad 0$$

$$\overset{\downarrow\ \downarrow\uparrow}{10011}\rightarrow \qquad\qquad 00$$

$$\overset{\downarrow\ \downarrow\uparrow}{1100}\rightarrow \qquad\qquad 100$$

$$\underbrace{110}_{\text{remainder}} \qquad \underbrace{0100}_{\text{output}}$$

■ ■

Since the multiplying circuit forms the output systematically in the shift register and the dividing circuit reduces the dividend systematically in the shift register, the multiplying and dividing circuits of Figures 10.1 and 10.2 can be combined to multiply by $h(x)$ and divide by $g(x)$ simultaneously as in Figure 10.4. These three basic circuits are all that is required for implementing the encoding and decoding of cyclic codes.

It is not necessary to view the encoder as a multiplier and the decoder as a divider. Because, as we shall see, the generator polynomial $g(x)$ divides $x^N - 1$, there must exist some polynomial $h(x) = (x^N - 1)/g(x)$. Multiplying modulo $x^N - 1$ by the polynomial $x^N - 1$ is equivalent to multiplying by unity. Hence, multiplying (or dividing) by $g(x)$ can be replaced with dividing (or multiplying) by $h(x)$. It is seen from Figures 10.1 and 10.2 that multiplying or dividing by a fixed polynomial $g(x)$ requires a shift register whose length corresponds to the degree of the fixed polynomial. It follows that encoding and decoding of a cyclic code requires shift registers of length $N - M$ or M, whichever is less. In this chapter all the codes considered are such that $N - M < M$, and all of the circuits shown will employ shift registers of length $N - M$.

It is sometimes convenient to force the code into a systematic format. When error-correction procedures are employed, the systematic format simplifies the decoder implementation. If the code is to be used for error detection only, however, there is no reason for the systematic format except for the possibility that it may fit into the overall system more conveniently. It is conceptually easy to imagine a system that transforms the information polynomial $f(x)$ of degree $M - 1$ into another polynomial $h(x)$ of degree $M - 1$ prior to the encoder and after the decoder forms the inverse transformation. If this transformation satisfies the equation

$$f(x)x^{N-M} + p(x) = h(x)g(x) \qquad\qquad (10.2)$$

where $g(x)$ is the generating polynomial of degree $N - M$ and $p(x)$ is a polynomial of degree $N - M - 1$, then the encoded polynomial is in systematic form for the original polynomial $f(x)$. The factor x^{N-M} is needed to

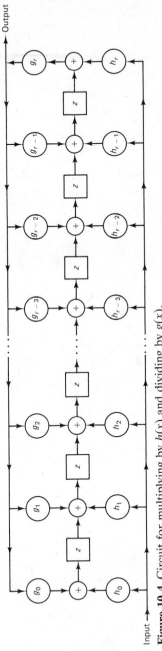

Figure 10.4 Circuit for multiplying by $h(x)$ and dividing by $g(x)$.

shift the information bits to the right or to transform the leading information bit, which is the coefficient of x^{M-1}, into the coefficient of x^{N-1} ($x^{N-M}x^{M-1} = x^{N-1}$). The coefficients of $p(x)$ correspond to the parity checks, and $p(x)$ is called the *parity check polynomial*.

EXAMPLE 10.5 ■

We wish to transform four information bits, 0101 [$f(x) = x + x^3$], into a seven-bit systematic code word

$$\underbrace{1000}_{p(x)}\underbrace{101}_{x^3f(x)}$$

where the parity checks are 100 [$p(x) = 1$]. We hope to use the method suggested by Equation 10.2, where $g(x) = 1 + x^2 + x^3$. Let us determine the intermediate polynomial $h(x)$ by dividing both sides of Equation 10.2 by $g(x)$: $[x^3f(x) + p(x)]/g(x) = h(x)$.

```
           1011
   1011 ) 1000101
          1011
          0111
          1011
         10110
         1011
```

Thus, if we transform $f(x)$ into $1 + x^2 + x^3$ and then multiply by $g(x)$, we get the desired systematic code word, 10000101. ■ ■

Observe that if we divide $x^3f(x)$ by $g(x)$, we obtain $h(x)$ with a remainder of 100:

```
           1011
   1011 ) 0000101
          1011
         01110
         1011
         00110
         1011
          100
```

If the remainder is always the correct parity check polynomial $p(x)$, we do not need $h(x)$, since we can simply append $p(x)$ on to $x^3f(x)$. We can see that this method will work by rearranging Equation 10.2 into

$$f(x)x^{N-M} = h(x)g(x) + p(x) \tag{10.3}$$

Thus, if we divide $f(x)x^{N-M}$ by $g(x)$, the answer is $h(x)$ (which can be suppressed) with a remainder of $p(x)$.

It follows that a systematic encoder can be implemented as shown in Figure 10.5. Initially gates 1 and 3 are enabled, gate 2 is disabled, and the

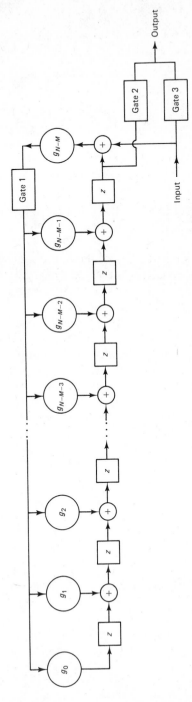

Figure 10.5 Implementation for a systematic encoder/decoder.

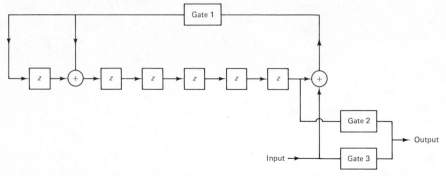

Figure 10.6 Systematic encoder/decoder for the [63, 57] generalized Hamming weight-3 code.

memory devices are set to zero. The input is fed directly into the output and simultaneously into a switching circuit that multiplies by x^{N-M} and divides by $g(x)$. This circuit is identical to Figure 10.3 with the output suppressed since only the remainder is desired. After the entire input is fed into the channel (and the switching circuit), the three gates change their states. This disconnects the feedback loop and feeds the parity check polynomial that has been developed in the shift register into the channel.

The identical circuit can be used as a decoder. Note that the premultiplication by $\dot{x}^{N-M}$, while not necessary, does not affect the ability to detect errors; it just shifts a detectable error pattern into another detectable error pattern. An implementation of a systematic encoder/decoder for a [63, 57] generalized Hamming code, which will be discussed later, is given in Figure 10.6.

So far we have not considered error-correction techniques. Our decoder gives the remainder, which we can use to correct errors by having a table of remainders and error patterns stored in the receiver. We seek more efficient methods, however. We shall postpone this discussion until after we have discussed how to find the generating polynomials and error-correcting capabilities of cyclic codes.

10.4 PROPERTIES OF CYCLIC CODES

As indicated earlier, a number of algebraic properties and theorems based on ideals can be used to find good cyclic codes. A rigorous discussion and proof of these theorems is beyond the scope of this text. We shall, however, state without proof some of the most relevant theorems [1].

We first note that if a polynomial $p(x)$ of degree M is not divisible by any polynomial of degree less than M (except for those of degree zero), it is called *irreducible*. The first two theorems enable us to find any cyclic code.

Theorem 10.1. Every polynomial that divides $x^N - 1$ generates a cyclic code.

Theorem 10.2. Every irreducible factor of $x^{2^K-1} - 1$ has degree K or less, and every irreducible polynomial $p(x)$ of degree K is a factor of $x^{2^K-1} - 1$.

EXAMPLE 10.6 ■

    We seek a cyclic code of block length 7. According to Theorem 10.2, we need to find the irreducible polynomials of degree 3. The polynomials $1 + x + x^3$ and $1 + x^2 + x^3$ are irreducible. All other third-degree polynomials are not. Using modulo-2 addition, we verify that

$$x^3 + x^2 + x + 1 = (x^2 + 1)(x + 1) \qquad x^3 + 1 = (x^2 + x + 1)(x + 1)$$
$$x^3 + x^2 + x = (x^2 + x + 1)x \qquad\qquad x^3 + x = (x^2 + 1)x$$

In agreement with Theorem 10.2,

$$x^7 - 1 = x^7 + 1 = (1 + x + x^3)(1 + x^2 + x^3)(1 + x)$$

Any one, or any combination, of these polynomials generates a cyclic code of length 7. We have already seen that $1 + x^2 + x^3$ generates the Hamming [7, 4] code of Example 10.1. The degree of the generating polynomial always corresponds to the number of parity checks K.

■ ■

    It is possible to find all the irreducible polynomials using a computer if the degree is not excessive. Peterson [1], for example, lists all irreducible polynomials of degree 34 or less, which enables us to find all cyclic codes of block lengths $2^K - 1$ for $K \le 34$.

    Let us now consider the error-detecting capability of cyclic codes. Cyclic codes are particularly useful for detecting multiple errors that are close together, called *burst errors*. A burst of length d is defined as a vector whose only nonzero components are among d successive components, the first and last of which are nonzero. For example, the error pattern 00000011010000 corresponds to a burst of length 4. In many situations the main source of errors is not stationary noise but the result of some sudden change in the system generating the data, such as a relay switching or a motor or engine turning on for a short time. In this environment detecting or correcting bursts is more important than detecting or correcting multiple random errors. The next two theorems indicate the burst-error-detection capability of cyclic codes, which represents one of the major advantages of cyclic codes.

Theorem 10.3. A code vector of an [N, M] cyclic code cannot be a burst of length K, where $K \le N - M$. Therefore, every [N, M] cyclic code can detect any burst of length $N - M$ or less.

EXAMPLE 10.7 ■

    The Hamming [7, 4] code is capable of detecting any burst of length $7 - 4 = 3$ or less. There are 23 such error patterns, including the seven single-bit error patterns. Note that such detection is not possible if single-bit errors are corrected. ■ ■

More knowledge of the structure of code words is needed in order to determine the capability of detecting multiple bursts or correcting burst errors. The problem of finding good cyclic codes is one of finding generating functions that specify either the minimum weight or the structure of the code.

10.5 GENERALIZED HAMMING AND FIRE CODES

These classes of cyclic codes can correct either single errors or, for the Fire codes, a single burst of errors. We shall see in Section 10.6 that the receivers are easy to implement.

In order to find the appropriate generating polynomials, we need some more group-theoretic definitions. An irreducible polynomial is said to be of *order e* if e is the smallest integer such that the polynomial divides $x^e - 1$. A polynomial of degree M is said to be *primitive* if $e = 2^M - 1$, that is, if $p(x)$ divides $x^N - 1$ for no N less than $2^M - 1$. The order of all the irreducible polynomials can be determined (see Problem 10.7). We can now state the theorems that enable us to find the Hamming and Fire codes.

Theorem 10.4. Any cyclic code generated by a primitive polynomial $p(x)$ is a generalized Hamming code of weight 3.

Theorem 10.5. If $p(x)$ is a primitive polynomial, the polynomial $g(x) = (1 + x)p(x)$ generates a generalized Hamming weight-4 code.

EXAMPLE 10.8 ■
Assume that there are six primitive polynomials of degree 6. It follows that there are six generalized Hamming [63, 57] cyclic codes, all of which permit detection of any combination of two errors and all burst errors of length 6 or less. One of these primitive polynomials is $g(x) = 1 + x + x^6$. The systematic encoder of Figure 10.6 was generated by this polynomial. The polynomial $(1 + x)(1 + x + x^6) = 1 + x^2 + x^6 + x^7$ generates a Hamming weight-4 code that can be used to correct single errors and to detect any combination of two errors. A simple implementation that achieves this will be given in the next section. ■ ■

Theorem 10.6. A code generated by $g(x) = p(x)(x^c + 1)$, where $p(x)$ is an irreducible polynomial of degree M, that does not divide the polynomial $(x^c + 1)$ is a *Fire code*. The block size N is the least common multiple of c and e [the order of $p(x)$]. No code word is a vector that is the sum of a burst of length M or less and a burst of length d or less, where $d \leq c - M + 1$. (It is assumed that $c > M$.)

EXAMPLE 10.9 ■
Consider the Fire code generated by $g(x) = (1 + x + x^3)(x^9 + 1)$. The polynomial $1 + x + x^3$ is primitive; hence the order is 7. The block size

TABLE 10.2 DESCRIPTION OF THE STRUCTURE OF IMPORTANT
CYCLIC CODES

Code name	Minimum weight	Additional structure[a]
Generalized Hamming weight-3 code	3	None
Generalized Hamming weight-4 code	4	None
Fire codes	3	No code word is the sum of a burst of length M and a burst of length d $(d = c - M + 1)$

[a] That is, structure other than the fact that no code word is a burst of length $N - M$ or less.

N is the least common multiple of 7 and 9, or $N = 63$. This polynomial generates a [63, 51] code that can detect bursts of length 12 or less or two bursts one of length 3 or less and the other of length 7 or less. This code can be used to correct bursts of length 3 or less. A simple implementation of this decoder will be given in Section 10.6. ■ ■

A summary of some important cyclic codes and their structure is given in Table 10.2.

10.6 IMPLEMENTATION OF SINGLE-BIT OR SINGLE-BURST ERROR CORRECTORS

Let us now consider the possibility of correcting burst errors when the structure of the code makes it possible. Error patterns corresponding to bursts of length b can be expressed in polynomial form as $E(x) = x^m B(x)$, where $B(x)$ is any arbitrary polynomial of degree $b - 1$ or less and m is the location of the first error bit. Note that single random errors correspond to $E(x) = x^m$ and are equivalent to bursts of length 1. Understanding the theory of burst error correction is equivalent to understanding how the decoder circuit works. Figure 10.7 shows a switching circuit that can be used to correct all bursts of length b or less in addition to detecting other error patterns (provided, of course, that the code makes it theoretically possible to do so).

Let us first describe the mechanics of the burst error correcting decoder and then explain why it works. The input is fed into a temporary buffer storage of N memory devices in addition to being fed into a circuit like that of Figure 10.4, which generates the remainder. After the input is completely stored in the memory, the $K - b$ leftmost memory stages in the shift register ($K = N - M$) are tested for all 0s. If they are not all 0, the first information bit from the buffer storage is fed into the output and the switching circuit goes through a shift cycle. The test (for a correctable pattern) is made again and the procedure continues. If the test is ever satisfied, the gates change their

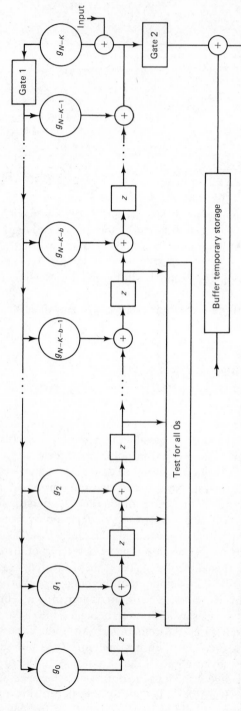

Figure 10.7 Systematic decoder for correcting all bursts of length b or less.

state and the bits in the shift register are added modulo 2 to the remainder information bits as they come out of the buffer storage, which presumably corrects the error. If the test is not satisfied before the entire input is fed into the output from the buffer storage, then an uncorrectable error has been detected.

Consider the case where the burst error occurs near to the beginning of the block, that is, the first error corresponds to the coefficient of x^{N-1-n}, where n is small ($n < N - M - b$). The initial remainder generated in the shift register corresponds to the remainder obtained by dividing $x^{N-M}E(x)$ by $g(x)$ (only the error pattern contributes to a nonzero remainder). For burst errors near the beginning of the block, the highest degree of the polynomial $x^{N-M}E(x)$ is less than the highest degree of $g(x)$ (i.e., $N - M - 1$). It follows that the remainder is the numerator itself, which is the error pattern. Cycling the shift register moves the error pattern to the right, and after n shifts the error pattern appears at the rightmost bits of the shift register just as the first incorrect bit is ready to emerge from the buffer. Modulo-2 addition of the error pattern corrects it. For $n \geq N - M$, the situation is much more complicated since the error pattern will not appear in the shift register, and it is questionable whether the circuit of Figure 10.7 will work. For this case, two factors must be considered. First, it must be proven that a false correctable pattern does not appear in the shift register, and, second, it must be proven that at the right time the true correctable pattern does appear in the shift register.

A false correctable error pattern can appear in the shift register only if the received polynomial can be regarded as the modulo-2 sum of a wrong but legitimate code word and a correctable error pattern. It follows that an error pattern $E(x)$ is correctable if and only if no nonzero code word can be regarded as the sum of two such patterns. An error pattern $E(x)$ is correctable and, simultaneously, an error pattern $D(x)$ is detectable if and only if no nonzero code word can be regarded as the sum of one of each of these patterns. This is precisely the structure of the Fire codes.

Let us now consider an error pattern that while theoretically correctable is not near the beginning of the block. Let $R(x)$ be the result of the parity check calculation or the initial remainder:

$$E(x) = g(x)S(x) + R(x) \tag{10.4}$$

Let the remainder after multiplication by x and division by $g(x)$ (one shift cycle) be $R_1(x)$. Then

$$xR(x) = g(x)S_1(x) + R_1(x) \tag{10.5}$$

Combining these equations results in

$$xE(x) = xg(x)S(x) + g(x)S_1(x) + R_1(x) \tag{10.6}$$

It follows directly from Equation 10.6 that the remainder in the shift register after one shift cycle is identical to the remainder obtained by shifting the error pattern to the right and dividing by $g(x)$. Thus, if one continues these shift cycles, the error pattern becomes the remainder and appears in the b rightmost stages of the shift register at precisely the right time to correct the error. This completes the argument.

In Example 10.8, we claimed the polynomial $g(x) = 1 + x^2 + x^6 + x^7$ generates a Hamming weight-4 code that can be used to correct single errors and simultaneously detect double errors. The receiver that accomplishes this is shown in Figure 10.8. In Example 10.9, we claimed that $g(x) = (1 + x + x^3)(1 + x^9)$ generates a Fire code that corrects a burst of length 3 while simultaneously detecting a burst of length 7 or less. The receiver that accomplishes this is shown in Figure 10.9. The encoder for these two codes is essentially identical to the decoders with the test for zero logic circuit and buffer storage removed (as in Figure 10.5).

Microprocessor technology makes the implementation of these decoders quite simple. It may be most efficient to implement the shift register and exclusive-or devices using hardware, but testing the memory for 0s and enabling the gates and the shift register can all be easily accomplished with microprocessors.

10.7 FURTHER COMMENTS ON CYCLIC CODES

Other cyclic codes, such as the Bose–Chaudhuri–Hocquenghem codes, can be used for multiple-random-error correction. We shall not consider this important class of codes, however, for a number of reasons. First, in order to indicate how to find the generating polynomials (even by rote or mechanical means), we would have to go much deeper into the algebra of polynomials. Second, the major effort in multiple-error-correcting codes involves devising correction algorithms that are more efficient than storing large tables of remainders and their associated error patterns. These algorithms are also beyond the scope of this text. Third, in the remaining sections of this chapter we shall discuss convolution codes that are expressly designed to correct multiple random errors. Although the correction algorithms considered appear to be designed for convolution codes, they do give insights into methods of multiple-error correction.

Our purpose in this last chapter is to indicate one method of efficient single- and burst-error-correction techniques and one method of efficient multiple-error-correction techniques.

10.8 DEFINITION OF CONVOLUTION CODES[2]

When the signal-to-noise ratio is quite small (0 dB or less) we anticipate many errors in a large block of binary pulses. The channel capacity theorem (see Section 9.6) tells us that to correct many of these errors the source rate must

---

[2] This development follows the work of Wozencraft and Jacobs [2].

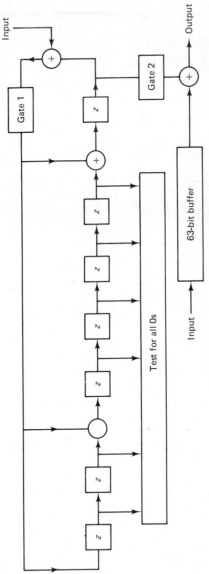

Figure 10.8 Single-error-correcting decoder for a systematic [63, 56] generalized Hamming weight-4 code.

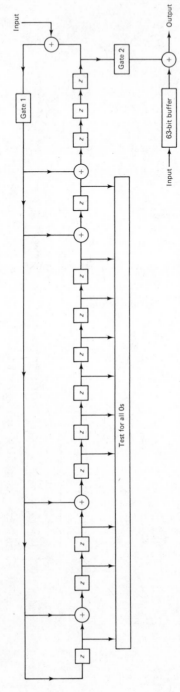

Figure 10.9 Decoder for correcting bursts of length 3 or less for a systematic [63, 51] Fire code.

be much smaller than the pulse rate. This means that the ratio of information bits M to total bits N must be small when the signal-to-noise ratio is small and we need to correct many random errors. In this situation, algebraic block codes become difficult to find and hard to decode efficiently. An alternative approach to the coding problem is to impose a hardware constraint (rather than an algebraic one) with the following two properties: First, the resultant codes must be decodable with a feasible computer algorithm; second, the total number of possible codes must be small enough to permit a trial-and-error search in order to find a good code. *Convolution codes* satisfy these properties and have been found to be very efficient when there are many errors.

According to Shannon's theory, the larger the block size, the better the code. The ultimate block size is the entire sequence of bits that one wishes to communicate. Let the block size be this entire sequence (L bits long), whatever L happens to be. Let us further assume that the convolutional encoder has the structure of Figure 10.10. Each bit that enters the shift register generates γ output bits. The last bit continues to generate output bits as it moves through the shift register. The total number of output bits is ($L + K - 1)\gamma$, and the resultant source rate becomes

$$R = \frac{L}{(L + K - 1)\gamma} \rightarrow \frac{1}{\gamma} \quad \text{bits/pulse} \quad (L \gg K) \tag{10.7}$$

As we indicated previously, however, such small source rates are necessary if one hopes to correct many random errors. If the size of the shift register K (called the *constraint span*) is modest, and if the number of connections γ is of the order 5, the total number of possible encoders is quite small considering the very large encoded block size ($L + K - 1$).

We shall argue in the rest of this chapter that near-optimal decoding algorithms that can correct many errors are quite feasible for these convolutional encoders. Using such algorithms enables us to find, by the simulation method, "good" convolution codes by trying different sets of connections. This method may be tedious (and a disadvantage relative to algebraic codes), but it is feasible. Another problem is that the error-correcting capability of

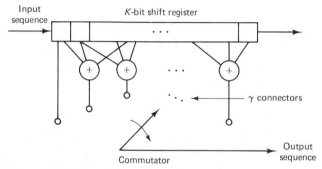

Figure 10.10 Convolutional encoder.

these codes can only be found by exhaustive simulation techniques. Neverthe-less, convolution codes have been found and empirically evaluated that approach the limits of the channel capacity theorem.

10.9 ERROR-FREE DECODING

In order to explain the *tree structure* of these convolution codes, let us first consider the unrealistic situation of no errors. We can identify a particular input vector with a branch of a tree, as in Figure 10.11. In this and all trees, we assume that a 0 branches up and a 1 branches down. We can construct a similar tree for the encoded block **y**; each node is associated with γ bits, with the exception of the last node, which is associated with $K\gamma$ bits.

EXAMPLE 10.10 ■

Consider the convolutional encoder shown in Figure 10.12, where $K = 4$ and $\gamma = 3$. Assume the shift register is initially set to zero. The code tree

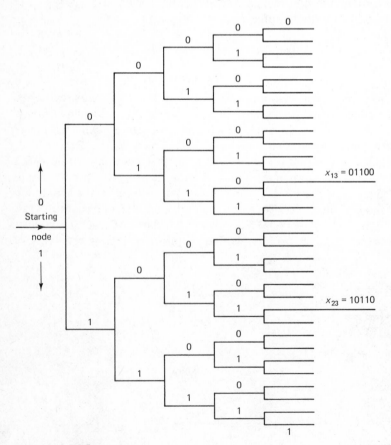

Figure 10.11 Input tree.

for this encoder is shown in Figure 10.12. The entire output sequence is indicated for only a few inputs, such as $x_{13} = 01100$ and $x_{23} = 10110$ of Figure 10.11. The encoded output corresponds to the same branch of the tree as the input. Thus identifying the branch corresponds to the decoding. We follow the encoded block

$y_{23} = 11101010011000100001000$

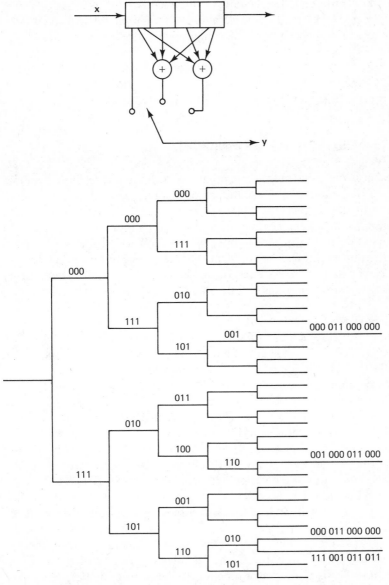

Figure 10.12 Convolutional encoder and code tree.

through the tree. Branching up corresponds to a 0 branching down to a 1. Thus $x_{23} = 10110$.   ■ ■

Of course, the number of branches grows exponentially with L, and for storing large data blocks the code tree in the receiver memory is not feasible. A more reasonable decoder would construct two encoders A and B and use the following algorithm:

1. Put a 1 into encoder A and a 0 into encoder B.
2. Compare the result of step 1 with the first γ bits of **y** to see which expression is correct; this determines whether the first bit of **x** is a 1 or a 0.
3. Reset the shift register of the "incorrect" encoder to that of the "correct" one.
4. Return to step 1.

10.10 TREE-SEARCHING TECHNIQUES

In the presence of errors (pulse errors that may or may not result in symbol errors) we might be tempted to use the same decoding algorithm with a slight modification of step 2: pick the closest or most nearly correct branch when neither output matches the input γ bits. As long as there are only one or fewer errors with each group of γ bits, this method will work. With small input signal-to-noise ratios, however, one eventually encounters a few errors in a row, causing an incorrect decision. Once the wrong path is chosen, a number of subsequent mistakes will be made even in the absence of errors.

EXAMPLE 10.11 ■

Suppose for the encoder of Example 10.10 that 111, 010, 100, 110, and 001000011000 were transmitted but received as $r = 111, 100, 100, 110$, and 001000011000. There are two errors in the second group of three pulses. Examining the code tree of Figure 10.12, we see that the second group 100 will be decoded as a 1 (i.e., 100 is closer to 101 than 010) and we have branched incorrectly. The next group 100 is closer to 110 than 001, and so the decision for a 1 is correct. The next group, however, or 110, is closer to 010 than to 101 and will be decoded incorrectly as a 0. Furthermore, the last group, 001000011000, is equally close to both choices, and a random decision will have to be made. Thus, the input 10110 will be decoded as $1110x$, where x is chosen arbitrarily. After the wrong branch, it will appear that seven more errors occurred.   ■ ■

Let us examine more closely how a wrong branch causes many subsequent errors. In Figure 10.13, the cumulative number of errors of a number of branches is plotted as a function of the decision nodes.

If we follow the correct branch, the cumulative number of errors increases in a linear fashion and follows the average number of errors we

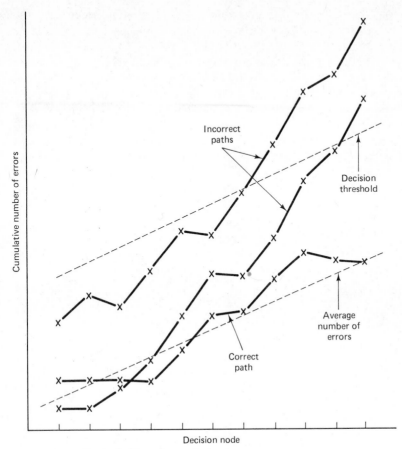

Figure 10.13 Typical plots of cumulative errors.

expect to encounter. If we branch incorrectly, we get a large number of subsequent errors, and the cumulative number increases faster. As shown in Figure 10.13, a wrong path might seem at first to be a better choice. If we followed all branches, however, the correct path would become evident, but this is not feasible.

The preceding discussion and Figure 10.13 suggest a tree-searching algorithm that can be implemented. The wrong branches will display an increasing error; therefore, we need an algorithm that discards unreasonable branches but stores the reasonable ones. For convolution codes that are "good," most branches are rapidly eliminated, and after a few decision nodes the number of possible branches stabilizes to some acceptable number. This is the basis for finding good codes experimentally. We can define the probability of error as the probability of eliminating the right branch. We must determine this probability theoretically or, by simulation, obtain the average number of computations per symbol and find the percentage of time that the buffer memory will overflow (which is a function of memory size). It is not uncommon with some convolution codes that the probability of buffer overflow is greater than the probability of error; more precisely, the probability of buffer overflow largely determines the probability of error (when the buffer overflows, the most unlikely branches must be discarded).

The threshold for discarding unreasonable branches could appear as the upper dashed line in Figure 10.13. If we know the probability p that a single pulse is in error, we can set this threshold to fix the overall probability of error (in the absence of buffer overflow). The number of errors in the correct path satisfies the binomial distribution.[3] If N is the number of decision nodes and γ the number of connections, the average number of errors is $N\gamma p$ and the variance is $N\gamma p(1 - p)$. A reasonable threshold T for maximum acceptable errors is $N\gamma p + k\sqrt{N\gamma p(1 - p)}$, where the parameter k determines the probability of error. Since the binomial CDF can be approximated by the Gaussian CDF (except for very large k), the probability of pulse error is approximately $Q(k)$. Because buffer memories are finite and overflow occurs, the probability of error is increased.

Typically, when the probability that a single pulse is in error is not known, the threshold can be based on the total number of errors $T_{\min}$ of the "best" branch. One possibility is

$$T = T_{\min} + k\sqrt{T_{\min}(1 - 1/N)\gamma}$$

Another is $T = T_{\min} + E$; when a branch has E more errors than the best branch, it is eliminated. Let us consider an example based on this last algorithm.

EXAMPLE 10.12 ■

Consider the convolutional encoder shown in Figure 10.14, where $K = 4$ and $\gamma = 5$. The first four decision nodes of the code tree are also shown. Let us assume that the first four input bits of $\mathbf{x}$ are 1, 1, 0, and 1. From the code tree we see that the encoded signal is $\mathbf{y} = 11111, 10101, 01101, 11011, \ldots$. Let us now assume a noise error pattern of $\mathbf{n} = 10010, 00111, 00000, 00100, \ldots$. Observe, using the bounds of Equation 9.19, that in a block of 20 bits at least 16 parity checks are needed to correct six errors, so $R = \frac{4}{20} = \frac{1}{5}$ (which is the rate of the convolutional code). The received signal $\mathbf{r} = \mathbf{y} + \mathbf{n} = 01101, 10010, 01101, 11111, \ldots$. We now consider the algorithm corresponding to $T = T_{\min} + 3$; that is, when a branch has at least three more cumulative errors than the best branch, it is eliminated. In the tree shown in Figure 10.15, we indicate the cumulative number of errors. A $\times$ indicates that the branch is eliminated from memory. We observe that at the second node, the best branch is not the correct branch, but all four possible branches are retained in memory. By the third node, the correct branch becomes the best branch, and by the fourth node there is only one other branch retained in memory. If we had to make a decision at this point, it would be the correct one. This encoder appears capable of achieving a very small

---

[3]See any classic text in probability, such as Feller [3].

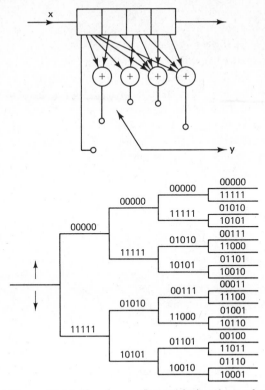

Figure 10.14 $K = 4$, $\gamma = 5$ convolutional encoder and code tree.

overall probability of error when $p \leq 0.3$ without a very large buffer memory. In order to verify this, however (or to find the range of values of p for which this algorithm usually works), simulation studies are needed. ■ ■

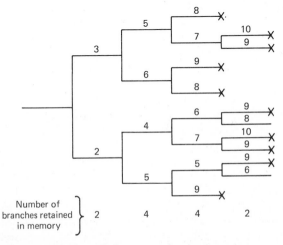

Figure 10.15 Cumulative errors in tree branches.

10.11 CONCLUSIONS

The last two chapters have been an introduction to the extensive subject of coding theory. The principal goal was to explain the underlying assumptions and basic techniques used. As seen in this chapter, the procedures are essentially algebraic and algorithmic and have little connection with classical communication techniques. The purpose of coding is to reduce the probability of error of standard binary communication techniques. These procedures are more flexible than other alternatives, such as orthogonal signaling. It is largely because of the convenience of coding techniques that binary communication signals play such an important role in modern discrete communications.

PROBLEMS

10.1. In Example 10.6 we saw that $g(x) = 1 + x + x^3$ is an irreducible polynomial of degree 3. It should therefore generate a $[7, 4]$ cyclic code that is different from the code implemented by the circuits of Figure 10.3. Determine the encoding and decoding (error-detecting) circuits for this code.

10.2. Determine the encoding and decoding (error-detecting) circuits that will implement a Hamming $[7, 4]$ code in systematic form that is equivalent to the Hamming $[7, 4]$ code implemented by the circuits of Figure 10.3.

10.3. Repeat Problem 10.2 except that your systematic code should be equivalent to the $[7, 4]$ code discussed in Problem 10.1.

10.4. Eight polynomials of degree 4 of the type $x^4 + a_3x^3 + a_2x^2 + a_1x + 1$ could be irreducible.
   (a) Prove that only the three polynomials $x^4 + x^3 + x^2 + x + 1$, $x^4 + x^3 + 1$, and $x^4 + x + 1$ are irreducible. (*Hint:* Argue first that one need only check whether the polynomials $x + 1$, $x^2 + 1$, and $x^2 + x + 1$ divide into the polynomials being tested.)
   (b) Show that the polynomial $x^{15} - 1$ factors in a manner that is consistent with Theorem 10.2.

10.5. Extend the results of Problem 10.4 and Theorem 10.1 as follows:
   (a) Argue that there are three $[15, 11]$ cyclic codes that can be generated.
   (b) What are the burst-error-detection capabilities of this code if they are used for detection only?
   (c) Design an encoder and error-detecting decoder (not necessarily systematic) for one of these $[15, 11]$ codes.

10.6. Prove that the $[7, 4]$ code implemented in Problems 10.1 and 10.3 is a Hamming code (i.e., $1 + x + x^3$ is primitive).

10.7. **(a)** Prove that two of the three $[15, 11]$ codes discussed in Problem 10.5 are Hamming codes (i.e., two of the irreducible polynomials are primitive while the third is of order 5).
   (b) Design a Hamming $[15, 11]$ encoder and decoder that is in systematic form. [*Hint:* Divide the polynomials into $x^{15} + 1$ and check the remainder at every step. If the polynomial is not primitive, the remainder pattern will prematurely match that of the divisor (ignoring the constant 1) but multiplied by x^m. One can then verify that the order is $15 - m$.]

10.8. **(a)** Determine the generating polynomial for a Hamming $[15, 10]$ weight-4 code.
　　(b) Design a decoder (as in Figure 10.8) that will correct single errors and simultaneously detect double errors.

10.9. Consider the irreducible polynomial of degree 4 that is not primitive but of order 5:
　　(a) Determine a generating polynomial for a $[45, 32]$ Fire code.
　　(b) Argue that this code can correct bursts of length 4.
　　(c) Design a decoder that will correct bursts of length 4 or less for a systematic $[45, 32]$ Fire code.

10.10. Consider now one of the primitive polynomials of degree 4:
　　(a) Argue that these polynomials can be used to generate $[45, 32]$ Fire codes with the same capabilities as the code of Problem 10.9.
　　(b) Using either of these primitive polynomials, find a $[120, 108]$ Fire code that can correct bursts of length 4 or less.
　　(c) Indicate one advantage and one disadvantage of this code relative to the $[45, 32]$ Fire code.

10.11. Consider the convolution encoder shown, where $K = 4$ and $\gamma = 4$:

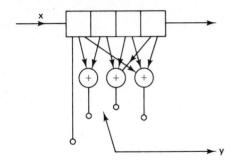

　　(a) Construct the code tree (four decision nodes deep) for this encoder.
　　(b) If the first four input bits are 1010, what are the first 16 encoded bits?

10.12. Consider the same encoder and input of Problem 10.11. Assume a noise error pattern of 1001, 0000, 0010, 0100.
　　(a) Determine the cumulative number of errors on each of the 16 branches (after the fourth node).
　　(b) Does the correct path have the smallest number of cumulative errors?
　　(c) Assuming that branches having three or more errors than the best branch are eliminated, what is the maximum number of branches that must be retained in memory?

REFERENCES

1. W. W. Peterson, *Error Correcting Codes*, Wiley, New York, 1961.
2. J. M. Wozencraft and I. M. Jacobs, *Principles of Communications Engineering*, Wiley, New York, 1965.
3. W. Feller, *An Introduction to Probability Theorem and Its Applications*, Wiley, New York, 1950.

Index